PREPARATION OF THIN FILMS

Joy
George

Cochin University of Science and Technology
Cochin, Kerala, India

Marcel Dekker, Inc.   New York • Basel • Hong Kong

Library of Congress Cataloging-in-Publication Data

George, Joy.
 Preparation of thin films / Joy George.
   p.   cm.
 Includes bibliographical references and index.
 ISBN 0-8247-8196-1
 1. Thin films. 2. Surface chemistry. 3. Vapor-plating.
 I. Title.
 QC176.83.G46   1992
 621.381'52--dc20   91-46019
                                                           CIP

This book is printed on acid-free paper.

MARCEL DEKKER, INC.
270 Madison Avenue, New York, New York 10016

Current printing (last digit):
10 9 8 7 6 5 4 3 2 1

PRINTED IN THE UNITED STATES OF AMERICA

In loving memory
of
my beloved daughter
SUMA ANIL MATHEW
Lecturer 1980–1986
Faculty of Home Science
B.C.M. College, Kottayam
Kerala, India

Preface

Applications of thin films in optics and electronics have made extraordinarily rapid progress in recent years. Consequently, the development of deposition techniques for the preparation of thin films with controlled, reproducible, and well-defined properties plays an increasingly important role in technological applications. Since the advent of the early thermal evaporation techniques, a wide variety of deposition methods for elemental and compound/alloy films has been developed.

Thin film deposition technology is still undergoing rapid progress, and although several books in the area of deposition technology have been published at regular intervals over the past 30 years, most describe only selected deposition processes and techniques. In writing this book, however, I have attempted to bring together in a single volume the various deposition techniques, mostly of inorganic compound thin films. Details presented in earlier books are included for continuity only, and over 1000 references are cited for more exhaustive details.

No excuse is made for the fact that several of the techniques described in the book are not of "industrial scale." Production-oriented preparation techniques are given second priority, the primary audience of the book being research scientists in the area of materials science and university postgraduate students.

Although I am fortunate in having contributed to the field of thin films during my professional career, part of the book's content is the result of the work of others, and references are made in the text at appropriate places.

In writing this book, I have found the continuing support and encouragement of my past and present Ph.D. research students to be most helpful and stimulating.

Thanks are due to my wife and children for the patience and understanding they have shown during the course of the work.

Joy George

Contents

Preface   *v*
Introduction   *viii*

1.   Vacuum Evaporation   1

2.   Sputtering   41

3.   Ion Beam and Ion-Assisted Deposition   101

4.   Reactive Deposition Techniques   141

5.   Ionized Cluster Beam Methods   207

6.   Chemical Methods of Film Deposition   223

7.   Epitaxial Film Deposition Techniques   303

8.   Other Methods of Film Deposition   335

9.   Summary   353

Index   *367*

Introduction

The real purpose of books is to trap the mind into doing its own thinking.
Christopher Morley

Any solid or liquid object with one of its dimensions very much less than that of the other two may be called a "thin film." The most commonly observed phenomenon associated with thin films, which attracted the attention of physicists as early as the second half of the seventeenth century, is the fascinating colors on a thin film of oil floating on the surface of water.

Thin films of metals were probably first prepared in a systematic manner by Michael Faraday, using electrochemical methods. In the earlier stages, scientific interest in thin solid films centered on antireflection coatings for lenses, multilayer interference filters, automobile headlights, and decorative coatings. Application of thin film technology has revolutionized the field of optics and electronics. The need for new and improved optical and electronic devices has stimulated the study of thin solid films of elements, as well as binary and ternary systems, with controlled composition and specific properties, and has consequently accelerated efforts to develop different thin film deposition techniques.

For example, since the discovery in 1986 of high-temperature superconducting (HTS) oxides, there has been much unprecedented research activity in the preparation of high-quality thin films. These materials have potential applications in superconducting electronics, and a wide spectrum of deposition methods, both high vacuum and chemical, have been used for the preparation of HTS films.

Thin film properties are strongly dependent on the method of deposition, the substrate materials, the substrate temperature, the rate of deposition, and the background pressure. Specific applications in modern technology demand such film properties as high optical reflection/transmission, hardness, adhesion, nonporosity, high mobility of charge carriers/insulating properties, chemical inertness toward corrosive environments, and stability with respect to temperature, stoichiometry, and orientation in single crystal films. The application and the properties of a given material determine the most suitable technique for the preparation of thin films of that material.

Although well-established conventional vacuum evaporation, D.C. sputtering, and simple chemical reaction techniques are able to produce reasonably good results, in certain cases the increase in the range of optical and semiconductor thin film applications requiring stringent and wide-ranging properties has led to the development of a variety of deposition techniques. These include electron beam evaporation, laser evaporation, magnetron sputtering, reactive and activated reactive deposition, ion beam deposition, molecular beam epitaxy, metal organic chemical vapor deposition, and plasma CVD. While most of these processes have reached a high level of development, the full potential of processes such as magnetron sputtering, ion beam deposition, or ionized cluster beam remains largely unexplored.

The book contains nine chapters. In Chapter 1, thin film preparation by vacuum evaporation is described, classifying these methods according to the mode used in heating the material. Vacuum evaporation by electron beam (EB), first introduced in the early 1960s, has certain significant advantages compared with conventional heating modes. These make it useful for the preparation of materials such as refractory metals or refractory metal oxide. EB evaporation is used very extensively for the preparation of thin films of a number of materials. The EB evaporation method of preparing thin films is covered in detail, giving the different electron gun structures as well as particular experimental setups used for the preparation of selected materials. This first chapter also discusses recent developments using CW and pulsed laser as the thermal evaporation source for the preparation of thin films of elements, ceramics, and semiconductors.

Chapter 2 covers the preparation of thin films by the conventional sputtering technique and its variants (e.g., triode sputtering, getter sputtering, radiofrequency sputtering, magnetron sputtering, ion beam sputtering). Magnetron sputtering has become a significant method for the deposition of metal and dielectric films since its development in the early 1970s, and this method is treated in great detail. Since a number of reviews are available on the general features of magnetron sputtering and the various target configurations, only the recently developed cathode configurations and the method of preparation of thin films are described. The ion beam sputtering deposition method is also discussed in some detail in this chapter.

The term "ion-assisted deposition" can cover a number of different techniques that involve the bombardment of the growing film by energetic ions. In the ion plating method, the ionization of the vapor generated by evaporation or sputtering takes place in inert gas (argon) glow discharge, and the ions of the coating material and the support gas impinge on the substrate. In ion beam deposition, the beam consists of the desired film material and is deposited in a high vacuum ($\approx$ 10-7–10-4 torr; the substrate is not immersed in a glow discharge). These are therefore grouped together and presented in Chapter 3, "Ion Beam and Ion-Assisted Deposition." Other ion-assisted techniques, such as that using a hot hollow cathode (HHC), are also included in this chapter.

Chapter 4, on reactive deposition techniques, presents reactive evaporation, reactive sputtering, reactive ion beam sputtering, and reactive ion plating methods. Activated reactive evaporation is also included. Chapter 5 is devoted to the ionized cluster beam and reactive ionized cluster beam techniques. Chapter 6 is on chemical methods of film deposition. The processes that have been arranged under chemical methods and covered in this chapter include electrodeposition, electroless deposition, anodization, thermal growth, and chemical vapor deposition (CVD), the CVD method covering a major portion of this chapter. In addition to general CVD, other variants such as laser CVD, photo CVD, and plasma-enhanced CVD methods of thin film preparation are dealt with in detail.

Molecular beam epitaxy (MBE) is essentially an ultra high vacuum evaporation process for growing epitaxial films by directing onto the clean heated substrate beams of atoms or molecules formed by thermal evaporation. Metal organic chemical vapor deposition (MOCVD) is a more recently developed technique that involves the reaction of mixtures of metal organic compounds at a substrate surface maintained at the proper reaction temperature. Both MOCVD and MBE, shown to be useful methods for producing high-quality semicondutor thin films for device applications, are described in Chapter 7. Two other epitaxial methods, liquid phase epitaxy (LPE) and hot wall epitaxy (HWE), are also discussed in this chapter.

The high degree of order in Langmuir–Blodgett (LB) film is one of the main points of interest in these films, and the LB technique is rapidly gaining worldwide attention focused on noncentrosymmetric multilayer films for potential pyroelectric, piezoelectric, and nonlinear optical device applications. Since this book is intended to cover the different methods of preparation of thin films, mostly of inorganic materials, only an introductory treatment of the preparation of LB films is given. This appears in Chapter 8, which also deals with the solution spray method.

Chapter 9 summarizes the different deposition techniques.

I

Vacuum Evaporation

> If you are out to describe the truth, leave elegance to the tailor.
>
> *Albert Einstein*

1.1 BASIC CONSIDERATIONS

Deposition of thin films by evaporation is very simple and convenient, and is the most widely used technique. One merely has to produce a vacuum environment in which a sufficient amount of heat is given to the evaporant to attain the vapor pressure necessary for evaporation, then the evaporated material is allowed to condense on a substrate kept at a suitable temperature.

A vast number of materials can be evaporated in vacuum and caused to condense on a substrate to yield thin solid films. Deposition consists of three distinguishable steps.

1. Transition of the condensed phase (solid or liquid) into the gaseous state.
2. Traversal by the vapor of the space between the vapor source and the substrate (i.e., transport of vapor from the source to the substrate).
3. Condensation of the vapor upon arrival at the substrate (i.e., deposition of these particles on the substrate).

Substrates are made from a wide variety of materials and may be kept at a temperature depending on the film properties that are required. When evaporation is made in a vacuum, the evaporation temperature will be considerably lowered and the formation of oxides and incorporation of impurities in the growing layer will be reduced. The pressure used for normal evaporation work is about 10^{-5} torr. This also ensures a straight line path for most of the emitted vapor atoms, for a substrate-to-source distance of approximately 10–50 cm in a vacuum system.

1

The majority of materials are evaporated from the liquid phase, while some are evaporated from the solid state. According to Knudsen [1], the maximum number of molecules dN_e evaporating from a surface area A_e during a time dt, is

$$\frac{dN_e}{A_e dt} = (2\pi \, m \, KT)^{-1/2} P_e \tag{1.1}$$

where P_e is the equilibrium pressure and m the mass of the molecule. The rate of free evaporation of vapor species m_e from a clean surface of unit area in vacuum is given by the Langmuir expression [2]

$$m_e = 5.83 \times 10^{-2} P_e \left(\frac{M}{T}\right)^{1/2} \text{g cm}^{-2} \text{s}^{-1} \tag{1.2}$$

where T is the temperature, M is the molecular weight of the vapor species, and P_e ($\approx 10^{-2}$ torr) is the equilibrium vapor pressure. In terms of the molecules, we may write the evaporation rate as follows:

$$N_e = 3.513 \times 10^{22} P_e \left(\frac{1}{MT}\right)^{1/2} \text{mol cm}^{-2} \text{s}^{-1} \tag{1.3}$$

However the rate of deposition of the vapor on a substrate depends on the source geometry, the position of the source relative to the substrate, and the condensation coefficient.

Taking an ideal case of deposition from a clean and uniformly emitting point source onto a plane receiver, the rate of deposition by Knudsen's cosine law varies as $\cos \theta / r^2$, where r is the radial distance of the receiver from the source and θ is the angle between the radial vector and normal to the substrate direction. If d_0 is the thickness of the deposit at the center vertically above the point source at a distance h and d at a distance l from the center then,

$$\frac{d}{d_0} = \frac{1}{[1 + (l/h)^2]^{3/2}} \tag{1.4}$$

If the evaporation takes place from a small area onto a parallel plane, then the thickness distribution is given by

$$\frac{d}{d_0} = \frac{1}{[1 + (l/h)^2]^2} \tag{1.5}$$

The substrate is bombarded not only by the particles of the evaporated substance but also by those of the residual gases. The residual gases in evaporation systems do have a profound influence on the growth and properties of the films. First there is the possibility of collisions between gas molecules and vapor molecules during the transit of the latter from the source to the substrate. The number of collisions depends on the mean free path (i.e., the average distance traversed between subsequent collisions). The number of atoms N from the total number N_0 traversing a distance l without having a collision is given by

$$N = N_0 \exp\left(\frac{-l}{\lambda}\right) \qquad (1.6)$$

where λ is the mean free path in the residual gas. Usually the films are deposited at a pressure of the order of 10^{-5} torr or less and only a negligible number of collisions between the residual gas and the vapor molecules will take place. As a result, the vapor molecules will exhibit straight line propagation.

Second, the film will be badly contaminated by the residual gases in the vacuum system. Such contamination can arise from gas molecules impinging on the surface of the substrate during deposition. The impinging rate of gas molecules is given by the kinetic theory of gases

$$N_g = 3.513 \times 10^{22} \frac{P_g}{(M_g T_g)^{1/2}} \text{ cm}^{-2}/\text{s} \qquad (1.7)$$

where P_g is the equilibrium gas pressure at temperature T_g.

Table 1.1 gives the mean free path (mfp) and other relevant data for air at 25°C for different pressures.

It can be seen that under the conditions of vacuum normally used (10^{-5} torr) and deposition rates of about 1 Å/s, the impingement rate of gas atoms is quite large,

Table 1.1 Mean Free Path and Other Relevant Data for Air at 25°C

Pressure (torr)	Mean free path (cm)	Collisions per second between molecules	Impingement rate (cm^{-2}/s)	Number of monolayers per second
10^{-2}	0.5	9×10^4	3.8×10^{18}	4400
10^{-4}	51	900	3.8×10^{16}	44
10^{-5}	510	90	3.8×10^{15}	4.4
10^{-7}	5.1×10^4	0.9	3.8×10^{13}	4.4×10^{-2}
10^{-9}	5.1×10^6	9×10^{-3}	3.8×10^{11}	4.4×10^{-4}

which means that a good amount of gas sorption will occur if the sticking coefficient of the gas atoms is not negligibly small. To obtain films with the minimum number of impurities, pressures in the region of ultra high vacuum ($<10^{-9}$ torr) must be used. Excellent treatments of the subject of vacuum evaporation are given by Holland [3] and Maissel and Glang [4].

1.2 METHODS OF EVAPORATION

To evaporate materials in a vacuum system, a container is required to support the evaporant and to supply the heat of vaporization while allowing the charge to reach a temperature high enough to produce the desired vapor pressure. To avoid contamination of the films deposited, the support material itself must have negligible vapor and dissociation pressures at the operating temperature. Rough estimates of the operating temperatures are based on the assumption that vapor pressures of 10^{-2} torr must be established to produce useful condensation rates. Materials commonly used are refractory metals and oxides. The possibilities of alloying and chemical reactions between the evaporant and the support materials must be taken into account while choosing a particular support material. The shape in which the support materials are used depends very much on the evaporant.

The important methods of evaporation are resistive heating, flash evaporation, electron beam evaporation, laser evaporation, arc evaporation, and radio frequency (rf) heating.

1.2.1 Resistive Heating

When the material to be evaporated is raised in temperature by electrical resistance heating, the most commonly used source materials (support materials) are the refractory metals tungsten, tantalum, and molybdenum, which have high melting points and very low vapor pressures. The simplest sources are in the form of wires and foils of different types as shown in Figure 1.1. Electrical connections to the wire or foil are made directly by attaching thin ends to heavy copper or stainless steel electrodes. The sources shown in Figure 1.1a and 1.1b are commonly made of thin tungsten/molybdenum wire (diameter 0.02–0.05 in.). Here the evaporants are fixed directly to the source in the form of wire (hung in the form of a slider in the case of hairpin). Upon melting, the evaporant wets the filament and is held by surface tension. Multistrand filaments are generally used because they offer greater surface area than single-wire elements. These sources have four main drawbacks: (a) they can be used only for metals or congruently evaporating alloys, (b) only a limited quantity of the material can be evaporated at a time, (c) the material to be evaporated should wet the resistive filament wire upon melting, and (d) once heated, these elements become very fragile and will break if not handled carefully. Dimpled foils (Figure 1.1c) fabricated from sheets of tungsten, tantalum, or molybdenum 0.005–0.015 in. thick, are the most commonly used sources when only

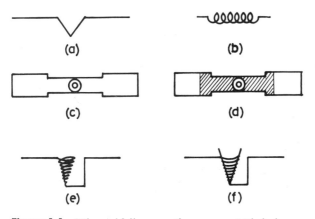

Figure 1.1 Wire and foil evaporation sources: (a) hairpin source, (b) wire helix, (c) dimpled foil, (d) dimpled foil with oxide coating, (e) wire basket, and (f) crucible with wire spiral heater.

small quantities of the evaporant are available or needed. All three refractory metals become brittle after being heated in a vacuum, especially if alloying with the evaporant takes place. Oxide-coated metal foils (Figure 1.1d) are also used as evaporation sources. Here Mo or Ta foils about 0.01 in. thick are covered with a thick layer of alumina, and operating temperatures up to 1900°C are possible. The power requirements of such sources are much above those of uncoated foils due to the reduced thermal contact between the resistively heated metal and evaporant. Wire baskets (Figure 1.1e) are used to evaporate small chips of dielectrics or metal, which either sublime or do not wet the source material on melting. Crucibles of quartz, glass, alumina, graphite, beryllia, and zirconia are used with indirect resistance heating (Figure 1.1f). A multiple Knudsen source [5] enables one to get uniform deposition of the films. Vapor sources of various designs have been described by many authors [6-8]. Even though new and more sophisticated techniques for the preparation of thin films have been developed, electrical resistive heating is still commonly used in the laboratory and in industry to prepare thin films of elements [9-12a], oxides [12b-14], dielectrics [15], and semiconductor compounds [16-19a].

For the preparation and deposition of high temperature superconducting (HTS) oxide thin films by evaporation, the constituent elements [19b] or compounds (e.g., BaF_2) [19c] are coevaporated by resistive heating and annealed in oxygen to obtain the superconducting phase. Azoulay and Goldschmidt [19d] have reported the preparation of Y–Ba–Cu–O films by layer-by-layer evaporation of Cu, BaF_2, and YF_3, using solely resistive evaporation from tungsten boats onto $SrTiO_3$ and postdeposition annealing..

Table 1.2 lists materials with appropriate temperatures required to produce vapor pressures of 10^{-2} torr and the suitable support materials

The main disadvantages of evaporation by simple resistive heating are (a) the reaction of the evaporant material with the support crucibles, (b) the difficulty in attaining high enough temperatures for the evaporation of dielectrics (Al_2O_3, Ta_2O_5, TiO_2, etc.), (c) low rates of evaporation, and (d) the dissociation of compounds or alloys upon heating.

1.2.2　Flash Evaporation

A common difficulty encountered in the preparation of thin films of multicomponent alloys or compounds that tend to distill fractionally* is that the chemical composition of the film obtained deviates from that of the evaporant. This difficulty is best overcome in flash evaporation. Here small quantities of the material to be evaporated are dropped in powder form onto a boat hot enough to ensure that evaporation takes place instantaneously. The temperature of the boat should be high enough to evaporate the less volatile material fast. When a particle of the material evaporates, the component with the higher vapor pressure evaporates first, followed by components with lower vapor pressure. In practice, the feed of material is continuous, and there will be several particles in different stages of fractionation on the boat. Moreover, since no material accumulates on the boat during evaporation, the net result of these instantaneous discrete evaporations is that the vapor stream has the same composition as the source material. If the substrate temperature is not high enough to permit reevaporation to take place, stoichiometric compound or alloy films will be formed. The powdered material can be fed into the heated support using different arrangements (mechanical, electromagnetic, vibrating, rotating, etc.) for material feeding.

Harris and Siegel [20], the first to utilize flash evaporation, used a motor-driven belt as the transport mechanism. An electromagnetically vibrating material feeder was used by Campbell and Hendry [21] for preparing thin films of Ni–Cr alloys. Flash evaporation techniques have been used to prepare semiconducting thin films of certain compounds from Groups III–V [22–24]. Ellis [25] used flash evaporation to prepare copper sulfide films. Platakis and Gatos [26] had developed a flash evaporation technique using a U-tube-type source, which permitted the preparation of compound semiconductor films with high structural and chemical homogeneity. Tyagi et al. [27] had used a compact and simple arrangement for flash evaporation that can be easily incorporated into a vacuum coating unit. A pitcher-shaped quartz crucible heated to 1200°C by a molybdenum filament basket was used as the source. PbS and PbS–Ag were used for the preparation of thin films. The results suggested that with this arrangement, dissociation and dopant separation could be

---

*Different components have different vapor pressures at any given temperature.

Table 1.2 Temperatures and Support Materials Used in the Evaporation of Various Materials

Material	Melting point (°C)	Temperature (°C) required to produce $V_p = 10^{-2}$ torr	Support materials		Remarks, if any
			Wire or foil	Crucible	
Ag	961	1030	W, Ta, Mo, Nb	Mo, Ta, C	Mo preferred
Al	660	1220	W, Ta	BN, graphite	Wets and creeps out of containers; reacts with graphite; use stranded W wire
Al₂O₃	2030	1800	W, Ta	W, Ta	Oxygen-deficient films from W
As	817	300		Al₂O₃, C, BeO	Sublimes; toxic
Au	1063	1400	W, Mo	W, Mo, C, Al₂O₃	
B	2300	2100		C	Does not alloy
Ba	725	610	Mo, W, Ta		
BaF₂	1280	1100	Mo, W, Ta		
Be	1280	1230	W, Mo, Ta	C	Toxic
Bi	271	670	W, Ta, Mo	Mo, Al₂O₃	
Bi₂O₃	817	1840		Al₂O₃	
Bi₂S₃	685 (decomposes)		W		Sulfur-deficient films
Bi₂Te₃	820		W, Ta, Mo		Reactive evaporation gives stoichiometric films
Ca	850	600	W	Al₂O₃	
CaF₂	1360	1280	Ta, W, Mo	Ta, W, Mo	
Cd	321	265	W, Ta, Mo, Fe, Ni, Nb	Mo, Ta, fused quartz	Sublimes; contaminates vacuum system
CdS	1750	670	W, Mo	Ta, W, graphite, quartz	Dissociates during evaporation
CdSe	1250	660		Mo, Ta, quartz, Al₂O₃	
CdTe	1041	570	W, Ta	Ta, Al₂O₃, graphite	

Table 1.2 Continued

Material	Melting point (°C)	Temperature (°C) required to produce $V_p = 10^{-2}$ torr	Support materials		Remarks, if any
			Wire or foil	Crucible	
Ce	804		W	Al$_2$O$_3$, C	
CeF$_3$	1324	1265	Mo, Ta, W	Mo, Ta, W	
CeO$_2$	1950	1810		W, Al$_2$O$_3$	Reacts with the source
Co	1492	1520	W, Nb	Al$_2$O$_3$, BeO	Alloys with W
Cr	1900	1400	W, Ta	Al$_2$O$_3$, BeO	
Cu	1083	1260	W, Ta, Mo, Nb	Mo, Ta, W, C, Al$_2$O$_3$	Mo boat preferred
Fe	1536	1480	W	Al$_2$O$_3$, BeO, W	Alloys with refractory metals
Ga	30	1130		Al$_2$O$_3$	
GaAs	1238			Ta, W	Decomposes; flash evaporation is simplest
GaP	1350		Ta, W	Ta, W	Flash evaporation is simplest
Ge	958	1400	W, Mo, Ta	W, Mo, Ta, C, Al$_2$O$_3$	Wets refractory metals
In	156	950	W, Fe, Mo	Mo, W, C, BeO, Al$_2$O$_3$	Mo boats preferred
InP	1070		Ta, W		Results in P-rich films
La$_2$O$_3$	2315	1850	W		
LiF	870	800		Mo, Ta	
Mg	650	440	W, Mo, Ni, Ta, Fe, Nb	Mo, Ta, C, Fe	Sublimes
MgF$_2$	1263	1130	W, Ta, Mo	Mo, Ta	Very little dissociation
Mn	1250	940	W, Mo, Ta, Nb	Al$_2$O$_3$	Wets refractory metals

Material					Remarks
Mo	2620	2510	Mo		Slow rate; EB evaporation preferred
Na₃AlF₆ (cryolite)	1000	865	Mo, Ta		
NaCl	801	670		Ta, W, C	
Ni	1453	1530	W, W foil lined with Al₂O₃		Alloys with refractory metals; EB evaporation preferred
Pb	327	715	W, Mo, Ni, Fe, Chromel	W, Mo, Ta, Fe, Al₂O₃	Does not wet W, Ta, Mo; toxic
PbCl₂	678	430	W	Al₂O₃	
PbF₂	855	700	W	Al₂O₃	
PbO	890	740			
PbS	1112	675	W, Mo	Quartz, Mo	Purest films from quartz; reacts with Mo
PbSe	1065	1150	W, Mo	Al₂O₃	
PbTe	917	1460	Ta, Mo	Al₂O₃, BeO	Alloys with refractory metals
Pd	1552		W	ThO₂, ZrO₂	Alloys with refractory metals; EB evaporation preferred
Pt	1769	2100	W		
Rh	1960	2040	W	ThO₂, ZrO₂	EB preferred
Sb	631	530	Mo, Ta, Ni	Ta, Al₂O₃, C, Mo	Toxic
Sb₂S₃	546	550	Mo, Ta		Dissociates during evaporation; two-source evaporation preferred
Se	217	240	Mo, Ta, Fe, stainless steel 304	Mo, Ta, C, Al₂O₃	Wets support materials; wall deposits contaminate vacuum system; toxic
Si	1410	1350		BeO, C. Mo	Tends to attack refractory oxides; films contaminated by SiO; EB gives purest films
Sn	232	1250	W, Ta, Mo	C, Al₂O₃, Ta	Wets Mo

Material			Support	Crucible	Remarks
SnO_2	1127	580	W	Ta, W	EB preferred
Ta_2O_5	1800	2000	Ta, W	Al_2O_3, Ta, C, Mo, glass	Contaminates vacuum system; wets all refractory metals; toxic; glass crucible with wire coil heaters preferred
Te	450	375	W, Mo, Ta		
Th	1750	2196	W	W	Wets W
Ti	1670	1750	W, Ta	C, ThO_2	
TiO_2	1850	2000	W, Ta		Decomposes to TiO and Ti
Ti_2O_3	1760	1850	W, Ta		
Tl	304		Ni, Fe, Ta, Nb	Ta, Al_2O_3, quartz	
V	1920	1850	W, Mo	Mo	Wets the supports
W	3390	3230	W		Wets Mo; alloys slightly with W
WO_3	1473	1140	W		EB preferred
Y_2O_3	1500		W	Ta, W, Pt	0.5 mm thick foil; cannot be used more than 3–4 times
Zn	419	345	W, Ni, Ta	Al_2O_3, C, Mo	Contaminates vacuum system; wets refractory metals; high sublimation rate
Zn_3P_2		1000	Mo		
ZnS	1830		Mo, Ta	Mo, Ta, C	Slight deviation from stiochiometry
ZnSe	1520	820	Mo, Ta	Mo, Ta, Al_2O_3	
ZnTe	1238	1000	Mo, Ta	Al_2O_3	
Zr	1850	2380	W	W	Deposits contain trace of W; EB preferred
ZrO_2	2715	2400	W	W	EB preferred

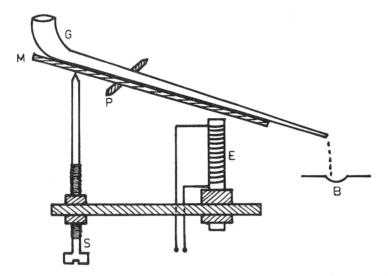

Figure 1.2 Schematic diagram of flash evaporator: M, mild steel plate; G, pipe-shaped glass tube; P, pivot; S, screw; E, electromagnet; B, heated molybdenum boat. (Data from Ref. 28.)

avoided. George and Radhakrishnan [28] used the flash evaporation technique for preparing Sb_2S_3 thin films, using an electromagnetically vibrating feeder similar to that used by Campbell and Hendry [21]. A schematic diagram of the setup is shown in Figure 1.2. The mild steel plate, which can be adjusted by screws on a pivot, is attracted by the electromagnet, and intermittent passage of the current through the coil makes the mild steel plate vibrate at a desired frequency. The pipe-shaped glass tube made of Corning glass serves to drop the powdered material onto a heated molybdenum boat. The whole system is enclosed in an aluminum cover and was used inside the vacuum system. The power to the winding of the electromagnet was fed from an astable multivibrator working at a frequency of 2 Hz. Gheorghiu et al. [29] prepared films of amorphous GaAs, GaP, and GaSb in an ultra high vacuum (10^{-7} Pa) by flash evaporation of the crystalline powder using an evaporation setup consisting of a Blazers vibrating feeder associated with various tungsten crucibles. The investigators were able to reduce the loss of powder and improve the constancy of the evaporation rate (controlled with a calibrated quartz microbalance) by inserting a cooled vibrating metallic foil between the feeder and a flat crucible 2 cm wide, which distributed the powder grains more uniformly over the whole crucible area.

Several reports are available on the preparation of semiconducting compounds by flash evaporation [30–34]. Sridevi and Reddy [30] used a simple flash evaporation technique and prepared thin films of $CuInSe_2$ to study their electrical and optical properties. The starting material in the form of fine powder was taken in a stainless steel hopper and vibrated with the help of a vibrator. A calling-bell type of plunger was used as the vibrator, the amplitude of which was varied with the help of a Variac. The powder dropped slowly onto a tantalum boat kept at about 1300°C and evaporated instantaneously. The lattice parameters determined agreed with those of the bulk. Flash evaporation has been used also for the preparation of epitaxial films of $CuInSe_2$, $LiInSe_2$, and $Li_xCu_{1-x}InSe_2$ [35–38].

Flash evaporation has been very widely used for the preparation of cermet (CE-Ramic plus METal) films, which are mixtures of metals and dielectrics. Their resistivity increases with the dielectric content, can be varied over a wide range, and possesses great stability at high temperature. Cermet films prepared by flash evaporation were reported as early as 1964. Braun and Lood [39] made thin film cermet resistors with a flash evaporator [20] using mixed powder of SiO and Cr with a starting material of 69 wt % Cr. Scott [40] used 77 wt % Cr, and Schabowska et al. [41] used 50 wt % Cr for preparing Cr–SiO films. The electrical and structural properties of flash-evaporated cermet Cr–SiO thin film (70 wt % Cr/30 wt % SiO) were reported by Milosavlgevic et al. [42]. A tungsten evaporation source was held at a constant temperature of 2000°C and the mixture was continuously dropped during the evaporation process. Taylor et al. [43] investigated the effect of composition and sheet resistance on the gage factor and temperature coefficient of resistance of thin Cr–SiO cermet films. Sintered cermet granules of 99.9% purity, granule diameters of 0.4–0.7 10^{-3} m, and ratios of mass of chromium to silicon monoxide of 70:30, 60:40, and 50:50 were used as the starting material. Schabowska and Scigala [44a] studied the electrical conduction in Cr–SiO cermet films with various metal concentrations (50, 60, and 70 wt % Cr) deposited by flash evaporation using a commercial evaporating and feeding system. The vapor source was a tungsten boat.

Tohge et al. [44b] have reported the preparation of amorphous chalcogenide films (Ge–Bi–Se) on glass substrates by flash evaporation using powders of the melt-quenched glasses or partially crystalline glasses as evaporation sources. Thin films of $Ge_{20}Bi_xSe_{80-x}$, with x up to 17 at. %, were prepared, and the films above x = 10 at. % exhibited n-type conduction even in the as-deposited form. These investigators found that annealing had only a very slight effect on the electrical conductivity and optical band gap.

High T_c superconducting oxide films have also been prepared by flash evaporation. Ece and Vook [44c] have reported the deposition of Y–Ba–Cu–O thin films on magnesium oxide substrates by dropping powders of sintered $YBa_2Cu_3O_{7-x}$ onto resistively heated tungsten. The as-deposited films were then annealed in oxygen for 60 minutes at 930°C, and these films exhibited superconductivity. Annealing at 945°C improved the texture of the films. Amorphous gallium arsenide (a–GaAs) thin films have been prepared by Manorama et al. [44d], recently using flash evaporation of finely powdered GaAs.

A serious drawback of the flash evaporation technique is the difficulty in preoutgassing the evaporant powder. Degassing the powder can be accomplished to some extent by vacuum storage for 24–36 hours prior to deposition. Otherwise, large quantities of gas may be released during evaporation. Also the expanding gases can cause "spitting" during evaporation.

1.2.3 Electron Beam Evaporation

Instead of the simple process of evaporation of any material by resistive heating, which often suffers from several disadvantages—for instance, the reaction of the material with the support crucible, low evaporation rates—the vaporizaion of materials can be accomplished by electron bombardment. Here a stream of electrons is accelerated through fields of typically 5–10 kV and focused onto the surface material for evaporation. The electrons lose their energy very rapidly upon striking the surface, and the material melts at the surface and evaporates. That is, the surface is directly heated by impinging electrons, in contrast to conventional heating modes. Because the material in contact with the support crucible remains solid, in effect the molten material is contained in a crucible of itself and the reactions are minimized. Direct heating allows the evaporation of materials from water-cooled crucibles, and these are very commonly used in electron beam (EB) evaporation. Such water-cooled crucibles are necessary for evaporating reactive and in particular reactive refractory materials, to avoid almost completely reactions with the crucible walls. This allows the preparation of high purity films because crucible materials or their reaction products are practically excluded from evaporation. By this type of heating any material can be evaporated, and the rate of evaporation varies from fractions of an angstrom per second to micrometers per second. The electron beam sources have been found to be versatile and reliable, and electron beam evaporators are used even for materials that can be quite easily and satisfactorily evaporated from an ordinary refractory metal boat. EB technology is rather expensive and complicated, and its use is not justified if the more easily controlled alternative electrical resistance heating is available. The method is of practical importance in certain cases requiring high purity films and in the absence of suitable support materials.

Electron beam guns can be classified into thermionic and plasma electron categories. In the former type the electrons are generated thermionically from heated refractory metal filaments, rods, or disks. In the latter type, the electron beams are extracted from a plasma confined in a small space.

In thermionic systems, which feature a simple work-accelerated electron gun structure,* there is a hot cathode in the form of a wire loop close to the evaporant, and the electrons converge radially on the work. The simplest is the pendant drop

---

*A work-accelerated structure is one in which the electric field is maintained between the cathode and the evaporant.

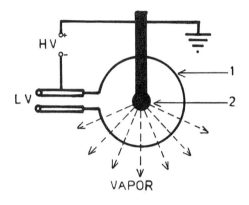

Figure 1.3 Work-accelerated electron gun structure—pendant drop configuration: 1, hot cathode; 2, pendant drop. (Data from Ref. 45.)

configuration (Figure 1.3) introduced by Holland [45]. The metal to be evaporated should be in the form of wire or rod centered within the cathode loop. The tip of the rod melts, and the evaporation takes place from the molten tip and is deposited on the substrates located below the source. Because the drop of molten metal at the tip is held by surface tension, this method is limited to metals with high surface tension and vapor pressures greater than 10^{-3} torr at their melting points. Careful control of the electric energy supplied is also necessary to avoid a temperature that too greatly exceeds the melting point.2.6

In another arrangement the cathode loop is below (Figure 1.4) and has water-cooled supports, the electron beam being focused electrostatically [46]. Yet an-

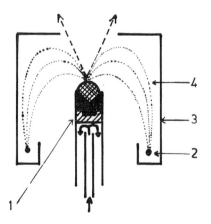

Figure 1.4 Work-accelerated electron gun structure—electrostatic focusing: 1, cold finger; 2, tungsten filament; 3, focusing electrode; 4, electron paths. (Data from Ref. 46.)

other arrangement, designed by Chopra and Randlett [47], is a demountable system, whose components can be moved relative to one another to make adjustments for focusing and beam current. The filament is shielded from the vapor by a grounded Ta shield, and this also acts to focus the electron beam electrostatically onto the evaporant placed in a water-cooled copper pedestal. Since the melting and evaporation of the material are confined to the surface, the water-cooled support pedestals pose no contamination problems. Configurations similar to this are relatively easy to implement and have been used to evaporate materials such as Si [48], Mo [49], and Ta [50].

Another class of thermionic system consists of self-accelerated electron guns that have a separate anode with an aperture through which the electron beam passes toward the work. An example [51] is shown in Figure 1.5. Here the electron beam is focused by a negatively biased filament having a conical anode and magnetic lens. Focal spots a few millimeters in diameter are used to evaporate the materials. These guns operate at higher voltages, offer more flexibility, and are most commonly used. Telefocus guns have been used successfully to evaporate refractory metals such as Nb [52], which require temperatures higher than 3000°C. The telefocus guns have adequate power density on the evaporant even if the distance between the gun and the crucible is large. The path of the electron beam is a straight line, and therefore either the substrate or the gun must be mounted off to the side, unless the

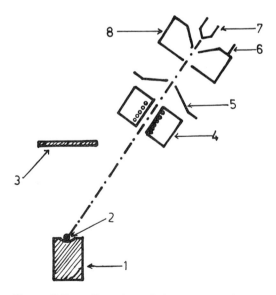

Figure 1.5 Self-accelerated electron gun—electrostatically and magnetically focused: 1, support; 2, evaporant; 3, substrate; 4, magnetic lens; 5, anode; 6, negative bias; 7, hot cathode; 8, filament housing. (Data from Ref. 51.)

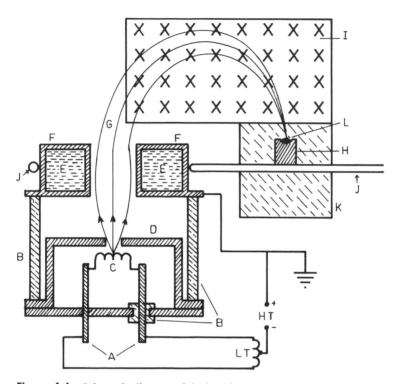

Figure 1.6 Schematic diagram of the bent-beam electron gun: A, L.T. electrodes; B, ceramic insulators; C, tungsten filament; D, shield; E, magnetic focusing coil; F, anode assembly; G, electron paths; H, copper hearth; I, extended pole pieces; J, water cooling tube; K. permanent U-magnet. (Data from Ref. 53.)

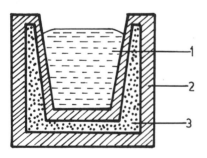

Figure 1.7 Water-cooled copper crucible: 1, evaporant; 2, copper crucible; 3, water cooling.

electron beam is bent through a transverse magnetic field [53]. In this bent beam electron gun designed by Banerjee et al. (Figure 1.6), the operating high tension (HT) voltage is 9 kV and currents up to 200 mA are used. By altering the HT voltage and the focusing current, the electron beam can be focused onto one or more supports situated between the pole pieces.

To accommodate the evaporant for electron beam evaporation, crucibles of different types are used, depending on the required level of evaporation rate. Water-cooled copper crucibles (Figure 1.7) have been found to be useful in a wide range of applications. Crucibles of this type are used for the evaporation of refractory materials such as tungsten as well as for the evaporation of highly reactive material (e.g., Ti). When high power losses are to be avoided or when the evaporation rate at a given power level has to be increased, crucible inserts (Figure 1.8) act as a heat barrier. The use of crucible inserts yields a more uniform temperature distribution over the molten pool and also a greater pool depth. The material selection depends on factors such as thermal conductivity, chemical resistivity to the hot evaporant, and high resistance to thermal shock. Ceramics based on Al_2O_3, graphite, titanium nitride, or boron nitride are used for crucible inserts.

Various electron beam evaporator devices with axial gun, magnetic focusing, and magnetic bending are now readily available commercially, to produce thin film for optical, electronic, and optoelectronic applications.

Heiblum et al. [54a], had designed and built a UHV compatible electron gun evaporator that could be incorporated into a molecular beam epitaxy (MBE) system. A schematic view of the EB gun head is shown in Figure 1.9. A circular line filament of thoriated tungsten is placed between a shield and a cage, both biased at a high negative voltage of 5–10 kV. The filament is heated by a 12 V, 30 A power supply and is floating (negative high voltage). Electron trajectories follow the elec-

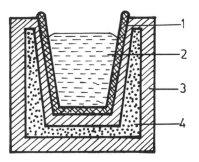

Figure 1.8 Water-cooled copper crucible with inserts: 1, crucible inserts; 2, evaporant; 3, copper crucible; 4, water cooling.

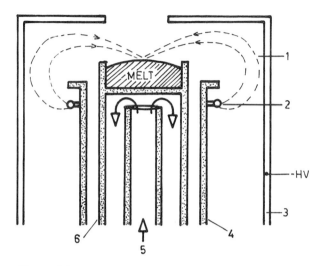

Figure 1.9 Schematic view of the EB gun head for UHV compatible evaporator in an MBE system: 1, electron trajectories; 2, filament; 3, cage; 4, shield; 5, water flow; 6, hearth. (Data from Ref. 54a.)

tric field lines established by the potential difference between the negative high voltage components and the grounded hearth. With a focusing mechanism, the relative position between the hearth and the rest of the gun head could be changed, and the investigators were able to focus the electron beam exactly on the top of the melt. In the degassing mode, with both cage and shield grounded and the filament held at 3 kV, the electrons, accelerated toward the shield and cage, heat and outgas them thoroughly. All the head parts were made from molybdenum, tantalum, and tungsten, which may have reduced considerably the outgassing due to heating. With this setup, Heiblum and his group were able to eliminate most of the problems arising during electron gun evaporation in an MBE system and to evaporate Mo and W onto GaAs at pressure of 2×10^{-9} torr and less.

Electron beam evaporation has been used for the preparation of thin films of a number of materials: MgF_2 [54b], Ga_2Te_3 [55], Nd_2O_3 [56], $Cd_{1-x}Zn_xS$ [57], Si [58], $CuInSe_2$ [59], InAs [60], $Co-Al_2O_3$ cermet [61], $Ni-MgF_2$ cermet [62], TiC and NbC [63a], V [63b], SnO_2 [63c], TiO_2 [64a], indium-tin oxide (ITO) [64b, 64c], Be [65a], Y and Si [65b], and $ZrO_2-Sc_2O_3$ [66a]. Electron gun evaporation has been used also to prepare HTS thin films [66b-e].

In the thermionic emission type of guns discussed, the chamber pressure must be limited to pressures of 10^{-4} torr or less for reasonable beam control and cathode element life. But this limitation of 10^{-4} torr maximum operating pressure does not occur with the plasma electron beam source [67], which utilizes ionizable gas at a pressure of 10^{-3} torr or higher. The plasma electron beam can be used practically for the same purposes as thermally emitted electron beams. There are two types of plasma electron beam gun: the cold hollow cathode guns and the hot hollow cathode (HHC) guns. HHC has the advantage of being a low voltage, high current arc generating device and, when used to produce metallurgical coatings, the typical operating range has an arc current of 50–200 A. The arc sends a beam of electrons to strike the evaporant material, and the beam is used to simultaneously evaporate and to ionize the material. Therefore this type of coating can be considered to be a variant of the ion plating, as discussed in Chapter 3.

1.2.4 Laser Evaporation

In laser evaporation, lasers are used as the thermal source to vaporize the evaporant materials, and preparation of thin films by laser evaporation is a high vacuum technique, where the source of power for evaporation is kept outside the vacuum system. The vaporized material is deposited onto substrates placed in front of the source material inside the vacuum chamber. This technique offers several advantages.

1. Lasers are clean and introduce minimal contamination from the heat source.
2. Film contamination from the support material is reduced because of the surface evaporation characteristics of the beam.
3. With the high power densities obtained by focusing the laser beams, high melting point materials can be vaporized at high deposition rates.
4. Because of the small beam divergence, the laser and the associated equipment could be kept far away, an attractive feature in radioactive areas.
5. Simultaneous or sequential multisource evaporation can be done easily by directing the laser beam with external mirrors.

It was Smith and Turner [68] who made a preliminary study and showed that many materials can be vaporized in a vacuum by a directed laser beam as the evaporation power source. They used a ruby laser external to the vacuum chamber and a lens focused the radiation from the ruby rod through a window in the bell jar onto the surface of the sample to be evaporated. Most of the films were evaporated from powdered materials placed in small inclined crucibles in the bell jar and deposited onto substrates placed 20 to 50 mm above the crucibles. Lateral motion of the lens allowed the focal spot to fall where desired on the surface of the material. The study cited reports several materials producing optically satisfactory films, including Sb_2S_3, $ZnTe$, MoO_3, $PbTe$, and Ge.

Although some studies using laser as a thermal source have been carried out since this initial report, special attention has been paid only in recent years to the preparation of thin films by laser evaporation.

Fujimori et al. [69] prepared carbon films using continuous wave (CW) CO_2 laser (80 W) as a heat source. The laser beam passed through a ZnSe window into the chamber and was reflected and focused by a Be–Cu concave mirror onto the

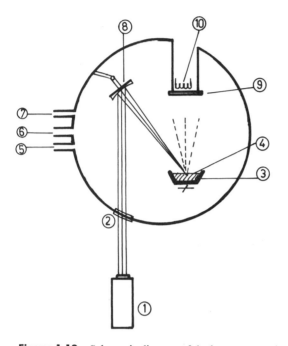

Figure 1.10 Schematic diagram of the laser evaporation system: 1, CO_2 laser; 2, ZnSe window; 3, Mo boat; 4, source material; 5, to pump; 6, to vacuum gages; 7, to mass filter; 8, concave mirror; 9, substrate; 10, infrared heater. (Data from Ref. 69.)

source material in a molybdenum boat. Graphite and diamond, both in powder form, were used as source material. Unlike carbon, these powders are easy to evaporate because of their low heat capacity. The laser beam was scanned across the source by the rotation of the concave mirror as well as by the linear drive of the molybdenum boat. The schematic of their setup is shown in Figure 1.10. With graphite powder as source material, these investigators observed that the films showed properties similar to those of graphite.

A new ceramic coating technique using a high power CO_2 laser as the heat source was developed by Mineta et al. [70], and films of Al_2O_3, Si_3N_4, and certain other ceramic materials were deposited on Mo plates (Figure 1.11). The CW CO_2 laser beam was guided into the vacuum chamber through a ZnSe lens and a KCl window and focused onto the periphery of the ring target, which had been preheated to 800°C. The films were deposited onto the substrate and kept 25–75 mm from the target surface, which had been preheated up to 300–600°C. The experiments revealed that hard and homogeneous films with high adhesion could be prepared with this method. Also the composition of the film did not differ much from that of the base material. This method would be particularly useful for preparing

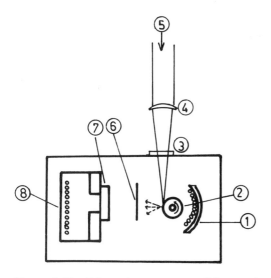

Figure 1.11 Schematic arrangement of the experimental setup for the ceramic coating technique using a CO_2 laser: 1, electric heater; 2, ceramic ring; 3, KCl window; 4, ZnSe lens; 5, laser beam; 6, shutter; 7, substrate; 8, electric heater. (Data from Ref. 70.)

thin films of various kinds of hard, high melting point, low vapor pressure materials. A pulsed Nd:YAG laser with its beam focused on a silicon plate was used by Hanabusa et al. [71] to deposit amorphous silicon films.

Pulsed laser evaporation can cause rapid heating and cooling of the source material with very high peak temperature, and instant evaporation occurs over small areas of the target. Power is delivered in the form of high power pulses creating flash evaporation conditions, and this leads to the important advantage of congruent evaporation of compound materials. There will be little or no fractionation of its constituents even if they have widely different vapor pressures, and this technique is suitable for the deposition of thin films of wide range of compounds and alloys. Also the original purity of the source material is maintained, eliminating crucible contamination because the source pellet/target becomes its own crucible. The average temperature of the irradiated targets and substrates is kept low, and therefore the deposition proceeds at low temperatures. During the evaporation of solid targets, the interaction of a pulsed laser with a solid target can produce high energy particle fluxes (electrons, ions, and neutral species), the energies of the particles depending on the material and laser power. These cause surface cleaning due to ion etching and also increase the number of nucleation sites for deposition will be increased, enhancing the process of epitaxial growth.

Yang and Cheung [72] have reported using the pulsed laser evaporation (PLE) technique for the preparation of SnO_2 film on GaAs and glass substrates using a

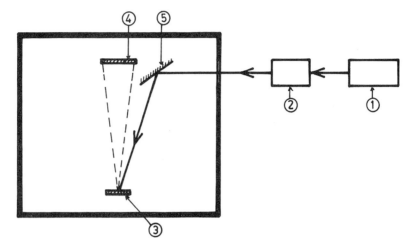

Figure 1.12   Schematic diagram of the pulsed laser evaporation system: 1, pulsed YAG laser; 2, scanner; 3, source; 4, substrate; 5, mirror. (Data from Ref. 72.)

high power pulsed laser. A train of pulses from an acousto-optically Q-switched Nd:YAG laser was scanned by a pair of galvanometric mirrors, directed into the vacuum chamber, and finally focused onto the surface of the target (SnO_2 pellets). The power densities of the laser pulses were in the 10^7 W/cm^2 range, pulsed laser frequency was 2000 Hz, pulse width was 200 ns, and the scan rates were in the range of 1–10 cm/s. Base pressure in the chamber was lower than 5×10^{-7} torr. The substrates were placed 2.5 in. above the surface. A schematic diagram of this pulsed laser setup is shown in Figure 1.12. Yang and Cheung had characterized these thin films of SnO_2 and had shown that the deposition rates were limited by the evaporation mechanism, which indicated that a 1.06 μm laser was not suited for SnO_2 evaporation. Also they observed congruent evaporation, but the films obtained were mixtures of SnO and SnO_2. Again with the setup above, "splashing" of the source material during evaporation could not be eliminated, and that also accounted for the poor quality of the films obtained by these investigators.

   With the setup shown in Figure 1.13, Sankur and Cheung [73] prepared highly oriented crystalline and transparent ZnO films on a variety of substrates by CO_2 laser evaporation of ZnO. The CO_2 laser was used in the pulsed mode, and the average laser power was varied by varying the energy per pulse or the pulse repetition rate. Pellets 1.25 cm in diameter, pressed out of 99.999% pure powder or hot sintered ZnO, were used as the source material. The average laser power was 10 W, and the power density on the pellet was in the 10^4 W/cm^2 range. The pressure was 10^{-7}–10^{-6} torr during evaporation. The films obtained were found to be nearly stoichiometric. The investigators' studies on the structural, optical, electrical, and acoustic properties had shown that this deposition technique could be used to pro-

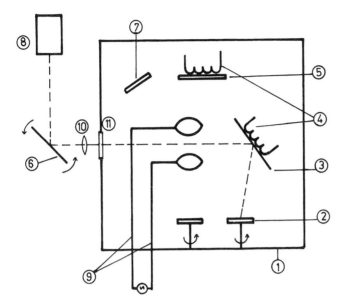

Figure 1.13 Experimental setup for laser deposition of ZnO films: 1, vacuum chamber; 2, sources and rotary pedestal; 3, mirror; 4, heaters; 5, substrate; 6, scanning mirror; 7, thickness monitor; 8, CO_2 laser; 9, glow discharge electrodes; 10, lens; 11, window. (Data from Ref. 73.)

duce device quality ZnO films with good reproducibility. Film grown at high background O_2 pressure (10^{-2} torr) or in an ac glow discharge at 10^{-2} torr were of poor crystallinity.

The PLE technique has been used to prepare thin films of Cd_3As_2 [74–77], as shown schematically in Figure 1.14. An acoustic-optically Q-switched Nd:YAG laser, having a wavelength λ of 1.06 μm, was used as the heat source. The laser pulse duration was 1.7×10^{-7} s, with a pulse repetition rate of 1 kHz. The vacuum chamber with a quartz window mounted on a computer-controlled X–Y stage enabled the target (Cd_3As_2 single crystal) to be moved relative to the position of the laser beam. The target could be rotated during vaporization to obtain uniform removal of target material. It was shown that the films grown onto room temperature substrates were amorphous [74] or polycrystalline [75], depending on the background pressure. The electrical quality of polycrystalline films improved markedly as the substrate temperature was increased from 295 to 433K, and at these deposition temperatures, the deposited films were stoichiometric [76]. The films deposited at about 430 K were highly oriented, with electrical characteristics approaching those of the bulk material; these conditions could be used for growing epitaxial films of Cd_3As_2 with electrical characteristics comparable to the bulk ma-

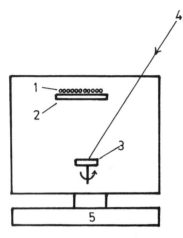

Figure 1.14 Schematic diagram of a PLE system: 1, heater; 2, substrate; 3, target; 4, Nd:YAG laser beam; 5, X–Y stage. (Data from Ref. 74.)

terial if grown on single crystals. No evidence of the material being splashed from the target was observed [77].

Baleva et al. [78,79], using the setup illustrated in Figure 1.15, prepared thin films of $Pb_{1-x}Cd_xSe$ films (x = 0, 0.02, 0.05) with a free generating Nd:glass laser having a pulse duration of 400 μs. The time interval between the pulses was 6 seconds. A rotating window and a plate with a hole allowed about 500 pulses to pass without significant decrease in laser energy and to fall on the target. The target holder could be rotated so that a constant material content could be evaporated from pulse to pulse. The pressure in the vacuum chamber was about 10^{-6} torr. The targets were in the form of pressed tablets (4 mm radius, 2 mm thick) made from the

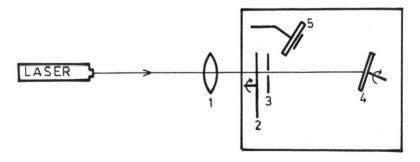

Figure 1.15 Experimental setup for laser deposition: 1, optical system; 2, rotating round window; 3, protecting plate; 4, target holder; 5, substrate holder. (Data from Ref. 78.)

alloy formed by melting stoichiometric mixture of 99.999% pure lead, cadmium, and selenium. Potassium chloride and BaF_2 with respective orientations of [100] and [111] were used as the substrates. Baleva et al. have studied the structural [78], electrical, and optical properties [79] of the films and have established that the following are most suitable technological conditions for the deposition of $Pb_{1-x}Cd_xSe$ films with the composition reproducing that of the source material:

Laser power density, 10^5 W/cm^2
Distance between substrate and target, 72.5 cm
Substrate temperature, 100-400°C

Ogale et al. [80a] have reported the deposition of iron oxide films on alumina substrates by pulsed ruby laser (694 nm, 30 ns) from a bulk α-Fe_2O_3 pellet. Iron oxide pellets synthesized from high purity (99.999%) γ-Fe_2O_3 powder served as the target, mounted in a vacuum chamber giving an ultimate pressure of 2×10^{-7} torr. The target was irradiated at an angle of 45°, and the evaporated material was deposited on polycrystalline Al_2O_3 and vitreous carbon substrates mounted in a heater assembly kept 4 cm from the pellet. The substrates were heated to 200°C during deposition to enhance film adhesion. Using a quartz lens, an energy density of about 10-12 J/cm^2 was obtained in the target, and the irradiation was carried out at a repetition rate of 3 pulses per minute. The target holder was rotated to change the laser irradiation position, to minimize texture effect. The films obtained were characterized using conversion electron Mossbauer spectroscopy, Rutherford backscattering spectroscopy (RBS), and scanning electron microscopy (SEM). The stoichiometry of the deposited films could be varied between FeO and Fe_3O_4 by controlling the oxygen partial pressure during deposition over a range from 5×10^{-7} to 10^{-4} torr. They also showed that by suitable thermal treatment, Fe_3O_4 could convert into α-Fe_2O_3.

Auciello et al. [80b] have prepared TiN films at room temperature on Si (100) substrates using an ArF (193 nm) laser. The TiN films obtained had the characteristic golden color, as well as good adhesion. In the experiment cited, the laser beam was directed at a stoichiometric bulk TiN target positioned at 45° to the beam direction, in an ultra high vacuum chamber. The laser-ablated material was deposited on (100) Si and C substrates at room temperature. The substrates were placed around a 5 mm hole on a holder about 3 cm from the target and parallel to it. A quartz crystal resonator, kept behind the substrate holder aligned with the hole (the evaporated material depositing simultaneously on the substrate and the crystal resonator), was used for in situ measurements of film thickness. Rutherford backscattering analysis without substrate interference was done in TiN films deposited on carbon substrates. A quadrupole mass spectrometer was located in a differentially pumped chamber and directed at the target for monitoring the species evolved from target and also the background gas atoms and molecules in the target chamber. By judicious predeposition conditioning of the target chamber and control of its atmosphere (e.g., nitrogen filling), the authors could reduce the oxygen content in the films to low levels. RBS and Auger electron spectroscopy (AES) studies showed good composition uniformity of the films. Studies by transmission electron microscopy (TEM) and X-ray diffraction (XRD) analysis yielded a lattice constant of 0.423 ± 0.003 nm.

Scheibe et al. [80c] have used an XeCl excimer (wavelength, 308 nm; pulse duration, 20–30 ns; energy per pulse, 100 mJ; pulse repetition frequency, 1–20 Hz) laser source with uniform flux distribution across the spot for the vapor deposition of thin films of gold and carbon. The schematic deposition of their experimental setup is shown in Figure 1.16. The lens and the diaphragm in front of the laser beam were selected to cut off the marginal beams from the laser source. Another lens projects the diaphragm on the target plane; and the resulting spot has a well-defined boundary and a very uniform power density inside. The target was placed at an angle of 45° with respect to the incident laser beam, in the vacuum chamber with a residual pressure of 0.01 Pa. A high frequency photocell is used to control the pulse shape and energy.

The target materials were of high purity. Polycrystalline gold and three kinds of carbon—highly oriented pyrolytic graphite, polycrystalline graphite, and amorphous carbon—were used as targets. To avoid the formation of deep craters, Si (111) single-crystal wafers, freshly cleaved KCl, and NaCl and ZnSe crystals were used as substrates. The substrate temperatures varied from 20 to 300°C.

The structural and IR optical properties of the films were investigated. The in-

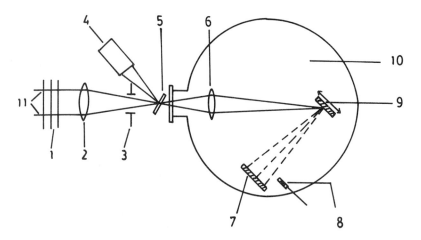

Figure 1.16 Schematic diagram of the experimental setup for vapor deposition using XeCl excimer: 1, glass attenuators; 2, lens; 3, diaphragm; 4, photocell; 5, beam splitter; 6, lens; 7, substrate; 8, probe; 9, target; 10, vacuum chamber; 11, laser. (Data from Ref. 80c.)

vestigators observed that a uniform power density distribution across the spot, without the hot spots, reduced particle emission from the target and produced films of better quality. Hard amorphous transparent carbon films prepared by this technique are suitable for IR optical elements such as antireflection coatings on ZnSe.

Pulsed laser evaporation has been used for the epitaxial growth of thin films. Dubowski [80d] has reviewed this topic on PLE and epitaxy of thin semiconductor films, discussing different aspects such as laser-induced damage and vaporization of solid targets, the nature and energy distribution of ejected particles, and deposition rate.

Pulsed laser evaporation for the epitaxial growth of $Cd_{1-x}Mn_xTe$ has been reported recently [80e]. The growth was carried out in a high vacuum system with a base pressure of less than 2×10^{-9} torr. Both XeCl and Nd:YAG lasers were used for the simultaneous evaporation of high purity targets ($Cd_{1-x}Mn_xTe$ with x = 0.073 and 0.56), and a CdXeCl excimer laser (x = 0.308 μm, triggering rate $\leq$ 80 Hz) vaporized the $Cd_{1-x}Mn_x$ targets; the repetition rate and pulse peak power of the Nd:YAG laser were adjusted to get a constant flux of Cd in the range of 10^{15}–10^{16} atoms/cm^2·s. The layers were grown on high quality (111) GaAs substrates kept at a fixed temperature in the range of 210–310°C. The quality of the layers were evaluated using reflected high energy electron diffraction spectroscopy (RHEEDS), scanning electron microscopy (SEM), electron-dispersive analytical X-ray (EDAX) spectroscopy, and low temperature PL and the films were of molecular beam epitaxy quality.

Yoshimoto et al. [81a] have reported recently the laser deposition of CeO_2 thin films on Si substrates using a specially designed UHV (ultra high vacuum) system. The UHV chamber could be evacuated to a base pressure of 8×10^{-10} torr. A CeO_2 disk (10 mm diameter, 2 mm thick) was used as the target, and an ArF excimer laser beam (193 nm) was focused on the target through a synthesized quartz window. The pressure in the chamber during deposition was 0.4–1 $\times 10^{-8}$ torr, and CeO_2 film deposition was carried out at temperatures between 600 and 800°C on Si (001), (111), and (110) substrates. In some of the experiments, the pressure was maintained at 1×10^{-7} torr (O_2 flushed into the chamber).

The deposited films were evaluated for crystallinity by in situ RHEEDS and ex situ X-ray diffraction. Some of the films were characterized by X-ray photoelectron spectrometer (XPS) analyzer and showed tetravalence of Ce in the as-grown film. The investigators observed that the orientation of the deposited CeO_2 films was strongly dependent on the surface state of the substrates. On a clean surface of Si (111), CeO_2 (111) films were found to grow epitaxially at 600 and 700°C. Koinuma et al. [81b] reported the layer-by-layer epitaxial growth of CeO_2 on Si (111) as verified by in situ RHEEDS studies during film deposition by UHV laser evaporation and showed that the use of $SrTiO_3$ as an interlayer on Si (001) was effective in facilitating the epitaxial growth of CeO_2 (001).

Other reports of epitaxial growth of thin films by laser evaporation have also been published very recently; materials include diamondlike carbon films [81c] and ferroelectric bismuth titanate [81d]. Laser evaporation has been used to prepare thin films of CdTe, Cd, and InSb [81e]; PbTe and doped PbTe [82a], polymers [82b]; lead chalcogenides [82c]; ceramic coatings from sintered Mg_2SiO_4 [82d]; Se [83]; BN [83b]; iron oxide and zinc ferrite [83c]; $BaTiO_3$ [83d], and SnO_x [83e].

Since the discovery of high temperature superconducting oxides, the preparation and characterization of HTS oxide thin films for application in new and fascinating areas of superconducting electronic devices has created a lot of interest. Moreover, the advantages of pulsed laser evaporation as a technique for the preparation of thin films have made this method for the preparation of high T_c superconducting ceramic thin films very popular, and it is now very widely used. A great many reports have been published on the pulsed laser evaporation of HTS materials [84]. Most of the work has been done on $YBa_2Cu_3O_{7-x}$ (YBCO), and a few reports of interest published very recently on YBCO films are discussed below.

Serbezov et al. [85a] reported in 1990 the preparation of $YBa_2Cu_3O_{7-x}$ thin film by nitrogen laser evaporation; the substrates were heated by a CW single-mode CO_2 laser, and the annealing of the deposited films was performed by the same CO_2 laser in an oxygen atmosphere. The schematic diagram of their setup is shown in Figure 1.17.

The evaporation part consists of an optical table with optical holders and a quartz lens for focusing the N_2 laser radiation onto the $YBa_2Cu_3O_{7-x}$ stoichiometric target. In addition, the apparatus includes a high speed mini motor with target, planar stainless steel mirrors to direct the CO_2 laser beam for substrate heating and annealing, and a cylindrical KCl lens to focus the CO_2 laser onto the film for local annealing. The temperature of the substrate is selected by adjusting the power of the CO_2 laser, as measured by a laser power meter.

The experimental data reported are as follows.

N_2 laser	337.1 nm
	5–10 mJ energy per pulse
	6 ns pulse duration
	Pulse repetition frequency, 5–30 Hz
CO_2 laser	Single-mode CW CO_2
	Wavelength, 10.6 μm
	Output power, 40 W
Substrates	Poly-Al_2O_3, sapphire, $SrTiO_3$, and monocrystal Si
Heating temperatures	550–750°C for Al_2O_3, sapphire, and $SrTiO_3$
Annealing temperatures	600–700°C for dielectric substrates
Time	10–15 minutes
O_2 pressure	1 atm
For Si	500–1000°C; time, several seconds to 10 minutes; pressure, 10^{-2} torr

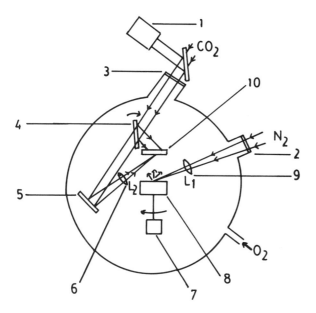

Figure 1.17 Schematic diagram of the setup for the preparation of YBa$_2$Cu$_3$O$_{7-x}$ thin film by nitrogen laser evaporation: 1, laser power meter; 2, quartz window; 3, KCl window; 4 and 5, planar stainless steel mirrors; 6, cylindrical KCl lens; 7, mini motor; 8, target; 9, spherical quartz lens; 10, substrate. (Data from Ref. 85a.)

The films were deposited in a vacuum of 10^{-5} torr. The rate of film growth was 1–2 Å/s. After annealing, the films remained in the O$_2$ atmosphere at 1 atmosphere pressure until T_s reached room temperature. The films were homogeneous in thickness and density over an area of 1 cm^2. Local superconducting planar structures were made by focusing the CO$_2$ laser onto the film surface by the cylindrical lens.

The salient features of interest in regard to this technique are as follows.

1. YBCO films of high quality were deposited on dielectric and Si substrates without buffer films.
2. The deposition was carried at relatively low heating and annealing temperatures.
3. An N$_2$ laser, the operation of which is technically simpler, was used as an evaporating source.
4. The investigators were able to obtain local regions of superconducting properties without destroying the film by using a CW CO$_2$ laser for local annealing instead of a conventional heater.

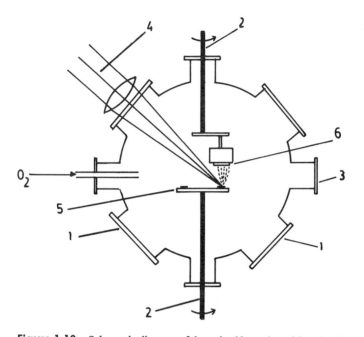

Figure 1.18 Schematic diagram of the pulsed laser deposition chamber: 1, quick-access door; 2, rotary/push/pull; 3, view port; 4, laser; 5, multiple target holder; 6, heated substrate holder. (Data from Ref. 85b.)

Substrates for the deposition of high temperature superconducting thin films must satisfy several criteria, including very low reactivity at the deposition temperature and the lowest possible dielectric constant. Successful substrate materials for the deposition of YBCO include ZrO_2, MgO, and $SrTiO_3$; the best films have been grown on single crystalline substrates, particularly $SrTiO_3$. For conductor applications, however, metal substrates and metal buffer layers may be required. Indeed, Russo et al. [85b] have reported the fabrication of metal buffer layers and superconducting Y–Ba–Cu–O thin films on metallic substrates using a versatile pulsed laser deposition chamber. Figure 1.18 is a schematic diagram of their deposition chamber. The chamber has two quick-access doors that allow rapid replacement of the substrates and targets. The target holder can accommodate four separate materials simultaneously and is capable of rotating each material to the laser beam. The individual targets are not rotated during deposition. The rotary-push–pull arrangement provides flexibility in adjusting the spacing between the target and substrate. Russo et al. deposited silver and platinum buffer layers on stainless steel, platinum, and several single-crystal substrates (MgO, $SrTiO_3$).

The laser sources used and their details are as follows.

Laser sources	XeCl (308 nm) excimer laser: 1 J per pulse; pulse repetition rate, 1 Hz KrF (248 nm) excimer laser: 650 mJ per pulse; pulse repetition rate of 5 Hz
Energy density on targets	3 J/cm^2 for YBCO 1–3 J/cm^2 for metal buffer layers
Pulse duration	25–30 ns for each laser

Russo and coworkers found that Ag buffer layers improved the resistive transition behavior for superconducting films on stainless steel and platinum. As-deposited YBCO films with T_c ($R = 0$) at 84 K were obtained on stainless steel with in situ laser-deposited Ag buffer layers.

Extending the work onto substrates used in semiconductor electronics, several workers have prepared YBCO films on Si, the primary material for semiconductor electronics. The main difficulty in using silicon as a substrate for HTS films is the redistribution of elements due to chemical reaction and/or the interdiffusional motions between the film and substrate that occur during annealing and can be detected by element depth profiling [85c]. High quality HTS films can be grown on epitaxial buffer layers deposited on silicon. A recent review on high temperature superconducting films on silicon has been published by Mogro-Campero [85d].

Hwang et al. [85e] have reported the preparation of in situ superconducting YBa$_2$Cu$_3$O$_{7-x}$ films by PLE with T_c ($R = 0$) up to 87 K and critical current density up to 6×10^4 A/cm^2 at 77 K on Si substrates with Mg$_2$Al$_2$O$_4$ and BaTiO$_3$ double buffer layers. The buffer layers were prepared [85f] by two different techniques: the first buffer layer (Mg$_2$Al$_2$O$_4$) by chemical vapor deposition (CVD) at 980°C, followed by the second buffer layer (BaTiO$_3$) by rf magnetron sputtering at 500°C. During deposition of the YBa$_2$Cu$_3$O$_{7-x}$ films, the substrate holder was kept at 650°C in 100 mtorr N$_2$O. After deposition, the sample in the chamber was cooled to room temperature in 200 torr of oxygen. Studies by TEM revealed that the Mg$_2$Al$_2$O$_4$ layer was heavily faulted. The subsequent layer (BaTiO$_3$) stopped most of the faults and provided a template for the HTS film growth. The superconducting properties of the film in Si substrates with the double buffer layers were much better than those of the films deposited directly on Si and on Si with ZrO$_2$ buffer layer. Hwang et al. [85g] observed that the films are still not as good as those of the films deposited directly on SrTiO$_3$ [85g], or LaGaO$_3$ and LaAlO$_3$ [85h].) The microstructure of the films was similar to that deposited directly on SrTiO$_3$ and exhibited a homogeneous, heavily faulted single-crystal-like structure free from secondary phase and grain boundaries.

Pulsed laser evaporation has been used to prepare high quality superconducting YBCO films at high deposition rates (145 Å/s) [86a]. An excimer laser (308 nm)

beam was made to strike a rotating target (1 in. diameter) at an angle of 45° such
that the evaporated flux fell on a 1 mm (100) $SrTiO_3$ substrate. The substrate, cemented by high quality silver paste on a heated holder, was rotated at 0.5 rpm during the deposition. An oxygen pressure of 150 mtorr was maintained during the
deposition. The film deposition rate could be varied from 1 to 145 Å/s by changing
the laser repetition rate from 1 to 100 Hz. After deposition the films were cooled to
200°C in 250 mtorr oxygen within 20 minutes. All the films were superconducting
without any further heat treatment and exhibited zero resistance at 90 K. The
method could be used to deposit films over large areas.

Another novel method for the preparation of YBCO thin films by 193 nm ArF
laser on (100) $SrTiO_3$ substrates, held at 700°C, has been reported [86b]. A cylindrical $YBa_2Cu_3O_{7-x}$ target was rotated along its axis so that a fresh surface was
always exposed to the laser beam. Target-to-substrate distance was 50 mm. Another salient feature of the system was that the substrates were heated by radiation
from a tantalum heater mounted in a quartz glass housing, isolated from the growth
chamber and separately pumped. This setup prevented oxidation of the heater even
at high oxygen partial pressure and enabled the deposition to proceed up to a partial
pressure of 5 torr. The deposition was carried out at different oxygen partial pressures ranging from 0.01 mtorr to 5 torr. After deposition, oxygen was fed into the
chamber until a pressure of 300 torr, before the films were slowly cooled down.
Smooth and highly oriented films with T_c above 80 K and critical current densities
of 1.0×10^5 A/cm^2 at 77 K without magnetic field were obtained at 10–50 mtorr of
oxygen.

Ohkubo et al. [86c] have reported recently the preparation of (001) epitaxially
grown $YBa_2Cu_3O_x$ thin films ($x = 6$-7) by pulsed laser deposition without postannealing. Pulsed laser beams from an ArF excimer laser (wavelength, 193 nm; energy per pulse, 3–4 J/cm^2; repetition rate, 10 Hz; beam size, $\approx 0.5 \times 2$ mm^2) was
used to irradiate the $YBa_2Cu_3O_x$ pellet prepared by a solid state reaction with
Y_2O_3, $BaCO_3$ and CuO. The pellet had a T_c of about 90 K. The laser beams were
scanned so that the beams did not irradiate a certain area for more than 1 minute.
The substrate was $SrTiO_3$ (100), about 7×15 mm^2 in size, and the distance between the substrate and the target was approximately 3 cm. During deposition, oxygen gas was introduced into the chamber near the target. The depositions were
performed at an oxygen pressure of 13 Pa and a substrate temperature of 730 ±
10°C.

After the deposition, the oxygen pressure was adjusted to values between 0.013
and 27,000 Pa within 1 minute. As soon as the oxygen pressure was adjusted to
each value, the substrate heater (infrared lamp heated from the opposite side of the
substrate) was turned off.

Ohkubo et al. have observed that the value of x in as-deposited epitaxial films
depends strongly on the oxygen pressure during the rapid cooling following the
laser deposition: x increased from 6 to 7 as oxygen pressure increased from 0.013 to

27,000 Pa. They concluded that this change of x with oxygen pressure is caused by the oxygen infusion during cooling.

1.2.5 Arc Evaporation

Arc evaporation to prepare thin films was first tried in a conventional vacuum evaporator by Lucas et al. [87] to obtain thin refractory metal films. Electrodes of the metal to be evaporated were mounted on insulated supports in a vacuum system evacuated to a pressure of 10^{-5} torr; one of the rods was rotatable and the other was fixed. Using a standard welding generator, the voltage was applied between the electrodes, and the rotatable electrode was brought into contact with the fixed rod, held there until a hot spot appeared, and then moved away, thus drawing an arc. This resulted in the rapid deposition of the film of the electrode metals on the substrate placed close to the electrodes. Films of niobium, tantalum, vanadium, and stainless steel were obtained using this setup.

Kikuchi et al. [88] obtained thin films of various metals (W, Ta, Mo, Zr, Ti, Fe, Ni, Cu, Al, Cd, Zn, and Sn) on rock salt and carbon substrates by arc evaporation. Other reports of the preparation of thin film using vacuum arcs have also been published [89,90], but in the absence of continuous smooth arc discharges, the results are not easily reproducible.

Igarashi and Kanayama [91] developed a vacuum arc method for the evaporation of refractory metals. Their arrangement consisted of two electrodes in the form of rods facing each other and aligned along the axis of a helical filament (Figure

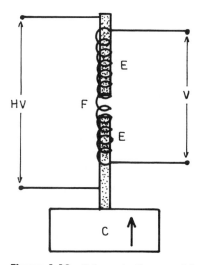

Figure 1.19 Schematic diagram of the electrode–filament arrangement for the vacuum arc method: C, gap control device; E, electrodes; F, helical filament. (Data from Ref. 91.)

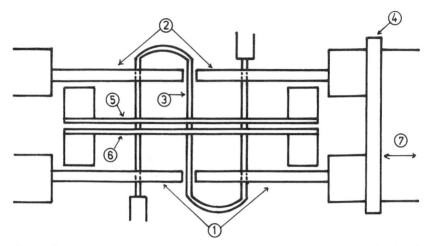

Figure 1.20 Schematic diagram of the electrodes–filament arrangement for the ultra high vacuum arc method: 1, Nb rods; 2, Ta rods; 3, tungsten filament; 4, insulator; 5, Ta shielding plate; 6, Nb shielding plate; 7, gap control system. (Data from Ref. 93.)

1.19). One of the electrodes could be moved linearly to adjust the interelectrode gap by a gear system driven from outside the vacuum chamber by a servo motor. The electrode coil system was set in the vacuum chamber. The arc-type discharge is easily started by applying between the electrodes an input voltage of only about 1.5–2.5 kV, while simultaneously bombarding the electrodes with the electrons emitted from the filament. This achieved a steady state discharge of long duration, and the discharge power caused the electrodes evaporate. Thus thin films of conductive refractory materials (W, Ta, C) were prepared. The gas supply that sustains the plasma state came from the melting surface of the two electrodes. Igarashi and Kanayama extended this technique to various other refractory metals (Mo, Nb, Ti, and Ni). The same investigators also succeeded in producing tungsten single-crystal films (500–1000 Å) using ultra high vacuum [92]. In a later paper, [93], they reported the preparation of high quality, single-crystal Nb and Ta films by the ultra high vacuum arc method. Using the setup shown in Figure 1.20, they were able to deposit Nb and Ta films alternately without breaking the vacuum. Rods of Nb and Ta 3 mm in diameter were placed facing each other for use as electrodes. Beneath the rods, 0.7 mm diameter wire served as the electron source. An alternating potential was applied between the rods, and then a current of 15 A was passed through the tungsten wire. This resulted in an electron bombardment over both the electrodes, and a discharge between the electrodes was attained smoothly. After the start, an autosustained continuous and stable discharge was achieved between the rods. The substrates were placed above the gap between the rods and were heated if necessary by a tantalum heater kept above the substrates.

1.2.6 Radiofrequency Heating

Radiofrequency heating has been used for vacuum deposition by several investigators [94–96]. By suitable arrangement of rf coils, levitation and evaporation can be achieved, thereby eliminating the possibility of contamination of the film by the support crucible [94]. The evaporant can also be contained in a crucible, which is surrounded by the rf coils. Thompson and Libsch [97] used boron nitride crucibles for evaporation of aluminum by induction heating. Ames et al. [98] prepared Al films using a specially designed boronitride–titanium diboride crucible to avoid the well-known problem of molten aluminum migrating upward out of the crucible. The upper portion of the crucible has reduced thickness so that the coupling in this region was strong enough to completely vaporize the migrating aluminum layer. The lower portion of the crucible was thick enough to permit adequate coupling to evaporate the metal and prevent excessive coupling, to minimize turbulence and spattering. Preparation of thin films by rf heating is limited because of the coupling needed between the coil and the evaporant and also the difficulty in positioning the coil and the samples properly in a vacuum system for effective coupling. Also the method requires relatively expensive and bulky rf heating equipment. Again, the evaporation rate is difficult to control. The method is not commonly used now as a technique for the preparation of thin films.

REFERENCES

1. M. Knudsen, *Ann. Phys.* 47:697 (1915).
2. I. Langmuir, *Physik,* 2, 14:1273 (1913).
3. L. Holland, *Vacuum Deposition of Thin Films,* Chapman and Hall, London, 1961.
4. L. I. Maissel and R. Glang, *Handbook of Thin Film Technology,* McGraw-Hill, New York, 1970, Ch. 1.
5. K. L. Chopra, *Thin Film Phenomena,* McGraw-Hill, New York, 1969, p. 15.
6. C. E. Drumheller, *Transactions, 7th American Vacuum Society Symposium,* Pergamon Press, Oxford, 1960, p. 306.
7. W. G. Vergara, H. M. Greenhouse, and N. C. Nicholas, *Rev. Sci. Instrum.,* 34:520 (1963).
8. L. I. Maissel and R. Glang, *Handbook of Thin Film Technology,* McGraw-Hill, New York, 1970, p. 1.43.
9. (a) M. N. Mahadasi and M. S. Al Robace, *J. Sol. Energy Res. (Iraq),* 3:1 (1986). (b) B. P. Rai, *Phys. Stat. Sol. (a),* 99:Pk 35 (1987).
10. (a) J. George, B. Pradeep, and K. S. Joseph, *Phys. Stat. Sol. (a),* 100:513 (1987). (b) Siham Mahmond, *J. Mater. Sci.,* 22: 3693 (1987).
11. (a) A. Kikuchi, S. Baba, and A. Kinbera, *Thin Solid Films,* 164:153 (1988). (b) K. Rajanna and S. Mohan, *Thin Solid Films,* 172:45 (1989).
12. (a) F. Volklein, *Thin Solid Films,* 191:1 (1990). (b) D. F. Bezuidenhont and R. Pretorius, *Thin Solid Films,* 139:121 (1986).

13. (a) C. Kaito and Y. Saito, *J. Cryst. Growth*, 79:403 (1986). (b) P. Singh and B. Baishya, *Thin Solid Films*, 148:203 (1987).
14. (a) M. A. Jayaraj and C. P. G. Vallabhan, *Thin Solid Films*, 177:59 (1989). (b) F. Lopez and E. Bernabeu, *Thin Solid Films*, 191:13 (1990).
15. J. D. Targove and A. R. Murphy, *Thin Solid Films*, 191:47 (1990).
16. (a) K. Suzuki, Y. Ema, and T. Hayashi, *J. Appl. Phys.*, 60:4215 (1986). (b) M. Isai, T. Kukunaka, and M. Ohshita, *J. Mater. Res. (USA)*, 1:547 (1986).
17. (a) S. Chaudhuri, A. Mondal, and A. K. Pal, *J. Mater. Sci. Lett.*, 6:366 (1987). (b) V. Damodaradas, N. Soundararajan, and M. M. Pattabi, *J. Mater. Sci.*, 22:3522 (1987).
18. (a) R. D. Gould and C. J. Bowler, *Thin Solid Films*, 164:281 (1988). (b) V. J. Rao, D. V. R. Raju, and P. B. Kadam, *Thin Solid Films*, 176:207 (1989).
19. (a) W. Z. Soliman, M. M. El-Nahas, and K. M. Mady, *Opt. Pura Apl.*, 22:115 (1989). (b) A. Narayana, N. Ochi, K. Takeuchi, H. Ito, and Y. Okabe, *IEEE Trans. Magn.*, MAG-25, 2549 (1989). (c) R. Feenstra, L. A. Boatner, J. D. Budai, D. K. Christen, M. D. Galloway, and D. B. Poker, *Appl. Phys. Lett.*, 54:1063 (1989). (d) J. Azoulay and D. Goldschmidt, *J. Appl. Phys.*, 66:3937 (1989).
20. L. Harris and B. M. Siegel, *J. Appl. Phys.*, 19:739 (1948).
21. D. S. Campbell and B. Hendry, *Br. J. Appl. Phys.*, 16:1719 (1965).
22. J. L. Richards, P. B. Hart, and L. M. Gallone, *J. Appl. Phys.*, 34:348 (1963).
23. J. L. Richards, in *The Use of Thin Films in Physical Investigations* (J. C. Anderson, Ed.), Academic Press, New York, 1966, p. 71.
24. E. K. Muller, *J. Appl. Phys.*, 35:580 (1964).
25. E. G. Ellis, *J. Appl. Phys.*, 38:2906 (1967).
26. N. S. Platakis and H. C. Gatos, *J. Electrochem. Soc.*, 123:1410 (1976); U.S. Patent 4,080,926 (March 28, 1978).
27. R. C. Tyagi, Rajendrakumar, and V. C. Sethi, *J. Phys. E*, 9:938 (1976).
28. J. George and M. K. Radhakrishnan, *J. Phys. D*, 14:899 (1981); M. K. Radhakrishnan, Ph.D. thesis, University of Cochin, 1981.
29. A. Gheorghiu, T. Rappeneau, S. Fisson, and M. L. Theye, *Thin Solid Films*, 120:191 (1984).
30. D. Sridevi and K. V. Reddy, *Indian J. Pure Appl. Phys.*, 24:392 (1986).
31. K. V. Reddy and J. L. Annapurna, *Pramana (India)*, 26:269 (1986).
32. B. S. V. Gopalan and K. R. Murali, *Mater. Chem. Phys.*, 15:463 (1986).
33. D. Sridevi and K. V. Reddy, *Thin Solid Films*, 141:157 (1986).
34. (a) J. L. Annapurna and K. V. Reddy, *Indian J. Pure Appl. Phys.*, 24:283 (1986). (b) I. S. Atwal and R. K. Bedi, *J. Mater. Sci.*, 24:110 (1989). (c) W. Horig, H. Neumann, V. Savelar, J. Lagzdones, and B. Schuman, *Cryst. Res. Technol.*, 24:823 (1989). (d) F. Volklein, V. Baier, U. Diller, and E. Kessler, *Thin Solid Films*, 187:253 (1990).
35. B. Schuman, G. Georgi, A. Tempel, G. Kuhn, Nguyen Van Nam, H. Neumann, and W. Horig, *Thin Solid Films*, 52:45 (1978).
36. B. Schuman, A. Tempel, C. Georgi, and G. Kuhn, *Thin Solid Films*, 70:319 (1980).
37. A. Tempel, B. Schuman, S. Mitaray, and G. Kuhn, *Thin Solid Films*, 101:339 (1983).
38. S. Mitaray, G. Kuhn, B. Schuman, A. Tempel, W. Horig, and H. Neumann, *Thin Solid Films*, 135:251 (1986).
39. L. Braun and D. E. Lood, *Proc. IEEE*, 54:1521 (1966).
40. R. E. Scott, *J. Appl. Phys.*, 38:2652 (1967).

41. E. Schabowska, T. Pisarkiewicz, and Z. Porada, *Thin Solid Films*, 72:L7 (1980).
42. M. Milosavlgevic, T. M. Nenadovic, N. Bibic, and T. Dimitrigevic, *Thin Solid Films*, 101:167 (1983).
43. A. G. Taylor, R. E. Thurstans, and D. P. Oxley, *J. Phys. E*, 17:755 (1984).
44. (a) E. Schabowska and R. Scigala, *Thin Solid Films*, 135:149 (1986). (b) N. Toghe, K. Kanda, and T. Minami, *Thin Solid Films*, 182:209 (1989). (c) M. Ece and R. W. Vook, *Appl. Phys. Lett.*, 54:2722 (1989). (d) V. Manorama, P. M. Dighe, S. V. Bhorasker, V. J. Rao, P. Singh, and A. B. Belhekar, *J. Appl. Phys.*, 68:581 (1990).
45. L. Holland, British Patent 754,102 (1951).
46. B. A. Unvala and G. R. Booker, *Phil. Mag.*, 9:691 (1964).
47. K. L. Chopra and M. R. Randlett, *Rev. Sci. Instrum.*, 37:1421 (1966).
48. A. P. Hale, *Vacuum*, 13:93 (1963).
49. R. A. Holmwood and R. Glang, *J. Electrochem. Soc.*, 112:827 (1965).
50. R. W. Berry, *Proceedings of the 3rd Symposium on Electron Beam Technology*, Alloyd Electronic Corporation, Cambridge, MA, 1961, p. 359.
51. R. A. Denton and A. D. Greene, *Proceedings of the 5th Symposium on Electron Beam Technology*, Alloyd Electronics Corporation, Cambridge, MA, 1963, p. 180.
52. P. Fowler, *J. Appl. Phys.*, 34:3538 (1963).
53. A. Banerjee, S. K. Bartheval, and K. L. Chopra, *Rev. Sci. Instrum.*, 47:1410 (1976).
54. (a) M. Heiblum, J. Bloch, and J. J. O'Sullivan, *J. Vac. Sci. Technol.*, A3:1885 (1985). (b) K. W. Raine, *Thin Solid Films*, 38:323 (1976).
55. S. Sen and D. N. Bose, *Phys. Stat. Sol (a)*, 66:Pk. 117 (1981).
56. S. Vincet, V. S. Dharmadhikari, and A. Goswami, *Thin Solid Films*, 87:119 (1982).
57. C. M. Mbow, D. Laplaze, and A. Cachard, *Thin Solid Films*, 88:203 (1982).
58. M. Milosavljevic, C. Jaynes, and I. H. Wilson, *J. Appl. Phys.*, 57:1252 (1985).
59. R. Trykozko, R. Bacewicz, J. Filipowicz, *Prog. Cryst. Growth*, 10:361 (1984).
60. G. Burrafato, N. A. Mancine, S. Santagate, S. O. Trofa, A. Torrisi, and O. Puglisi, *Thin Solid Films*, 121:291 (1984).
61. G. A. Niklasson, *J. Appl. Phys.*, 57:157 (1985).
62. M. Mast, K. Gindele, and M. Kohl, *Thin Solid Films*, 126:37 (1985).
63. (a) A. Kaloyeros, M. Hoffman, and W. S. Williams, *Thin Solid Films*, 141:237 (1986). (b) A. Borodznik-Kulpa and C. Wesolowska, *Acta Phys. Pol. A*, 70:413 (1986). (c) D. Das and R. Banerjee, *Thin Solid Films*, 147:321 (1987).
64. (a) H. W. Lehmann and K. Frick, *Appl. Opt.*, 27:4920 (1988). (b) R. Oesterlein and H. J. Krokoszinski, *Thin Solid Films*, 175:241 (1989). (c) H. J. Krokoszinski and R. Osterlein, *Thin Solid Films*, 187:179 (1990).
65. (a) R. O. Adams and C. W. Nordin, *Thin Solid Films*, 181:375 (1989). (b) M. P. Seigal, W. R. Graham, and J. J. Santiago-Aviles, *J. Appl. Phys.*, 68:574 (1990).
66. (a) S. B. Qadri, E. F. Skelton, M. Harford, and P. Lubitz, *J. Appl. Phys.*, 67:2655 (1990). (b) M. S. Osofsky, P. Lubitz, M. Z. Harford, A. K. Singh, S. B. Qadri, E. F. Skelton, T. Elam, R. J. Soulen, W. L. Lechter, and S. A. Wolf, *Appl. Phys. Lett.*, 53:1663 (1988). (c) J. Steinbeck, A. C. Anderson, B. Y. Tsuar, and A. J. Strauss, *IEEE Trans. Magn.*, MAG-25:2429 (1989). (d) F. H. Garzon, J. G. Beery, D. R. Brown, R. J. Sherman, and I. D. Raistrick, *Appl. Phys. Lett.*, 54:1365 (1989). (e) F. C. Case, *J. Appl. Phys.*, 67:4365 (1990).
67. C. Q. Lemmond and L. H. Stauffer, *IEEE Spectrum*, 1964, 66.

68. H. M. Smith and A. F. Turner, *Appl. Opt.*, 4:147 (1965).
69. S. Fujimori, T. Kasai, and T. Inamura, *Thin Solid Films*, 92:71 (1982).
70. S. Mineta, N. Yasunga, N. Tarumi, E. Teshigawara, M. Okutomi, and M. Ikeda, *Bull. Jpn. Soc. Process. Eng.*, 18:49 (1984).
71. M. Hanabusa, M. Suzuki, and S. Nishigaki, *Appl. Phys. Lett.*, 38:385 (1981).
72. H. T. Yang and J. T. Cheung, *J. Cryst. Growth*, 56:429 (1982).
73. H. Sankur and J. T. Cheung, *J. Vac. Sci. Technol.*, A1:1806 (1983).
74. J. J. Dubowski and D. F. Williams, *Appl. Phys. Lett.*, 44:339 (1984).
75. J. J. Dubowski and D. F. Williams, *Thin Solid Films*, 117:289 (1984).
76. J. J. Dubowski and D. F. Williams, *Can. J. Phys.*, 63:815 (1985).
77. J. J. Dubowski, P. Norman, P. B. Sewell, D. F. Williams, F. Krolicki, and M. Lewicki, *Thin Solid Films*, 147:L51 (1987).
78. M. I. Baleva, M. H. Maksimov, S. M. Metey, and M. S. Sendova, *J. Mater. Sci. Lett.*, 5:533 (1986).
79. M. I. Baleva, M. H. Maksimov, and M. S. Sendova, *J. Mater. Sci. Lett.*, 5:537 (1986).
80. (a) S. B. Ogale, V. N. Koinkar, S. Joshi, V. P. Godbole, S. K. Data, A. Mitra, T. Venkatesan, and X. D. Wu, *Appl. Phys. Lett.*, 53:1320 (1988). (b) O. Auciello, T. Barnes, S. Chevacharoenkul, A. F. Schreiner and G. E. Maguire, *Thin Solid Films*, 181:65 (1989). (c) H. J. Scheibe, A. A. Gorbunov, G. K. Baranova, N. V. Klassen, V. I. Konov, M. P. Kulakov, W. Pompe, A. M. Prokhorov, and H. J. Weiss, *Thin Solid Films*, 189:283 (1990). (d) J. J. Dubowski, *Chemtronics*, 3:66 (1988). (e) J. J. Dubowski, *J. Cryst. Growth*, 101:105 (1990).
81. (a) M. Yoshimoto, H. Nagata, T. Tsukahara, and H. Koinuma, *Jpn. J. Appl. Phys.*, 29:L1199 (1990). (b) H. Koinuma, H. Nagata, T. Tsukahara, S. Gonda, and M. Yoshimoto, *Ex. Abstr. 22nd Int. Conf. Solid State Devices*, Sendai, 1990, p. 933. (c) J. A. Martin, L. Vazquez, P. Bernard, F. Comin, and S. Ferrer, *Appl. Phys. Lett.*, 57:1742 (1990). (d) R. Ramesh, K. Luther, B. Wilkens, D. L. Hart, E. Wang, J. M. Tarascon, A. Inam, X. D. Wu, and T. Venkatesan, *Appl. Phys. Lett.*, 57:1505 (1990). (e) J. J. Dubowski, *Proc. SPIE*, 668:97 (1986).
82. (a) M. Baleva and D. Dakoeva, *J. Mater. Sci. Lett.*, 5:37 (1986). (b) P. Jayaramareddy and M. Sivajuddin, *Bull. Mater. Sci.*, 8:365 (1986). (c) M. S. Sendova, *J. Mater. Sci. Lett.*, 6:285 (1987). (d) P. A. Dearnly and K. Anderson, *J. Mater. Sci.*, 22:679 (1987).
83. (a) S. G. Hansen and T. E. Robitaille, *Appl. Phys. Lett.*, 50:359 (1987). (b) G. Kessler, H. D. Baner, W. Pompe, and H. J. Scheibe, *Thin Solid Films*, 147:L45 (1987). (c) S. Joshi, R. Nawathey, V. N. Koinkar, V. P. Godbole, S. M. Chaudhari, and S. B. Ogale, *J. Appl. Phys.*, 64:5647 (1988). (d) R. Nawathey, R. D. Vispute, S. M. Chandhari, S. M. Kanetkar, and S. B. Ogale, *Solid State Commun.*, 71:9 (1989). (e) C. M. Dai, C. S. Su, and D. S. Chuu, *Appl. Phys. Lett.*, 57:1879 (1990).
84. (a) B. Roas, L. Schultz, and G. Endres, *Appl. Phys. Lett.*, 53:1557 (1988). (b) E. Fogarassy, C. Fuchs, J. P. Stoquert, P. Siffert, P. Perriere, and F. Rochet, *J. Less Common Mater.*, 151:249 (1989). (c) W. Ludorf, X.-Z. Wang, and D. Bauerle, *Appl. Phys. A*, A49:221 (1989). (d) D. P. Nortan, D. H. Lowndes, J. D. Budai, D. K. Christen, E. C. Jones, J. W. McCamy, T. D. Ketcham, D. St. Julien, K. W. Lay, and J. E. Tkaczyk, *J. Appl. Phys.*, 68:223 (1990).
85. (a) V. Serbezov, S. Benacka, D. Hadgiev, P. Atanasov, N. Electronov, V. Smatko, V. Stribik, and N. Vassilev, *J. Appl. Phys.*, 67:6953 (1990). (b) R. E. Russo, R. P. Reade, J.

M. McMillan, and B. L. Olsen, *J. Appl. Phys.*, 68:1354 (1990). (c) A. Mogro-Campero, B. D. Hunt, L. G. Turner, M. C. Burrell, and W. E. Balz, *Appl. Phys. Lett.*, 52:584 (1988). (d) A. Mogro-Campero, *Supercond. Sci. Technol.*, 3:155 (1990). (e) D. W. Hwang, R. Ramesh, C. Y. Chen, X. D. Wu, A. Inam, M. S. Hegde, B. Wilkens, C. C. Chang, L. Nazar, T. Venkatesan, M. S. Matsubara, Y. Miyasaka, and N. Shohata, *J. Appl. Phys.* 68:1772 (1990). (f) S. Miura, T. Yoshitake, S. Matsubara, Y. Miyasaka, N. Shohata, and T. Satoh, *Appl. Phys. Lett.*, 53:1967 (1988). (g) D. M. Hwang, T. Venkatesan, C. C. Chang, L. Nazar, X. D. Wu, A. Inam, and M. S. Hegde, *Appl. Phys. Lett.*, 54:1702 (1989). (h) G. Koren, A. Gupta, E. A. Geiss, A. Segmuller, and R. B. Laibowitz, *Appl. Phys. Lett.*, 54:1054 (1989).

86. (a) X. D. Wu, R. E. Münchausen, S. Foltyn, R. C. Estler, R. C. Dye, C. Flamme, N. S. Nogar, A. R. Gracia, J. Martin, and J. Tesmer, *Appl. Phys. Lett.*, 56:1481 (1990). (b) T. Hase, H. Izumi, K. Ohata, K. Suzuki, T. Morishita, and S. Tanaka, *J. Appl. Phys.*, 68:374 (1990). (c) M. Ohkubo, T. Kachi, and T. Hioki, *J. Appl. Phys.*, 68:1782 (1990).

87. M. S. P. Lucas, H. A. Owen, Jr., W. C. Stewart, and C. R. Vail, *Rev. Sci. Instrum.*, 32:203 (1961).

88. M. Kikuchi, S. Nagakura, H. Ohmura, and S. Oketani, *Jpn. J. Appl. Phys.*, 4:940 (1965).

89. A. S. Gilmour, Jr., and D. L. Lockwood, *Proc. IEEE*, 60:997 (1972).

90. J. W. Robinson and M. Ham, *IEEE Trans. Plasma Sci.*, PS-3:222 (1975).

91. Y. Igarashi and M. Kanayama, *Appl. Phys. Lett.*, 28:481 (1976).

92. Y. Igarashi and M. Kanayama, *J. Appl. Phys.*, 52:7208 (1981).

93. Y. Igarashi and M. Kanayama, *J. Appl. Phys.*, 57:849 (1985).

94. E. A. Roth, E. A. Margerum, and J. A. Amick, *Rev. Sci. Instrum.*, 33:686 (1962).

95. J. A. Turner, J. K. Birtwistle, and G. R. Hoffman, *J. Sci. Instrum.*, 40:557 (1963).

96. J. Van Audenhove, *Rev. Sci. Instrum.*, 26:383 (1965).

97. F. E. Thompson and J. F. Libsch, *Sci. Solid State Technol.*, p. 50, December 1965.

98. I. Ames, L. H. Kaplan, and P. A. Roland, *Rev. Sci. Instrum.*, 37:1737 (1966).

2

Sputtering

The man who says that he will work when the world has become perfect and then he will enjoy bliss, is as likely to succeed as the man who sits beside a river and says "I shall cross when all the water has run into the ocean."

Swami Vivekananda

It has long been known [1] that when a surface is bombarded with high velocity positive ions, it is possible to cause the ejection of the surface atoms. This process of ejecting atoms from the surface by the bombardment of positive ions, usually inert, is commonly known as (cathode) sputtering. The ejected atoms can be made to condense on a substrate to form a thin film.

When a charged particle bombards the target surface, apart from the ejection of neutral atoms of the surface material, charged atoms and electrons are also emitted from the surface. The ejected neutral target atoms condense into thin films on the substrate. This process can be realized by having positive ions of a heavy gas such as argon strike the target surface, the target being connected to a negative voltage supply and the substrate holder forming the anode and facing the target. The sputtering yield, the most important parameter for characterizing the sputtering process, is defined as the number of atoms ejected from the target surface per incident ion. The sputtering yield depends on the bombarded material, its structure and composition, the characteristics of the incident ion, and also the experimental geometry. The sputtering yield has been determined for a number of metals bombarded with ions over a broad range of ion energy. The interested reader is referred to data compiled by Maissel [2] and Vossen and Cuomo [3]. The data on sputtering yield should be taken only to give a rough indication of the rate of deposition that might be expected from a given material.

Various theories have been put forward to account for the mechanism of cathode sputtering. Von Hippel [4] attributed sputtering the creation of high local temperature by the bombarding individual ions, leading to the evaporation of the target material from this microscopically small region. This hot spot evaporation theory has been replaced by what is often called knock-on sputtering, originally suggested by Stark [5], developed by Kingdon and Langmuir [6], and subsequently exposed to numerous developments. The elementary event here is an atomic collision cascade. According to this collision theory of sputtering, further developed by Keywell [7], the incident ion knocks atoms in the target from their equilibrium positions, thus causing these atoms to move in the material and to undergo further collisions, finally causing the ejection of atoms through the target surface. The concensus is that this mechanism is the most universal one and is applicable to the cathode sputtering process in general.

As early as 1877, the sputtering technique was used to coat mirrors [8]. But with the advent of improved vacuum technology, sputtering was replaced by the much faster process of vacuum evaporation where even materials like Ta, W, and Al_2O_3 could be evaporated using electron beam evaporation. After a lapse of several decades, the application of sputtering to the deposition of thin films started gaining considerable importance. With a better understanding of the sputtering processes and the development of rf sputtering and other variants of sputtering systems, sputtering has become one of the most versatile techniques in thin film technology for preparing thin solid films of almost any material. Some of the main advantages of sputtering as a thin film preparation technique are (a) high uniformity of thickness of the deposited films, (b) good adhesion to the substrate, (c) better reproducibility of films, (d) ability of the deposit to maintain the stoichiometry of the original target composition, and (e) relative simplicity of film thickness control. Item (d) is important for alloys and compounds.

A number of works exist on the broad subject of sputtering and sputtering process of thin film deposition [9-16].

2.1   GLOW DISCHARGE DC SPUTTERING

Several systems have been employed for the deposition of thin films by sputtering, and the simplest arrangement is the glow discharge d c sputtering system. A basic glow discharge dc sputtering system is shown in cross section in Figure 2.1. A plate of the material to be deposited (target) is connected to negative voltage supply and the substrate facing the target is mounted on the adjacent anode. Next, dc voltages of the order of 1-5 kV are applied across the electrode (current density, 1-10 mA/cm^2). A glow discharge is initiated by introducing a neutral gas such as argon into the vacuum chamber to provide a pressure of 10^{-1}-10^{-2} torr. When the glow discharge is started, the positive ions strike the target plate, removing mainly neutral atoms from the surface of the target, and these eventually condense as a thin film on the substrate. There are also electrons emitted from the target (cathode) by ion bombardment, and these are accelerated toward the substrate platform, where they collide with the gas atoms. These electrons help to maintain the glow dis-

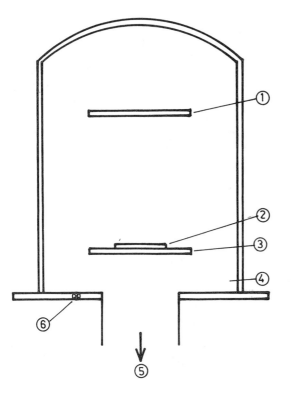

Figure 2.1 Glow discharge dc sputtering system: 1, cathode (target); 2, substrate; 3, anode; 4, vacuum chamber; 5, to vacuum pumps; 6, gas inlet.

charge, as the ionized gas atoms in turn bombard the target and release more secondary electrons. If the gas pressure is too low or the cathode–anode spacing too small, the secondary electrons cannot undergo sufficient ionizing collisions before they hit the anode. On the other hand, if the pressure and/or separation is too large, the ions generated are slowed by inelastic collisions, and when they strike the target, they will not have enough energy to produce secondary electrons. In effect, the operation of a sputtering system requires the generation of a sufficient number of secondary electrons to replace these lost to the anode or to the walls of the vacuum chamber.

The collision of the ejected atoms with the gas atoms in the plasma will cause the scattering of the former and reach the anode with random directions and energies. The probability of the sputtered material getting lost without striking the substrate thus increases with the cathode–substrate spacing. At constant pressure and constant voltage, the deposition rate is low at large distances from the cathode, the thickness distribution over the substrate showing a maximum at the center. Maissel [17] has suggested that the optimum conditions for ensuring uniformity of deposition are obtained when the cathode–anode distance is about twice the length of the Crookes dark space, and a plane cathode of about twice that of the substrate is used.

Sputtering is basically a low temperature process, and less than 1% of the total applied power is used for the ejection of the sputtered material and secondary electrons. A considerable amount of energy is dissipated as heat at the cathode by the ions that strike it, and the cathode gets hot. The maximum temperature attained and the rate of rise of temperature depend on the glow discharge conditions. Although the sputtering yield for most materials increases with temperature, it is not generally advisable to set the cathode temperature rise beyond a tolerable level during sputtering because of the possible problems with outgassing. It is desirable to cool the cathode, and several cooling methods have been reported [18,19].

In a practical sputtering system, the self-sustained dc discharge cannot be sustained at a pressure below 10 mtorr because there are insufficient ionizing collisions. For more details about the phenomena of self-sustained glow discharge, the interested reader is referred to the literature [20,21]. The most serious drawback of the self-sustained glow discharge as a technique for the sputter deposition of films is contamination of the deposited film by the inert gas used to produce the discharge. But the concentration of this trapped inert gas in the film can be lowered if the sputtering is done at low pressures. Another advantage that might be expected from sputtering at low pressure is the higher mean energy of the sputtering particles when they strike the substrate, resulting in better adhesion of the films. For the sputtering system to operate at pressures lower than 10–20 mtorr, there should be a source of electrons other than the secondary electrons emitted from the main cathode, or the ionizing efficiency of the available electrons should be increased. A sufficient level of ionization can be achieved at lower pressures by the use of an auxiliary high frequency discharge. Gawehn [22] has reported a system in which a high frequency coil surrounds the chamber to induce the formation of the plasma. Vratny [23] used a different technique, feeding the radiofrequency power directly to the cathode superimposed on the d c bias. In addition to the advantages of sputtering at low pressure, these methods give higher deposition rates at normal pressure, particularly at low voltages.

The ionization efficiency of the available electrons can be enhanced by the application of a magnetic field. One effect of the magnetic field is to cause electrons that are not moving parallel to it to move in helical paths around the magnetic lines of force. This means that the electrons must travel a larger path to advance a given distance from the cathode, effectively enhancing their ionization efficiency for a given linear distance traveled.

A bibliography of work on d c diode sputter deposition reported recently is given in Table 2.1.

Table 2.1 DC Diode Sputtering: A Bibliography

Target	Sputter gas	Remarks	Ref.
$ErRh_4B_4$, arc-melted target	Ar (Research grade)	Films were characterized by X-ray deffraction, SEM, and Auger electron spectroscopy.	91
Nb_3Ge	Ar (99.994% pure)	Nb_3Ge films with high critical temperature were obtained.	92
TaB_2-Cr-Si-Al, Fe-Cr-Si, Ta-Cr-Si-Al (composite targets)	Ar	Pressure, 60×10^{-3} torr.	93
Ni (60 mm dia.)	Ar	Metallization for MOS preparation.	94
Tantalum silicide		Film characteristics were analyzed by X-ray diffraction, backscattering, electron microprobe, and electron transmission microscopy.	95
Barium ferrite (sintered target)		c-Axis oriented barium ferrite films, easy direction of magnetization normal to the film plane.	96
Zr_2Rh	Ar	Target was 2.5 cm in diameter, 99.99% pure Rh foil spot welded to 99.98% pure Zr sheet. Substrate temperature 350–650°C. High quality superconducting Zr_2Rh film have been obtained.	97
Bi_2Te_3	Ar	Bi_2Te_3 films showed nonstoichiometric composition; Bi, 44.5 at %; Te, 55.5 at %.	98
PbTe	Ar	Stoichiometric films were obtained.	98
Titanium	Ar + N_2	Below a critical nitrogen pressure (4×10^{-2} Pa), α-Ti hexagonal phase formed.	99
Graphite	Ar	α-Carbon passivating layer was prepared. Appears to be a promising dielectric material for use in silicon solid state electronics.	100
Composite target of Ti pieces, fixed symmetrically on a graphite disk	Ar	Commercial grade titanium and nuclear grade graphite.	101

Table 2.1 (Continued)

Target	Sputter gas	Remarks	Ref.
Sintered TiC, (5.08 × 10^{-2} m dia., 3.2 × 10^{-3} m thick	Ar	TiC powder plus 1% paraffin wax compacted and sintered.	101
In 5.4 wt % Sn	Ar	Annealed in air at various temperatures to obtain transparent conducting ITO films.	102
Y–Ba–Cu–O, compacted and sintered	Ar	Compacted and sintered powder of various compositions were used as targets. Pressure, 30 mtorr. Substrates, sapphire and Al_2O_3. Substrate temperatures, RT to 500°C.	103
$Tl_{2.3}Ba_2Ca_2$ Cu_3O_x	Pure Ar	Substrates: MgO (100) and $SrTiO_3$ (100). The as-deposited films were heat treated under air or O_2 with an inclusion of a TlBaCaCuO powder compact to make it superconducting. Single- phase $Tl_2Ba_2Ca_2Cu_3O_{10}$ films have T_c onset at 125 K and T_c ($R = 0$) at 116 K. Film morphology and micro-structure studied by SEM and TEM.	140
$YBa_2Cu_3O_{7-x}$	Pure O_2 or an O_2–rich mixture of O_2 and Ar	Substrates: $SrTiO_3$, MgO, Al_2O_3, and oxidized Si. Substrate temperature, 500–800°C. After deposition the films were cooled in an oxygen atmosphere. Best films were obtained on $SrTiO_3$ at about 650°C.	141
$YBa_2Cu_3O_7$	Pure oxygen	Substrates (100) and (110) surface of $SrTiO_3$. Sibstrate temperature, 500–800°C. Films were superconducting without postannealing. For films grown around 700°C, zero resistance was obtained at 90K. Epitaxial films could be grown on $SrTiO_3$ (100).	142
Silver, palla-dium, and Ag-Pd	Pure argon	Epitaxial deposition of silver and palladium films onto rock salt single-crystal (100) face. Monophasic alloy film prepared by simultaneous sputtering of silver and palladium.	143

2.2 TRIODE SPUTTERING

As already mentioned, an alternative method to increase the ionization and sustain the discharge at low pressures is to supply additional electrons from a source other than the target cathode. Triode sputtering involves the injection into the discharge of electrons from an independent source by means other than the discharge itself. A hot cathode, which emits electrons through thermionic emission, is used to inject electrons into the discharge system. The thermionic cathode is usually a heated tungsten filament, which can withstand ion bombardment for long periods. The anode must be biased positively with respect to the substrate. But if the anode is at the same potential as the substrate, some of the electrons from the thermionic emitter will be deflected toward the substrate and will then be collected at the substrate, resulting in gross inhomogeneities in plasma density at the target.

A schematic representation of a triode sputtering unit is shown in Figure 2.2. The filament is placed in an elbow shown at the bottom of the bell jar, to protect the filament from any sputtered material. The plasma is confined between the anode and the filament cathode by a magnetic field provided by the external coil. Sputtering occurs when a high voltage negative potential with respect to the anode is applied to the target. Ions bombard the target as in diode glow discharge, and the target material is deposited on the substrate. The density of ions in the plasma can be controlled by adjusting either the electron emission current or the voltage used to accelerate the electrons. The energy of the bombarding ions can be controlled by varying the target voltage. Thus in a supported discharge like the triode discharge sputtering system, the ionization density is kept high by supplying extra electrons at the right energies from an additional electrode for efficient ionization. This method allows operation at much lower pressures ($\leq 10^{-3}$ torr) than in a conventional diode glow discharge system. The main limitation of this technique is the difficulty in producing uniform sputtering from very large flat targets. Also the supported discharge is hard to control, for reproducible results [24].

After the early works reported [25,26] on this type of sputtering, Sun et al. [27,28] made a study of the properties of sputtered tungsten films as a function of several deposition parameters (substrate temperature, deposition rate, etc.). The plasma discharge was maintained by using a thermionic cathode and was confined in a water-cooled aluminum tube. A voltage of +90 V with respect to the ground was given for the auxiliary anode, and the anode current was approximately 6 A. Tungsten films were deposited at an Ar pressure of about 10^{-3} torr. In a triode sputtering study reported by Lee and Oblas [29], 12 different metal targets were used for sputtering in the triode mode at a pressure of 2×10^{-3} torr. The configuration of the sputtering system is shown schematically in Figure 2.3. Here the cathode was mounted vertically with a grounded shield 12.7 cm behind the target and parallel to the substrate holder (grounded). The auxiliary anode and the filament assembly placed between the substrate and target increase the ionization of argon. At the ar-

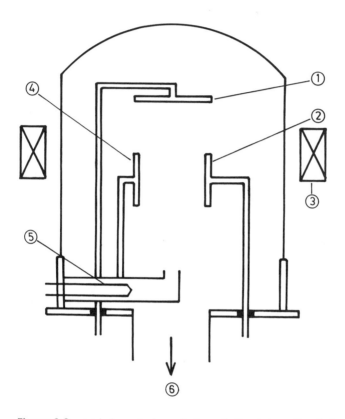

Figure 2.2 A triode sputtering unit: 1, anode; 2, substrate; 3, coil; 4, target; 5, filament; 6, to vacuum pumps.

gon pressure used for sputtering, the mean free path is large compared to the cathode fall distance; therefore most of the ions arrive at the cathode surface with the full applied potential of the cathode. Lee and Oblas [29] studied the effect of the target material on the argon content of the various sputtered films deposited under similar conditions. Adams et al.[30] prepared Al-Ag alloy films by triode sputtering, and films with a wide variety of compositions were produced and evaluated for use as solar reflectors. Patten and Boss [31] used d c triode sputtering of nickel in krypton plasma with different concentrations of oxygen to the study the effects of O_2 additions on the growth of colomnar shadowing defects in sputtered nickel. Sonkup and coworkers [32,33] prepared both n-type and p-type films of GaAs on single-crystal GaAs semi-insulating substrates [32,33], using a supported plasma technique similar to that described by Edgecumbe et al. [25], and studied the electrical properties of these sputtered films. Doping of the film to various densities

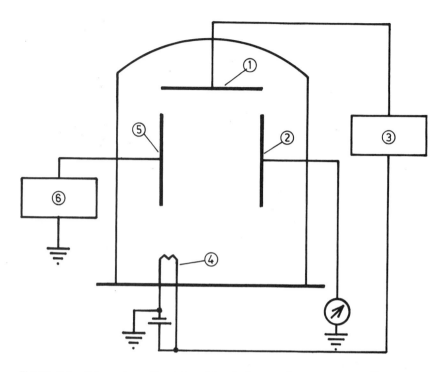

Figure 2.3 Schematic configuration of the triode sputtering system: 1, auxiliary anode; 2, substrate-anode; 3, power supply; 4, filament; 5, cathode; 6, high voltage supply. (Data from Ref. 29.)

was done by cosputtering a silicon or magnesium target of ultra high purity. The GaAs target was d c sputtered at the rate of 3 Å/s, while the silicon or magnesium targets were sputtered at a rate controlled by the application of a pulsed voltage of adjustable duty cycle to the target. The GaAs target was water cooled to minimize As sublimation. Ziemann et al. [34], who studied the plasma potential in a triode d c discharge as a function of the applied anode voltage, the injected electron current, and the gas pressure, related their results to the experimentally determined impurity content of the corresponding films obtained.

In ordinary diode sputtering systems, the substrate is immersed in the plasma and the deposited film surface is bombarded with ions and atoms. This can cause interfacial mixing, which in turn will result in the loss of the sharpness of the interfacial layers. If the layers are thin enough, destruction of the multilayers themselves will result, leaving a homogeneous film. To avoid these problems, Sella and Vien [35a] used a low energy d c triode sputtering system for the deposition of multilayers and obtained well-defined layers with sharp interfaces. In their setup

(Figure 2.4), the two targets are mounted back to back and can be rotated through 180° in a diffusion-pumped vacuum system. The substrate is fixed and water cooled. The argon pressure during sputtering is maintained at 10^{-3} torr. For controlling film thickness the investigators used a new method of monitoring, based on the dependence of the deposition rate on the target current, the other sputtering parameters being kept constant.

Gallias et al. [35b] have used a d c triode system to deposit tantalum films onto SiO_2 or Si substrates. Thick films (700 nm) deposited onto Si (111) or SiO_2/Si substrates (T_s 500°C) were polycrystalline. But the layers were amorphous when deposited at 50°C that is, without substrate heating.

Not much work has been reported on $ZrTe_x$ (x = 3 or 5), despite the interest of these materials due to their quasi-one-dimensional crystal anisotropy. However Caune et al. [35c] reported for the first time the preparation of $Zn\ Te_x$ films by sputtering a binary target using a triode sputtering system. Two half-disks, 100 mm in diameter, one 99.99% pure zirconium and the other of 99.95% pure tellurium, formed the targets. To prevent overheating of the low melting point tellurium, the halves were joined and soldered on the same water-cooled copper holder, and the applied cathode voltage never exceeded 1000 V. To regulate film composition, the surface ratio of the two cosputtered elements was adjusted by a mobile hemicircular mask, the sputtering rates of the two components under the working conditions being known.

Details of the sputtering conditions used in this experiment are as follows:

Sputter gas	Pure Ar
Residual pressure before sputtering	10^{-7} torr
Working pressure	10^{-3} torr (Ar)
Target potential	-1000 V
Target-to-substrate distance	100 mm
Ionic density	1.8 mA/cm²
Deposition rate	$\approx$ 13 mm/min
Substrate temperatures	$ZrTe_3$, $ZrTe_5$, 275°C
Target surface T_e/Zr ratio	$ZrTe_3$ = 0.7, $ZrTe_5$ = 2.3

Various kinds of substrate (e.g., polyimide, optical glass, silicon, molybdenum) were used, and the substrates rotated to get homogeneous films. The morphology of the films and their characteristics were studied using microprobe X-ray and standard X-ray diffraction methods. Maniv [24] compares d c diode and triode sputtering.

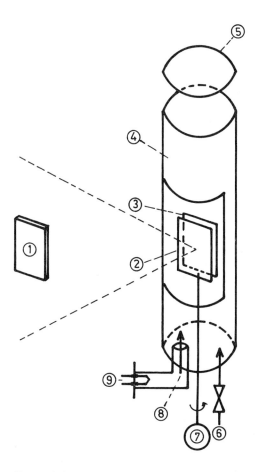

Figure 2.4 Schematic arrangement of the low energy dc triode sputtering system: 1, substrate; 2, target 1; 3, target2; 4, screen; 5, anode; 6, argon inlet; 7, motor; 8, electron beam; 9, filament. (Data from Ref. 35a.)

2.3   GETTER SPUTTERING

A depositing film is an active getter of impurities; thus a sputtering gas, when made to pass over an area of freshly deposited film, is cleansed of its impurities. The chemical cleanup occurs when the residual gas molecules impinging on an atomically clean metal surface, are chemisorbed or form a chemical compound. Molecules such as O_2, H_2, H_2O, N_2, and CO interact strongly with a large number of metals. However the rare gases are merely physisorbed with small interaction energies and are not readily trapped.

In a sputtering system consisting of a very large cathode and an equally large substrate holder, the material that is deposited at the center of the assembly will be free from the impurities present in the original gas or that generated through the outgassing from the walls. This is because the impurity atoms in the sputtering gas have made many collisions with the depositing film material before reaching the central area and therefore have a high probability of being removed. But in practical getter sputtering systems, the cathode is of finite size and the walls of the sputtering chamber are arranged very close to both cathode and substrate, without interfering with the glow discharge. In the simple getter sputtering system shown in Figure 2.5, two cathodes of the material to be sputtered are located symmetrically with respect to a rather close-fitting protective cylinder. The gases enter the bottom of the cylinder, and the maximum gettering action of the active admixtures occurs at the lower cathode. Films deposited on the cylindrical enclosure also getter reactive gases and bury them. After sputtering for a few minutes when all the reactive gases have been gettered, the shutter is removed and the sputtered material from the second cathode is allowed to deposit on the substrates.

Theuerer et al. [36–38] developed different versions of a getter sputtering system and had prepared thin films of niobium and tantalum. Cook et al. [39] have described a system operating satisfactorily. Here also a shutter is provided over the substrate to facilitate the reduction, to a large extent, of the impurities emitted from the walls of the chamber when the glow is first switched, as well as the impurity gases initially present at or near the surface of the cathode. Here the shutter is opened only after a period of time has elapsed after the glow has been switched off and the internally generated impurities have been reduced to considerable extent.

Preparation of thin films by getter diode sputtering has not been reported extensively since these early reports. Hong et al. [40] and Bacon et al. [41] used dc getter sputtering to deposit amorphous GdCo films with no bias voltage applied to the substrate during deposition, using the experimental apparatus similar to that described in the early work of Theuerer and Hauser [36]. These GdCo film show perpendicular anisotropy. The argon impurity ion is 1.0 at. % or less, which is far less than those obtained in the biased rf diode sputtered GdCo films; in addition, the distribution of argon is uniform in the films [41]. Hong et al. [42], using d c getter diode sputtering, prepared amorphous Tb(FeCo) film from arc-cast alloy targets.

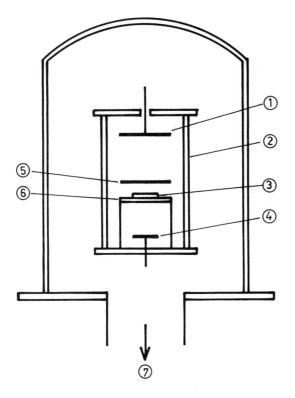

Figure 2.5 Schematic diagram of a getter sputtering system: 1, cathode; 2, protective cylinder; 3, substrate; 4, auxiliary cathode; 5, shutter; 6, anode; 7, to pump.

Freshly prepared $Tb_{(18.5)}(FeCo)_{81.5}$ (in at. %) showed a perpendicular anisotropy. The investigators studied and compared the aging characteristics of amorphous Tb(FeCo) film prepared from both dc getter diode and magnetron sputtering.

The main disadvantage of getter sputtering is that the cylindrical chamber acts as a wall, giving rise to wall losses and large nonuniformities in film deposition near the target edges.

2.4 RADIOFREQUENCY SPUTTERING

The sputtering techniques described earlier assume that for the deposition of thin films by sputtering, the target should be a conductor. In a conventional dc sputtering system, if an insulator is substituted for the metal target, a surface positive charge is built up on the front surface of the insulator during ion bombardment. This charge buildup can be prevented by simultaneously bombarding the insulator with both ion beam and electron beam particles [43]. But a sputtering scheme for depositing insulator films, first suggested by Anderson et al. [44], was subsequently developed into a practical method by Davidse and Maissel [45]. Here an rf potential is applied to the metal electrode placed behind the dielectric plate target. At rf potentials, the electrons oscillating in the alternating field have sufficient energies to cause ionizing collisions, and the discharge will be self-sustained. The high voltage at the cathode, which is essential in a dc glow discharge for the generation of secondary electrons, is no longer required here for the discharge to maintain itself. Since the electrons have much higher mobilities than ions, many more electrons will reach the dielectric target surface during the positive half-cycle than ions during the negative half-cycle, and the target will become self-biased negatively. The negative dc potential on the insulator target surface repels electrons from the vicinity of this surface, creating a sheath enriched in ions in front of the target. These ions bombard the target, and sputtering is achieved. This positive ion sheath is the counterpart of the Crookes dark space in a dc sputtering system. At frequencies less than 10 kHz, such an ion sheath will not be formed, and 13.56 MHz frequency is generally used for rf sputtering. It should be noted that since the applied rf field appears between the two electrodes, an electron escaping from the interelectrode space as a result of random collision will no longer oscillate in the rf field. Therefore these electrons will not get sufficient energy to cause ionization, hence will be lost to the glow. But if a magnetic field is applied parallel to the rf field, it will constrain the electrons without being lost to the flow, thus improving the efficiency of the rf discharge. So a magnetic field is more important in rf sputtering than it is for the dc case. A grounded metal shield is placed close to the other side of the metal electrode to extinguish the glow on the electrode side and to prevent sputtering of the metal electrode. Thin films of quartz, aluminum oxide, boron nitride, mullite, and various glasses have been prepared using rf sputtering [45].

Metals can also be rf sputtered if the rf power supply is coupled capacitively to a metal electrode. Then dc current will not flow in the circuit, and this allows the buildup of a negative bias on the metal electrode that is necessary for sputtering, the self-biasing effect caused by the difference in electron and ion mobilities. With this method, films of many metals can be deposited at rates of about 1000 Å/min for an rf power of approximately 1 kW at roughly 5 μ pressure [46].

The physical appearance of an rf sputtering system is almost like that of a dc system. The most important difference between rf and dc systems is that the former requires an impedance matching network between power supply and discharge chamber. It is also important that in an rf system, adequate grounding of the substrate assembly be ensured to avoid undesirable rf voltages, which can develop on the surface. A comparison between dc and rf sputtering system is given in Reference 24.

The use of rf sputtering for the deposition of thin films is of great interest because it enables more economical deposition onto substrates of large areas. Since the early reports of Davidse and Maissel [45], thin films of the number of materials have been prepared by rf sputtering. An intensive bibliography of the recent reports, with details of the deposition parameters, is given in Table 2.2.

2.5 MAGNETRON SPUTTERING

Magnetron sputtering technology has made significant progress since its development, in early 1970s, for the high rate deposition of metal, semiconductor, and dielectric films. In comparison to conventional diode sputtering, magnetron sputtering, apart from obtaining high deposition rates at lower operating pressures, makes it possible to obtain high quality films at low substrate temperatures. The basic principle of all magnetically enhanced sputtering technique was discovered by Penning [47] more than 50 years ago and was subsequently developed by Kay and others [48-51], resulting in the sputter gun [52] and cylindrical magnetron sources [53]. The planar magnetron structure was introduced in 1974 by Chapin [54], although the basic principle of a planar magnetron device had been demonstrated in 1959 by Kesser and Pashkova [55]. Since then several magnetron cathodes in dc and/or rf sputtering have been developed to deposit thin films for the fabrication of semiconductor and optical devices, and the technology in general has been reviewed extensively with regard to deposition theory, current-voltage characteristics, sputtered film structure, cathode geometry, and so on [56-60].

The magnetron effect can be described as a closed drift path of crossed electric and magnetic fields for electrons in a plasma discharge. For a simple planar magnetron cathode (the behavior of the various cathodes is similar), the arrangement consists of the planar cathode (target) backed by permanent magnets that provide a toroidal field, with field lines forming a closed path on the cathode surface. The difference in the mobilities of the ions and the electrons causes a positive ion sheath to be developed close to the target cathode, floating at a negative potential relative to the plasma. Because of the field due to the ion sheath at the cathode, ions are extracted from the plasma and accelerated to strike the target, resulting in the sputtering of the target material. The secondary electrons produced, upon entering the region of crossed electric (E) and magnetic (B) fields, are trapped in orbits that permit long travel distances close to the cathode. In the zones of efficient electron trapping, the electron density reaches a critical value, at which the ionization probability due to the trapped electrons is at a maximum. This means that a higher rate of secondary electron production by high energy positive ions is not necessary for effective sputtering.

Most magnetron sources operate in the pressure range from 1 to 20 mtorr and a cathode potential of 300-700 V. The sputtering rates are primarily determined by the ion current density at the target, and the deposition rates are affected by factors such as applied power, source-substrate distance, target material, pressure, and sputtering gas composition.

Table 2.2 Deposition Parameters of Various Materials Prepared by Rf Sputtering

Material	Target	Sputter gas	Sputtering power/ density	Deposition rate	Pressure	Substrate and substrate Temp.	Remarks if any	Ref.
PbTiO$_3$	Powder mixture of Pb$_3$O$_4$ and TiO$_2$ pressed on a quartz plate; Excess of Pb$_3$O$_4$	90% Ar + 10% O$_2$	3 W/cm^2	40Å/min	2×10^{-1} Torr	Pt foil, 300–350°C	Films at substrate temperatures up to 550°C have also been prepared.	104
Mumetal and permalloy	77 Ni–14 Fe–5 Cu–4 Mo and 82 Ni–18 Fe (wt %); 3–4 in. diameter	High purity bottled Ar further purified using a titanium chem-ad-sorption purifier		0.5–3.0 μm/h		Corning 7059 borosilicate glass	Bias voltage, 0 to –400 V. Uniform aligning field of 60 Oe was applied parallel to the film plane.	105
MoS$_2$	MoS$_2$ (25 cm dia)	Ar	2 W/cm^{-2}		10 m torr		Target-to-substrate distance, 2.5 cm.	106
Si	Semiconductor grade pieces of single crystals (20 cm dia, 5 mm thick)	Ar (99.999% pure)	0.5–1.3 kW 13.56 MHz	10–30 nm/ min	5 m torr	Ni and Ta 1 cm × 1 cm × 0.1 cm platelets 50–100°C	Bias voltage range, 0–300 V.	107
C	Electrographite Ek 45 (20 cm dia, 5 mm thick)	Ar (99.999% pure)	0.5–1.3 kW	10–30 nm/ min	5 m torr	Ni and Ta (1 cm × 1 cm × 0.1 cm) platelets 50–100°C	Bias voltage range, 0–300 V.	107

Material	Target	Gas	Power	Rate	Pressure	Substrate	Comments	Ref.
SiC	Hot pressed powder compact (20 cm dia, 5 mm thick)	Ar (99.999% pure)	0.5–1.3 kW	10–30 nm/min	5 mtorr	Ni and Ta (1 cm × 1 cm × 0.1 cm) platelets, 50–100°C	Bias voltage range, 0–300 V.	107
SiC	Hot pressed, very dense powder compact with 0.5% Al_2O_3 (20 cm dia., 5 mm thick)	Ar (99.999% pure)	0.5–1.3 kW	10–30 nm/min	5 mtorr	Ni and Ta (1 cm × 1 cm × 0.1 cm) platelets, 50–100°C	Bias voltage range, 0–300 V.	107
ZnSe	99.999% pure powder compressed and sintered to form a rigid disk.	Spectroscopically pure Ar		20–200 nm/h	1.7×10^{-2} torr	Cut 3° off (100) toward (110) Si-doped n-type GaAs, 160–360°C	Ar admitted through chamber containing freshly sublimed film of Ti. Single-crystal films have been prepared.	108
PLZT	Hot pressing ceramic powder (8 in. dia.)	Ar/O_2 = 5:1	9.5 W/cm²	1 m/h	4.5×10^{-3} torr	Al_2O_3, sapphire and glass, 500–550°C	Substrate bias + 80 V. Excess PbO in the target, 20%.	109
Ekonol	Disk (80 mm dia.) 6 mm thick	Ar	25–100 W	1–1.5 nm/min	5–70 mtorr	Glass, Mo metal foil, NaCl or KBr disks	Pressure > 5 m torr for mechanically stable films. Sputtered films appear to be an amorphous hydrocarbon derivative of the bulk polyester with a resistivity comparable to that of bulk.	110
Polymer metal	Polymer target surface with various sizes of metallic foil	Ar	25–100 W		5–70 mtorr	Glass, Mo metal foil, NaCl or KBr disks		110

Table 2.2 (Continued)

Material	Target	Sputter gas	Sputtering power/ density	Deposition rate	Pressure	Substrate and substrate Temp.	Remarks if any	Ref.
ZrB$_2$ and TiB$_2$	Vacuum hot pressed to yield compact disks (5 cm dia.) Density 80% for TiB$_2$ and 92% for ZrB$_2$ of the bulk value.	Ar (99.999% pure)	800 W	100–150 Å/min	2–3 × 10^{-3} torr		Negative bias, 75–80 V.	111
CuInSe$_2$	Hot pressed (3 in. dia.) CuInSe$_2$; starting materials Cu, In, and Se (99.999% pure)	Ar	50 W		30 mtorr	Glass, Mo Au–Cr/ alumina, ITO/ quartz, Cu	Substrate-to-target distance, 42 mm.	112
GeTe	Hot pressed (6 in. dia.) starting composition, Ge$_{51.3}$Te$_{48.7}$	Ar		5–6 Å/s	6 × 10^{-3} torr	Corning 7059 and Al$_2$O$_3$, RT and > 250°C.	RT gave amorphous films, > 250°C gave crystalline. Actual composition in the film, Ge$_{57.9}$Te$_{48.1}$.	113
Ta$_2$O$_5$	10 cm diameter, 5 mm thick Ta$_2$O$_5$	Mixture of Ar and O$_2$		4–10 nm/min	8 × 10^{-2} torr	Glass coated with In$_2$O$_3$ films	Target-to-substrate distance, 5 cm.	114
BaTiO$_3$	3.5 in. diameter, 99.7% pure BaTiO$_3$	5 or 10% O$_2$ mixture with Ar	400 W	55 Å/min	9 × 10^{-3} torr	99.8% pure Pt foil, 0.025 mm thick, 340–930°C		115

Material	Target	Sputtering gas	Power density	Deposition rate	Pressure	Substrate, temperature	Remarks	Ref.
CdS	99.999% pure cerac target (5 in. dia.)	Ar (99.9995% pure)	0.24–2.82 W/cm²		20–30 mtorr	Glass 60–300°C	Substrate bias voltage, –200 to 0 V.	116
α-Si α-Si:H	4 in. diameter, 99.999% pure Si target	Ar and 15% H₂/85% Ar in Si:H	200 W	80–120 Å/min	2 mtorr	Standard microscope slides and KBr crystals 100–300°C		117
Doped α-Si:H	Doped Si wafers	Ar + H₂	180 W		20 mtorr	Glass, 250°C	Target acted as dopant source.	118
Crystalline and amorphous SiC and SiC:H	Ceramic SiC	Ar and Ar/H₂ mixtures	1.2 W/cm² at 30 mtorr to 0.6 W/cm² at 3 mtorr		3 mtorr to 30 mtorr			119
$Bi_2YFe_{3.8}$, $Al_{1.2}O_{12}$, and Bi_2, $GdFe_{3.8}$, $Al_{1.2}O_{12}$	Sintered disks (70 mm dia.) of each composition	Ar/O₂, 9:1	6.1 W/cm²	6 nm/min	5.3×10^{-2} torr	Corning glass and fused quartz, RT–500°C	Best film quality is obtained when T_s is just below 440°C.	120
SiO₂	Stoichiometric target (99.999% pure), 5 in. diameter	Ar			1.4×10^{-2} torr	50–200°C	Target voltage was varied from 500 to 1500 V.	121
GdFe and CoCr		Ar	100 W		20 mtorr	Glass		122

Table 2.2 (Continued)

Material	Target	Sputter gas	Sputtering power/ density	Deposition rate	Pressure	Substrate and substrate Temp.	Remarks if any	Ref.
ZnO	Pressed disks of 99.999% pure ZnO powder (60 mm dia., 5–6 mm thick)	Ar/O$_2$ = 1:1	50–500 W		1–10 mtorr	Corning 7059 glass 300–700 K	Magnetic field was applied parallel to the target-substrate axis.	123
MoS$_2$	MoS$_2$ (99.9% pure)	Ar	400 W	235–500 (nm/min)	(8–13) × 10^{-3} torr	304 stainless steel, also precoated with Cr$_3$Si$_2$	Target-to-substrate spacing 4.1 cm; substrate bias, –50 to –100 V.	124
ZnO	ZnO target (6 in. dia.)	Ar/O$_2$ gas mixture	150–350 W		10 mtorr	GaAs 100–300°C	Dependence of growth rate on rf power and substrate temperature was investigated.	125
MgO	MgO (99.5% pure)	Ar (99.9997% pure)	400 W 11.3 W/ cm^{-2}		1.06– 2.66 Pa	Pure copper (OFHC) stainless steel (Fe + 25% Cr + 21.2% Ni), RT and 350°C	Studied the electrical insulating properties. Breakdown voltage found to depend on coating thickness, gas composition during sputtering, and substrate material.	144
GdTbFeCo	Iron base mosaic target; nominal area of composition, (Gd$_1$Tb$_1$)$_{30}$ (Fe$_4$Co$_1$)$_{70}$	Ar			10–50 mtorr	Corning 0211 glass	Both target and substrate table water cooled. Films sputtered using moderate argon pressures and substrate bias voltages are dense, smooth, morphologically featureless,	145

Material	Target	Gas	Intensity	Pressure	Rate	Substrate	Comments	Ref
Gold	Gold	(Ne,Ar)-O_2	10^7 W/m^2			Si (111)	and strongly resistant to oxidation. Results indicate a large increase in resistivity; also a decrease in reflectivity of half the value of metallic gold. Possible formation of a second phase with no long-range crystallographic order coincident with Au-O band formation.	146
$BaTiO_3$ (BaSr)TiO_3	Polycrystalline $BaTiO_3$ and (BaSr)TiO_3	O_2		50 Pa		Platinum, stainless steel, Al_2O_3, MgO plates, 700–1100 K	Systematic studies on the piezoelectric properties of $BaTiO_3$ thin films.	147
α-Si_{1-x} Sn_x:H	Silicon target with stripes of 99.999% pure tinfoil at the center.	Ar/H_2 mixture (24% H_2)		P_{Ar} 5mtorr	2–4 Å/s	Si and Corning glass 220 or 250°C	Mole fraction of x (Sn) varied up to 0.51. Alloy compositions measured using X-ray microprobe analysis. Addition of Sn moves the conduction band edge, thereby closing the optical bandgap.	148
Al-Ni-Si alloy	Homogenous Al-Ni-Si (at. %: Al, 65; Ni, 15; Si, 20)	Ar		Varied: 2–5 Pa	0.1 nm/s	Carbon/KBr/ Al alloys/ glass RT	Different substrates were used, depending on the analysis to be taken. Target biased between 800 and 1200V. Structure and properties of the films studied.	149

Table 2.2 (Continued)

Material	Target	Sputter gas	Sputtering power/ density	Deposition rate	Pressure	Substrate and substrate Temp.	Remarks if any	Ref.
ITO	4 in. hot pressed (90% In_2O_3 + 10% SnO_2) ceramic	Ar or Ar + O_2	Varied	Varied	Varied	Glass, quartz and freshly cleaved NaCl crystals 20°C	Effects of target substrate separation on growth characteristics and films properties were studied.	150
ZnO	ZnO (99.999% pure)	O_2/Ar mixture (28% O_2)	100 W	0.3 Å/s	2×10^{-3} torr	Corning glass 7059 and slides of c-plane sapphire crystal, 100–400°C	Structural properties of the films were investigated by X-ray diffraction; morphology studies by SEM. The films were in general optically clear and quite homogeneous and uniform.	151
ZnTe	Polycrystalline ZnTe (99.999% pure)	Ar (99.999%)	100 W	0.8 μm/h	Ar pressure, 0.15 torr	GaSb (100) wafers, 120–370°C	Films were characterized by optical reflectance spectrum, RHEEDS, and ion beam channeling measurements. Deposition temperature > 320°C required for films of high crystalline quality.	152
SrTiO₃	SrTiO₃: thickness 6 mm; diameter, 75 mm	Ar and O_2 (99.999% pure)	Rf operated at 1 kV		8×10^{-3} mbar	Al_2O_3 500°C	Films prepared with Ar/O_2 ratios of 1:0, 5:1, and 10:1. Polycrystalline films were obtained on substrate heated to 500°C. This is independent	153

Material	Target	Gas	Power	Deposition rate	Pressure	Substrate	Remarks	Ref.
CuGa Se$_2$	Compound target with 2% Se excess (at. %)			140 Å/min at 60°C 85 Å/min at 400°C	20 mtorr	Glass slides 60–400°C	of Ar/O$_2$ ratio. Films characterized by XRD, SEM, and EDAX. resistivity measurements with temperature 100–450K were made. Compositional, structural, optical, and electrical properties were strongly influenced by growth temperature.	154
Y$_2$O$_3$	Y$_2$O$_3$ sintered (purity > 99.99%)	Ar and O$_2$				Glass substrates coated with ITO 0.2 μm thick, 100°C	Oxygen content, rf power density, and pressure were varied. Deposition rates and dielectric properties of the films were studied.	155
Cobalt rich Co-Fe	Co plate (99.9% pure) bearing Fe chips (99.9% pure)	Ar and N$_2$	5.1 W/cm^2	70-60 nm/min	Total (Ar + N$_2$) gas pressure, 9 × 10^{-3} torr	Corning glass	A constant magnetic field of 200 Oe was applied parallel to the film to induce in-plane uniaxial anisotropy. Deposition rate changed with increasing N$_2$ partial pressure from 0 to 7.2 × 10^{-4} torr. Relationship between crystal structure and the magnetic properties studied.	156
InSb	Polycrystalline InSb, 3 in. diameter	Ar (99.9995% pure)	200 W	200 Å/min	5 × 10^{-3} torr	Sapphire (10 × 10 mm^2 area)	X-ray diffraction and SEM results showed that the InSb layer on sapphire (001) is epitaxial, with (111) parallel to the substrate surface.	157

Table 2.3 Deposition Conditions of Various Materials Prepared by Magnetron Sputtering

Material	Target	Sputter gas	Sputtering power/density	Deposition rate	Pressure	Substrate and substrate Temperature	Type	Remarks	Ref.
BaTiO$_3$	BaTiO$_3$ ceramic (70 mm dia.)	Ar/O$_2$, 80:20	80 W	90Å/min, on average	1×10^{-3}, 2×10^{-2} torr	Pt and fused quartz 500–700°C	Rf planar	Distance between substrate and target, 4 cm.	126
CdSe	Hot pressed CdSe (99.99% pure)	Ar	500 W	0.63 m/min		Sil wafers Corning 7059 glass	Rf planar	Deposition rate increased linearly with average power.	127
α-Si:H	High purity polysilicon	High purity Ar/H$_2$ mixtures	100–300 W			Pyrex glass unheated substrates 320 K	Rf planar	Source-to-substrate distance, 12 cm. deposition rate measured for a wide range of H$_2$ and Ar pressure and applied rf power.	128
PZT	Pressed commercial grade with excess PbO, 5–20 wt % (0.3–0.5 cm thick, 4 cm dia.)	100% O$_2$ and Ar + O$_2$ at various pressures	300 W	0.5–0.7 m/h with 100% O$_2$ at 40 mtorr	10–100 mtorr	Pt, conducting glass, polished Si (111) and fused quartz 100–650°C	Rf planar	Target-to-substrate distance, 5 cm.	129
ZnO	Commercially available sintered ZnO (99.999% pure, 80 mm dia.)	Ar	32–85 W	2.5–25 nm/min	1×10^{-2} to 6×10^{-2} torr	Corning 7059 glass	Rf planar		130

Material	Target	Gas	Power	Deposition rate	Pressure	Substrate	Sputtering	Remarks	Ref.
LiNbO$_3$	100 mm diameter sintered powder synthesized from Li$_2$O$_3$ and Nb$_2$O$_5$ (99 and 99.9% pure, respectively)	Ar/O$_2$	100 W, 13.56 MHz	0.2–0.3 m/h		Fused quartz plates, 0.2 mm thick, cooled by water or liquid N$_2$	Rf planar	No noticeable difference in the properties of films on substrates cooled by water and liquid N$_2$.	131
Mo	Mo (99.9% pure)	99.996% pure Ar and Ne		6 nm/s	> 10 torr Ar	3 in. Si wafer (100)	Circular dc magnetron	Substrate holder with rf bias; source to-substrate distance, 10 cm	132
MoSe$_2$	Pressed MoSe$_2$ powder (10^{-1} m dia.)	Ar	2.5×10^4 W/m^{-2}	10–25 nm/ min	1.5×10^{-2} to 5×10^{-2} torr	Silica glass disks -70–150°C	Rf planar	Target-substrate separation, 7 cm.	133
SnO$_2$	Hot pressed SnO$_2$ disk (10 cm dia., 99.99% pure)	Ar (99.999% pure) and O$_2$	50 W	12 nm/min	5×10^{-3} torr	Unheated Pyrex glass	Rf planar	Target-substrate separation, 7 cm.	134
Si-Cr alloys	Si and Cr (99.99% pure)	Ar			2.5×10^{-3} torr	Corning 7059 glass plates or Si wafers coated with SiO$_2$ films	S-gun	Cosputtering, continuous variation of Si/Cr ratio.	135
SiO$_2$	High purity SiO$_2$	30% O$_2$ + 70% Ar	500 W		5×10^{-3} torr	Si crystal 200°C	Rf planar		136
92.2 wt % Cu-7.8% Sn and 26% Cu-74% Sn	Alloy targets (90 mm dia.) prepared by melting 99.99% pure Sn in vacuum	Ar	8 W/cm^2	1.5 nm/s	4.5×10^{-3} torr	Stainless ferritic steel	Planar disk like	Temperature gradient along the length of the substrate in the range 1220–330 K.	137

Table 2.3 (Continued)

Material	Target	Sputter gas	Sputtering power/density	Deposition rate	Pressure	Substrate and substrate Temperature	Type	Remarks	Ref.
Y-Ba-Cu-O	$YBa_{1.86}Cu_{2.86}O_y$ Powder mixture of Y_2O_3, BaO_2, and CuO cold pressed into a disk (3 mm thick, 50 mm dia.) and sintered in oxygen-enriched atmosphere	Ar		52 nm/min at center of discharge to 4 nm/min at radius of 6.5 cm	9 mtorr	Fused quartz	Dc magnetron	Target-to-substrate distance, 4 cm. Superconducting films formed by sputtering of the single oxide target followed by annealing in oxygen.	138
Y-Ba-Cu-O	Y-Ba-Cu-O compacted and sintered Y-Ba-Cu-O powders of various compositions	Ar		0.05-0.3 nm/s	5 mtorr	Sapphire and Al_2O_3	Dc magnetron	Substrate temperature, RT to 500°C. Structure of the film studied by X-ray diffraction analysis.	103
$Ti_{1-x}B_x$	99.98% pure Ti disk (150 mm dia.) + 99.9% B single-crystal particles (≈ 5 mm) placed in it	Ar	1 kW	7-15 $gmin^{-1}cm^{-2}$	3.8×10^{-3} torr	Mo substrate 870 K	Rf planar	Composition varied by varying B. Distance between target and substrate, 80 mm.	139
Y-Ba-Cu-O	$(Y_{0.25}Ba_{0.75})_{0.66}$ $CuO_{3-\delta}$; $(Y_{0.29}Ba_{0.71})_{0.70}$ $CuO_{3-\delta}$, $(Y_{0.33}Ba_{0.66})_{0.66}$ $CuO_{3-\delta}$	Ar + O_2, $O_2/(Ar + O_2) < 0.75$	200 W	2-150 Å/min	0.4-13 Pa	$(00\bar{1}2)$ sapphire, (100) $SrTiO_3$, (100) MgO	Planar rf magnetron	Composition of prepared films strongly depended on the sputtering gas. Superconductivity of the	158

Material	Target	Method	Substrate	Pressure	Rate	Gas	Power	Remarks	Ref.
CdZnSO and ZnSO	Mixed ZnO and CdS or ZnS	Rf magnetron	Glass					films was influenced by substrates, and the sputtering and annealing conditions. Target contained 40% ZnO and 60% CdS. The lattice parameters of the films obtained were very near those of semiconductor CuInSe2.	159a
Bi(Pb)-Sr-Ca-Cu-O	Sintered disk; composition: Bi$_{1.7}$ Pb$_x$Sr$_2$ Ca$_{2.5}$Cu$_{3.75}$O$_y$ ($x = 1.5$, 1.75, 2.0, 2.25, and 2.5)	Dc magnetron	Polished MgO (100) 400°C	0.01 torr	20-30 nm/min	Ar + O$_2$ (10%)		After deposition, the films were cooled to room temperature in oxygen at 1 torr, then annealed in air for 3 hours at 850°C. Superconducting properties were sensitive to Pb content.	159b
Al$_2$O$_3$	Sintered Al$_2$O$_3$ (99.9%)	Rf magnetron	Two types of iron-based alloy (Fe-C-Ti), Ambient	40×10^{-3} torr	0.2 nm/s	Ar 99.999%	5 W/cm^2	One type of substrate contained 0.14% carbon and 0.7% titanium; the other type, 0.47% carbon and 0.7% titanium. Al$_2$O$_3$ deposited specimens were annealed at 900-1100 K in high	160

Table 2.3 (Continued)

Material	Target	Sputter gas	Sputtering power/density	Deposition rate	Pressure	Substrate and substrate Temperature	Type	Remarks	Ref.
								vacuum. Adherence of Al_2O_3 films on first type was better than that of second type.	
$YBa_2Cu_3O_{7-x}$	Single stoichiometric target $YBa_2Cu_3O_{7-x}$ 3 in. diameter	Ar	100 W	1 Å/s	10 mtorr	Silicon, SiO_2/Si, Si_3N_4/Si 100°C	Rf magnetron	As-deposited films were amorphous and insulating. In situ high temperature X-ray diffraction study to examine crystallization and film substrate reactions of amorphous $YBa_2Cu_3O_{7-x}$ thin films on different substrates were made.	161
WO_3	WO_3 powder (99.99%) target	Ar/O_2 mixture containing various amounts of O_2	Rf power, 100 W		30 mtorr	MgO (100) single crystal cleaved surface, sapphire ($11\bar{2}0$) cut (the A plate), (0001) cut (the C plate), and ($1\bar{1}02$) cut (the R plate) single crystals 300–500°C	Rf magnetron	Films orientation depended on substrate and on oxygen concentration in the spttuering gas.	162

Material	Target	Power	Rate	Pressure	Reactive gas	Substrate, temperature	Method	Remarks	Ref.
ErBa$_2$Cu$_3$O$_{7-x}$			2 nm/min	80–100 mtorr	Ar/O$_2$ 1:1	MgO single crystal, 650°C	Rf magnetron	As-grown films were single-crystal-like and consisted of subgrains 0.1–0.2 μm. A annealed films were epitaxial with either the a or b axis of the ErBa$_2$Cu$_3$O$_{7-x}$ unit cell along the <100> direction of the MgO substrate.	163
Pb-doped Bi–Sr–Ca–Cu–O	Single sintered target					MgO, 400°C	Dc magnetron	After deposition films postannealed at 850°C in air for 3 hours; films highly c-axis-oriented, zero resistivity at 115 K. Composition of sputtered films same as that of target except for a slight decrease in Bi concentration.	164a
Molybdenum	Mo (99.9%)	44–152 W	28–125 nm/min	1.2×10^{-2} to 6×10^{-3} mbar		Cast iron, 75 × 25 × 2 mm	Dc magnetron	Films characterized in terms of thickness, crystalline structure, adhesion, and roughness. Dependence of film adhesion on in situ cleaning by ion bombardment and bias voltage have been investigated.	164b

Table 2.3 (Continued)

Material	Target	Sputter gas	Sputtering power/density	Deposition rate	Pressure	Substrate and substrate Temperature	Type	Remarks	Ref.
a-Si$_{1-x}$ C$_x$ 0 ≤ x ≤ 1.0	Graphite disk and Si chips on the disk	Ar	270 W	80 to 7 nm/s as x increased from 0 to 1.0	5×10^{-3} torr	(111) Si and Corning 7059 glass RT	Rf magnetron	Carbon content in film was controlled by varying the number of silicon chips. Atomic structure of amorphous films studied by TEM, IR absorption, electron spin resonance spectroscopy, and electrical conductivity.	165
Gd-Ba-Cu-O	GdBa2Cu3-O$_{7-x}$	O$_2$-Ar				Si 740–770°C	Magnetron	By in situ treatment at 400°C in pure oxygen atmosphere, the orthorhombic phase with full superconducting transition at 84 K was obtained.	166
Lead zirconate titanate	0.1 m diameter PZT metal target	100% O$_2$	50 W	0.12 m/h	5–6 Pa	Single-crystal Si wafers, conducting glass, alumina, and optically polished fused quartz disk	Dc magnetron	Structural, electrical, and optical properties were investigated. Influence of growth conditions on surface morphology examined by SEM.	167a

70

Material	Target	Sputter gas	Power	Deposition rate	Pressure	Substrate	Method	Comments	Ref.
WB_x (27 at. % B)	Compound target (99.8% purity)	Ar	1.3 W/cm^2	14 nm/min	Ar pressure varied: 0.5–2.8 Pa	Si and GaAs	Rf magnetron	Heat treatment at 200°C for 15 minutes for Si or GaAs was done before deposition of metal films. Substrate bias voltage varied to obtain desired film properties. Electrical resistivity, deposition rate, and intrinsic film stress determined under various conditions.	167b
$Tl_2Ca_2Ba_2Cu_3O_x$	$Tl_2Ca_2Ba_2Cu_3O_x$ powder target, 2–3 mm thick	Ultra high purity Ar	250 W	30 Å/min	5 mtorr	(100) $SrTiO_3$	Rf magnetron	Powder target was prepared from stoichiometric amounts of high purity BaO, CaO, CuO, and Tl_2O_3 powders. Target was enriched with about 20% excess Tl_2O_3 to compensate for loss of Tl. As-deposited films heat treated in two steps: (a) sintering in air at 850°C for 15 minutes in an excess Tl partial pressure; (b) annealing in an oxygen flow of 500 standard cubic feet per	168

Table 2.3 (Continued)

Material	Target	Sputter gas	Sputtering power/ density	Deposition rate	Pressure	Substrate and substrate Temperature	Type	Remarks	Ref.
								minute at 750°C for 30 minutes. Zero resistance temperature of 107 K obtained in the films.	
Y–Ba–Cu–O–Ag	Y–Ba–Cu–O–Ag composite target containing 10 wt % of silver			100 Å/min		Polycrystalline YSZ, (001) YSZ, and (001) SrTiO₃, no extra heating	Planar type rf magnetron	Sputtered films heat treated at 900°C for 20 minutes in Ar, followed by oxygenation in flowing O₂. Temperature decreased at 50°C/h rate up to 500°C, where the film remained for 5 hours, and furnace cooled. Films had very smooth surfaces compared to postannealed HTSC film without silver.	169

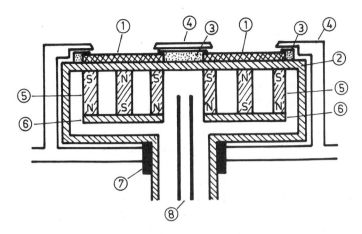

Figure 2.6 Cross-sectional schematic diagram of the planar magnetron cathode with multiple targets: 1, target; 2, backing plate; 3, target holder; 4, ground shield, 5, permanent magnet; 6, pole piece; 7, ceramic insulator; 8, cooling water inlet. (Data form Ref. 61.)

The basic operation of a magnetron cathode is somewhat altered in the case of rf magnetron sputtering. In the application of an rf potential to an insulator (or, via a blocking capacitor, to a metal electrode) in magnetron sputtering, the difference in the mobility of the ions and electrons results in a negative self-biasing of the cathode, which in turn gives the necessary potential for sputtering.

Quite a large number of papers have been published during the past few years on the preparation of thin films, including HTS oxide films, by magnetron sputtering. Table 2.3 lists the deposition conditions for a selected number of materials.

The conventional method, employing a planar cathode for depositing thin films on large substrates with uniform thickness, has the limitation of the size of the target in accordance with that of the substrates. It is expensive and difficult to produce large, high purity targets without cracks. To overcome this difficulty, a planar magnetron cathode with three targets was proposed by Serikawa and Okamoto [61], who successfully deposited silicon films using 0.5 wt % phosphorus-doped silicon as the target material. A cross-sectional schematic diagram of their setup is shown in Figure 2.6. Permanent magnets are placed under the targets and the targets are mechanically clamped to the backing plate with target holders, which are fastened with hollow screws. The outside edges of the targets are covered with a ground shield. The space between the ground shield and the high voltage cathode is about 3 mm (less than the cathode dark space) so that the ground shield can obstruct the arc discharge and prevent current losses.

Serikawa and Okamoto [61] had shown that phosphorus-doped Si films deposited from phosphorus-doped targets using their setup had excellent uniformity in thickness and resistivity. Table 2.4 gives the characteristics of the magnetron cathode.

Kobayashi et al. [62] developed a new sputtering method to achieve high rates of deposition of refractory metal silicide films. They employed a planar magnetron sputtering cathode with two electromagnetic coils coaxially wound in the magnetic yoke. A multiple ring target, consisting of three target pieces, was used. A center silicon disk, a molybdenum ring, and an outer silicon ring were arranged concentrically to form the single target plate. The glow ring was generated concentrically with the target plate, and the diameter of the ring could be changed by controlling the currents in the two electromagnetic coils. The glow ring diameter was controlled so that each target pieces got selectively sputtered and was adjusted to obtain the film of required composition and thickness. The schematic of the setup is shown in Figure 2.7. This single-wafer, single-cathode deposition technique permitted precise control of film composition and its distribution over a 4 in. wafer with a high deposition rate (100 nm/min).

Magnetic materials present difficulties in the preparation of thin films by magnetron sputtering, and Chang et al. [63] have developed a dc magnetron sputtering system to achieve a high rate of sputtering of nickel. They have shown that a minimal or sufficient magnetic field strength of about 300 gauss (G) at the Ni target surface, which can be generated with a strong magnet in the cathode assembly, is crucial in forming a stable plasma, and that the sputtering equipment for magnetic material deposition must contain an adjustment mechanism in its cathode design to control the magnetic field strength. A cross-sectional view of the sputtering system is shown in Figure 2.8. The cathode assembly includes the nickel target bonded onto a copper backing plate, a water cooling system, and a permanent magnetic assembly. The position of the magnetic assembly relative to the target can be adjusted and the magnetic field strength from 300 to 800 G can be controlled by ad-

Table 2.4 Characteristics of the Magnetron Cathode

System	rf planar (13.56 MHz)
Target	0.5 wt % phosphoros-doped Si (diameter, 4 in.; thickness, 5 mm)
Substrate	4 in. diameter Si wafers with 1 μm thick SiO_2 films
Separation between cathode and substrate	80 mm
Substrate temperature	100°C
Gas	Argon
Pressure	> 5 × 10^{-3} torr

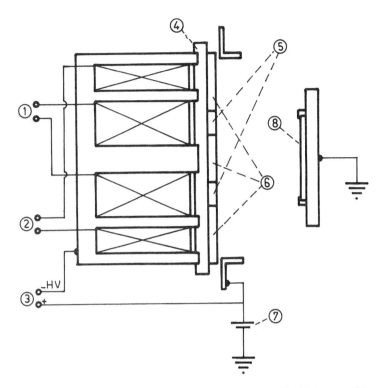

Figure 2.7 Schematic diagram of the sputtering setup for high rates of deposition of refractory metal silicide films: 1, inner coil; 2, outer coil; 3, sputtering power supply; 4, Cu backing plate; 5, molybdenum; 6, silicon; 7, anode bias; 8, wafer. (Data from Ref. 62.)

justing the spacing between the magnets and the target. A shield is provided to block low angle incident ions.

A conventional planar magnetron sputtering system, with a hollow cathode arc electron source added, has been reported by Cuomo and Rossnagel [64a]. A schematic diagram of this setup appears in Figure 2.9. This is a triode device in which the cathode is a magnetron cathode, and the hollow cathode electron source serves as a second cathode. The electron source is placed adjacent to the magnetron cathode such that it is near the edge of the cathode but still in the primary magnetic field. The positioning of the hollow cathode in the magnetic field is critical, and the electrons emitted from the hollow cathode produce additional gas ionization and result in the increase in plasma density at constant voltage. Deposition rates of up to 10 times were recorded over the conventional magnetron at the same voltage. The operating pressure could also be reduced significantly, and the magnetron device

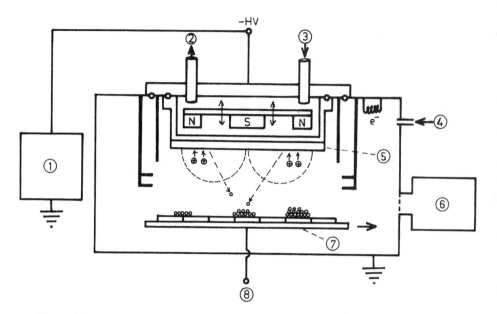

Figure 2.8 Cross-sectional view of the magnetron sputteing system: 1, dc power supply; 2, water outlet; 3, water inlet; 4, argon inlet; 5, nickel target; 6, vacuum pump; 7, substrate pallet; 8, substrate bias. (Data from Ref. 63.)

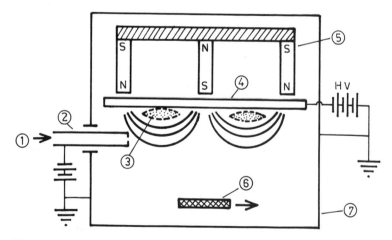

Figure 2.9 Schematic diagram of the planar magnetron device: 1, argon inlet; 2, hollow cathode; 3, plasma, 4, cathode; 5, magnet assembly; 6, sample; 7, vacuum chamber. (Data from Ref. 64a.)

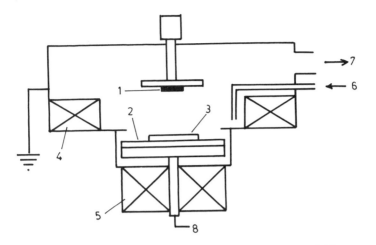

Figure 2.10   Schematic diagram of the compressed magnetic field magnetron sputtering system: 1, substrate; 2, Bi metal disk; 3, $SrCa_{1.26}Cu_{1.63}O_x$ disk; 4, compressing coil; 5, magnetron coil; 6, gas inlet; 7, to vacuum pump; 8, to rf power supply. (Data from Ref. 64e.)

could be operated in the 0.2–0.6 mtorr pressure range and at full power as low as 5 $\times 10^{-5}$ torr, in argon. With this low pressure sputtering, direct line-of-sight deposition is possible, and this allows the magnetron–target distance to be increased. The increased distance gives better film uniformity and also keeps the samples from being bombarded by ions or electrons, as a result of the proximity of the plasma. The increased distance also allows the operation of multiple processes simultaneously, such as multiple magnetron deposition, combined evaporation/deposition, or ion beam bombardment during deposition. Cuomo and Rossnagel [64a] prepared Ta/Au layers using their setup as just described.

A high deposition sputtering technique called compressed magnetic field (CMF) magnetron sputtering was developed by Hata et al. [64b, c]. Here there are two coils, one around and the other under the target, to confine the leakage lines to the target surface and to produce a leakage magnetic flux on the target surface, respectively. Using this technique, good quality hydrogenated amorphous silicon (a-Si:H) has been prepared using He gas with H_2 gas added [64d]. Yoshimoto et al. [64e] have recently reported using this type of sputtering technique to prepare HTS thin films of the Bi-Sr-Ca-Cu-O system. The schematic diagram of their sputtering system is shown in Figure 2.10. The target consisted of a double disk: a small sintered Sr-Ca-Cu-O disk (composition $Sr_1Ca_{1.26}Cu_{1.63}O_x$) with a diameter of 38 mm and thickness 2 mm, placed on a large Bi disk (diameter, 50 mm; thickness, 2mm). Sputtering was carried out by applying an rf field between the target and the substrate.

The details of the sputtering parameters reported are given below:

Sputter gas, Ar/O$_2$ 8:2 mixture
Pressure, 5 mtorr
Rf power, 100–230 W
Optimum power, 200 W
Substrate, MgO (100)
Substrate temperature, 600°C (without intentional heating)

The film after deposition was cooled to room temperature. The bismuth content in the film was controlled by changing the sputtering area of the binary target. The composition of the deposited films could be varied by controlling the currents in the magnetron and compressing coils. When the compressing and magnetron coils were 4.5 and 1.0 A, respectively, the composition of the as-deposited films was nearly 2.6:2.0:2.3:3.9. (Bi/Sr/Ca/Cu: this is close to the high T_c phase ratio of 2:2:2:3 for Bi/Sr/Ca/Cu). The films were deposited under these conditions with a deposition of about 130 Å/min. Films annealed between 840 and 890°C for 0.5 to 15 W exhibited superconductivity with T_c onset and T_c ($R = 0$) at 110 and 76 K, respectively.

2.6 FACING TARGETS SPUTTERING (FTS)

Although some of the magnetron-type sputtering systems already described could attain high deposition rates and low substrate temperatures, no magnetron apparatus can sputter-deposit magnetic materials at high rates. Hoshi et al. [65] developed a new sputtering apparatus with two facing targets to prepare thin films of magnetic materials at high deposition rates without greatly increasing the substrate temperature. This facing target sputtering apparatus has been used to prepare magnetic thin films of iron, nickel, and permalloy [66]. Here two disk targets of the same size are arranged in parallel, as shown schematically in Figure 2.11a. When a magnetic field is applied perpendicular to the target surface, it confines the high energy electrons in the space between the facing target planes. The confinement of the electrons promotes the ionization of the gas, resulting in a higher rate of sputtering. The arrangement of the substrate and the targets makes the electron bombardment of the substrate almost negligible. The temperature of the substrate is then not unduly raised. The schematic of the sputtering arrangement used by these authors is shown in Figure 2.11b.

A magnetic field up to 1200 G was applied perpendicular to the target surface by an external coil. The targets were nickel or iron (99.9% purity) disks, 60 mm in diameter and 3 mm thick. For deposition of Mo permalloy films, composite targets consisting of disks of ion and nickel and chips of molybdenum (Figure 2.11c) was used. The composition of the film was controlled by changing the diameter of the iron disks and the number of the Mo chips. The distance between the targets was kept at 50 mm. Substrates (Corning 7059 glass: 20 mm × 3 mm × 1 mm) were held vertically at a distance of 40–70 mm from the common axis of the target. Sputtering was done at a pressure of 5×10^{-4} to 8×10^{-2} torr, and a stable glow discharge could be maintained up to a discharge current of 1.5 A. Hoshi and coworkers had studied the crystal structure, composition, and the surface morphology of the films obtained, and had shown that with this setup they could deposit magnetic films on substrates at a temperature below 180°C with a deposition rate as high as 50 times that of the conventional d c diode type of sputtering.

The FTS approach has been used by Naoe and coworkers to prepare Co-Cr thin films for high density perpendicular magnetic recording media [67,68a]. The same groups recently reported the preparation of Fe/Ti multilayer films of various thickness of Fe and Ti [68b] and also TbFeCo [68c] thin films with large Kerr rotation angle by the FTS method. FTS has been used also to prepare HTS thin films. Hirata and Naoe [68d] have reported the preparation of YBCO thin films at low substrate temperatures. MgO <110> and Sr Ti O3 <110> were used as substrates, and the substrate temperature varied from room temperature to 500°C. The X-ray diffraction studies showed that the as-deposited films crystallized in the tetragonal phase. The composition of the films was found to be same as that of targets. Since the substrate is in a plasma-free position, resputtering does not happen, and consequently there is no difference in composition between the film and the target. High T_c (≈ 85 K) films of extremely smooth surfaces have been obtained at about 410°C.

2.7 ION BEAM SPUTTERING

One main disadvantage of the slow discharge system of sputtering is the high background pressure, which results in the inclusion of gas molecules in the sputtered films. In ion beam sputter deposition the ion beam generated at an ion source is extracted into the high vacuum by an extraction voltage and directed to a target of the desired material, which is sputtered and deposited on a nearby substrate. A simple schematic of ion beam sputtering is shown in Figure 2.12. Apart from the low background pressure, which results in less inclusion of gas and less scattering of the sputtering particles during transit, ion beam sputtering allows greater isolation of the substrate from the ion production process, unlike the glow discharge sputtering. As noted earlier, in glow discharge sputtering, the target, the substrate, and the depositing thin film are in the plasma atmosphere during deposition. Also, the directionality of the beam allows the investigator to vary the angle of incidence of the beam on the target as well as the angle of deposition on the substrate. Other advantages of ion beam sputtering over conventional sputtering processes are as follows.

1. The narrow energy spread of the ion beams enables us to study the sputter yield as a function of ion energy.
2. The process allows accurate beam focusing and scanning.

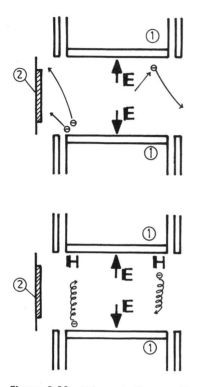

Figure 2.11 Schematic diagram of facing targets sputtering system. (a) Target configuration: 1, target; 2, substrate; top without magnetic field; bottom, magnetic field applied. (b) Sputtering system: 1, iron column; 2, ground shield; 3, magnet coil; 4, target; 5, substrate; 6, to vacuum pump; 7, dc high voltage source; 8, gas inlet. (c) Composite target (schematic diagram): 1, nickel target; 2, iron target; 3, molybdenum; 4, substrate. (Data from Ref. 66.)

3. Changes in target and substrate materials are allowed, keeping beam characteristics constant.
4. Independent control of ion beam energy and current is possible.

Again the target and the substrate are independent of the acceleration electrode, and therefore the damage due to collision of ions is much less than in conventional sputter deposition. Also the influence of the residual gas is much less because the chamber can be kept at a low pressure, the ion source being independent of the chamber. Also, ion beam sputter deposition has become useful in the field of epitaxial growth of semiconductor films. It is possible to deposit a wide variety of materials, with the condensing particles having a kinetic energy exceeding 10 eV, in a high vacuum environment. This will lead to high surface diffusion rates, even at low substrate

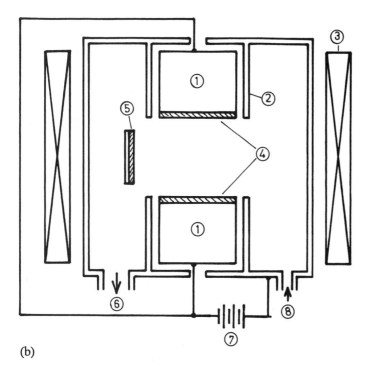

(b)

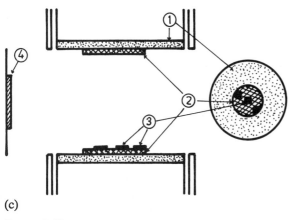

(c)

Figure 2.11 (Continued)

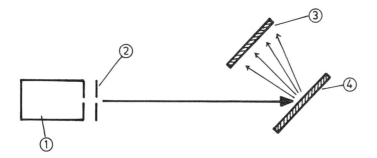

Figure 2.12   A simple ion beam sputtering system: 1, ion source; 2, extraction electrode; 3, substrate; 4, target.

temperatures, a condition favorable for epitaxial diffusion. The main disadvantage is that the bombardment target area is small and the deposition rate generally is low. Also ion beam sputter deposition is not suited for depositions of uniform thickness over large substrate areas.

The two most generally used types of ion source for ion beams sputter deposition are the Kaufman source and the duoplasmatron. For more information on these and other ion sources, interested readers may refer to Reference 69 and the references therein.

The ion beam sputtering deposition technique has been used to prepare metallic, semiconductor, and dielectric films. The following list offers a representative survey of the materials prepared: Au, Cu, Nb, W, SiO_2, TiO_2 [70]; Si, GaAs, InSb [71]; Mo, Ti, Zr, W, Cr [72]; Ta [73]; Ni, Al, Ni_3Al, Au [74]; Au, Ag, Co, Pt, Ni, Mo, AlN, Si_3N_4, Cr_3C_2, Ta_5Si_3, W, Cr, ZrO_2 [75]; Si:H [76]; Si [77]; ZnO [78a]; amorphous "diamondlike" carbon [78b]; $Co_{100-x}-Cr_x$ (x = 17–23) [78c]; (Co_{90} $Cr_{10})_{100-x}$ TM_x, where TM denotes vanadium, niobium, molybdenum, or tantalum, and x is 0–20 at. % [79a]; ZnS [79b]; rare earth Fe-Co [80a]; Al_2O_3 [80b]; Co-Cr [80c]; ZnO:Al [81a]; Cu/Ni and Fe/Ni multilayer [81b]; ITO [81c]; and YBCO [81d, e].

Described below are some recent reports on the preparation of thin films by ion beam sputter deposition.

Kitabatake and Wasa [82] used the setup shown schematically in Figure 2.13 for the deposition of carbon films. The effects of hydrogen ion bombardment were also studied. The 99.999% pure graphite disk target (100 mm diameter) was bonded to the water-cooled holder. An electron bombardment ion source with an ion beam diameter of 25 mm was used. The incident angle of the ion beam was about 30° to the target, and the substrate was placed near the target as shown in Figure 2.13. At

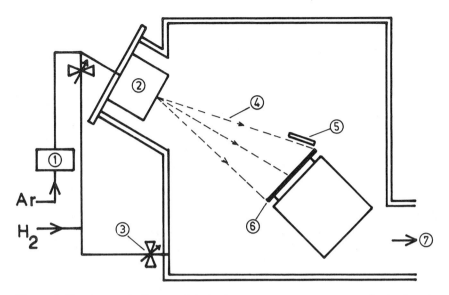

Figure 2.13 Schematic diagram of the ion beam sputter system apparatus for the deposition of carbon films: 1, mass flow controller; 2, ion source; 3, needle valve; 4, ion beam; 5, substrate; 6, target; 7, to vacuum pump. (Data from Ref 82.)

this position the ion beam sputtered the target and also grazed the substrate. The sputtering conditions are given in Table 2.5. The substrates were either Si (111) or fused quartz plates. Three types of film were deposited under different conditions to study the effect of hydrogen ion bombardment.

1. Pure argon fed through the ion source: the argon ions grazed the surface of the target.

2. Hydrogen additionally fed through the needle valve directly to the vacuum

Table 2.5 Sputtering Conditions for the Preparation of Carbon Films

Target	Graphite plate, 100 mm diameter
Ion source energy	1200 eV
Ion beam current	60 mA
Source-to-target distance	250 mm
Gas pressure	5×10^{-5} 2×10^{-4} torr
Film growth rate	300–400 nm/h

chamber: hydrogen molecules impinged on the surface of the substrate during deposition. The grazing ions were only argon.

3. Hydrogen additionally fed through the ion source: the grazing ions were a mixture of argon and hydrogen (H^+ and H_2^+). Kitabatake and Wasa had observed that the properties of the deposited carbon films were influenced by hydrogen ion bombardment. The hydrogen ion bombardment activated the growth of diamond on the deposited carbon films.

Hydrogenation profoundly affects the physical properties of amorphous carbon films. Of the methods of preparing thin films of hydrogenated amorphous carbon, ion beam sputter deposition with hydrogenation of the sputter gas enables one to change systematically the level of hydrogenation of the films and study their effects. Jansen et al. [83] used an ion beam 2 in. diameter, extracted from a 6 in. Kaufman-type ion source, for preparing carbon films from a high purity carbon disk; they studied the effect of hydrogenation on the properties of amorphous carbon films deposited by ion beam sputter techniques. A schematic diagram of their setup is shown in Figure 2.14. The system background pressure was about 10^{-7} torr, and the pressure during deposition was maintained at a total pressure of 3×10^{-4} torr by introducing hydrogen-argon mixtures of varying composition. At the maximum beam hydrogenation of 90%, the films contained 35-40 at. % hydrogen. The ion beam energy was 1200 eV under typical operating conditions, and the deposition rate was of the order of 0.1 μm/h.

Gulino [84] prepared thin films (300-3000 Å thick) of ZnTe by sputtering a pressed powder ZnTe target with an argon ion beam, using a commercially available sputtering system. The schematic arrangement of the ion beam sputtering setup is shown in Figure 2.15. The deposition conditions are given in Table 2.6.

Takeuchi et al. [85] developed a simple and compact ion beam source and used it to sputter-deposit copper and ruthenium oxide films. The schematic diagram of their ion beam sputter deposition system is shown in Figure 2.16. The target (a copper plate and a compressed RuO_2 powder disk) was placed 5.5 cm away from the acceleration electrode (A). The following procedure was adopted for the sputter deposition of thin films.

The chamber was evacuated to a pressure of 2×10^{-5} torr. The argon gas was introduced into the ion source, and the pressure in the chamber adjusted at $(3-4) \times 10^{-4}$ torr. A d c voltage of 600-700 V was applied between cathode and anode to ignite the discharge, and the gas flow was controlled to maintain a pressure of 2×10^{-4} torr in the chamber. An acceleration voltage (1-6 kV) was applied to extract and accelerate the argon ions. The argon ions bombarded the target, and the material deposited on the glass substrate to form thin films. During the sputter deposition the discharge voltage and discharge current were kept at 350 V and 200 mA, respectively. The properties of these sputter-deposited films were investigated, and it was observed that the crystallinity of the copper films increased with the acceleration voltage, whereas RuO_2 films were amorphous irrespective of the accelera-

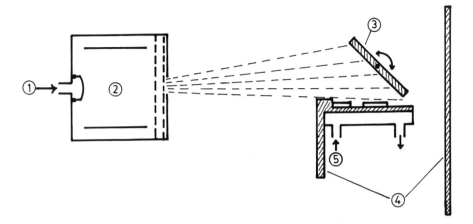

Figure 2.14   Schematic diagram of the ion beam sputter apparatus for the deposition of amorphous carbon films: 1, H_2 + Ar inlet; 2, ion source; 3, graphite target; 4, graphite shields; 5, H_2O or $LiquidN_2$. (Data from Ref. 83.)

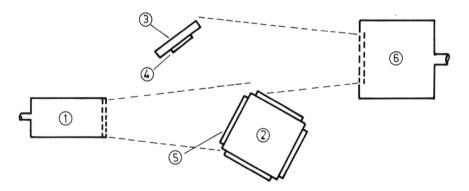

Figure 2.15   Schematic Diagram of the sputtering setup for the preparation of thin films of ZnTe: 1, 2.5 cm ion source; 2, target holder; 3, substrate holder; 4, substrate; 5, ZnTe target; 6, 15 cm ion source. (Data from Ref. 84.)

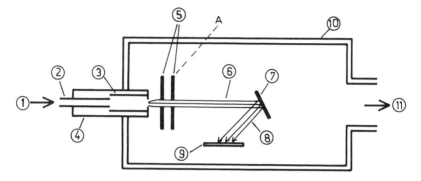

Figure 2.16 Schematic diagram of the ion beam sputtering system for the deposition of Cu and RuO_2 films: 1, Ar gas inlet; 2, anode; 3, cathode; 4, ion source; 5, acceleration electrodes, 6, Ar ion beam; 7, target; 8, sputtered materials; 9, substrate; 10, chamber; 11, to vacuum pump. (Data from Ref. 85.)

tion voltage. The resistivities of both copper and RuO_2 films decreased with increase in acceleration voltage.

T. Toshima et al. [86], who have prepared amorphous CoZr films by using low energy ion beam sputtering, investigated the magnetic properties of the films obtained by varying the accelerator voltage and the beam current of the sputtering system. They showed that the films are amphorous, with a lower Zr content, when are prepared with lower acceleration voltage or lower ion beam current. Deposition parameters are given in Table 2.7.

Schewebel et al. [87] have grown silicon homoepitaxial thin films by ultra high vacuum ion beam sputter deposition. They have reported the deposition characteristics and also the characteristics of the films obtained. The beginning of single-crystal growth was observed to occur at deposition temperatures as low as 250°C. Above 700°C, films of high crystalline and morphological quality were obtained.

Table 2.6   Ion Beam Sputtering Deposition Conditions for ZnTe

Target	Pressed powder disk, 12.7 cm diameter, 0.64 cm thick
Ion beam energy	1000 eV
Ion beam current density	3.5 mA/cm^2
Source-to-target distance	20.3 cm
Substrate-to-target distance	15.3 cm
Deposition rate	≈ 70 Å/min

Table 2.7   Ion Beam Sputtering Deposition Parameters for CoZr Amorphous Films

Target	Diameter, 127 mm
	Alloy targets: Zr content of 2.6, 3.4, and 4.8 at. %
	Co targets: Zr chips, $(5 \times 5 \times 1$ mm, diameter
Substrate	Corning 0211 glass: thickness, 0.5 mm (substrate rotated for uniform deposition)
Beam diameter	100 mm
Substrate-to-target distance	≈ 95 mm
Angle between target and ion beam	45°
Angle between substrate and ion beam	45°

Boron doping of the films could be achieved by sputtering a doped target. Films deposited at 710°C with thickness as low as 0.5 μm had mobilities equal to room temperature bulk mobility.

NiFe films with excellent magnetic properties and thickness uniformity have been prepared using a newly developed high rate ion beam sputtering system [88a]. The films were deposited with and without ion assist. Sputtering rates up to 1000 Å/min and thickness uniformity of ±3% over a 6.5 cm diameter area have been obtained. Two ion beam sources were used: 3 and 10 cm diameter Kaufman-type ion guns. Sputter deposition was carried out with the 10 cm diameter gun at 525 W beam energy at a pressure of 5×10^{-5} torr. Four NiFe targets, 20 cm in diameter, are arranged in a rotating cylinder, and the argon ion beam first bombards the one facing the ion beam from a selected target angle. The selected target can wobble about a vertical axis and move slowly up and down during sputtering; the substrate surface is kept parallel to the ion beam. A permanent magnet is placed behind the substrate and rotates with it during deposition to orient the deposited NiFe film. An ion source having a diameter of 3 cm is used to bombard the substrate before deposition for surface cleaning/ion assist during deposition. Substrate materials were Si, Al_2O_3, and glass wafers with a diameter of 6.6 cm. Argon and nitrogen can be fed into the chamber or the ion source. Nitrogen was fed into the chamber during Ar/N_2 sputtering to study the effect of nitrogen on the film properties. The investigators have shown that these ion-sputtered NiFe films have excellent magnetic, electrical, and microstructural properties, which make them suitable for thin film head applications.

Tustison et al. [88b] have shown that good quality epitaxial Fe films can be successfully grown by ion beam sputtering on GaAs substrates in the (100) orientation.

The ion beam (argon) from an ion source strikes the target surface, and the film is deposited on a heated (temperature-controlled) substrate mounted opposite the target. The best substrate temperature is 300°C. The deposition rate is approximately 0.3 nm/s. The single crystalline character of the films has been verified by means of X-ray and magnetic measurements. Magnetic measurements have shown that the crystalline anisotropy field of the films obtained is almost the same as that of bulk single crystals.

Krishnaswamy et al. [88c] have reported a preliminary study on ion beam sputter deposition as a technique for the epitaxial growth of both CdTe and HgCdTe films on single-crystal substrates of CdTe for use in infrared detector fabrication. A specially designed UHV dual ion gun system (Figure 2.17) was used for this purpose. Ion beam sputter deposition takes place from two independently cooled (liquid N_2) targets by two 3 cm Kaufman-type ion sources. The substrate holder can be rotated during deposition, and the target-to-substrate distance can be varied and adjusted for each run. The deposition parameters are given in Table 2.8. For $Hg_{1-x}Cd_xTe$ films, the investigators used single crystal pieces of composition $x =$ 0.2. Supplementary Hg was provided by ion beam sputtering from a cryogenically cooled source of solid mercury. High quality epitaxial CdTe films have been obtained at temperatures down to 140°C. HgCdTe epitaxial films with twin free films have been grown reproducibly at temperatures between 30 and 100°C. This technique can be effectively and conveniently used for the deposition of CdTe and HgCdTe thin films at growth rates from 1 to 3 μm/h or probably higher, and it is well suited to the low temperature growth of infrared and heterojunction superlattice structures.

Nagakubo et al. [89a] have prepared multilayer films of various thickness of Fe and Al by ion beam sputtering using a Kaufman-type ion source with two grids. Pure argon (99.999%) at a pressure of 1.5×10^{-4} torr was introduced through the ion source as the working gas. Planar targets Fe and Al (99.9) were sputtered by the uniform beam of argon ions (acceleration voltage, 500 V; extraction current, 3 mA). For preparing Fe/Al multilayer films, the films were deposited on water-cooled glass slides by repeating clockwise and counterclockwise rotation of the target holder alternately. The thickness of the films was controlled by adjusting the deposition times of the two metals. The dependence of magnetic properties and crystal structure on the thickness of the Fe and Al layers was investigated.

Kagerer and Koniger [89b] have reported the ion beam sputter deposition of thin films of platinum and manganine for the fabrication of thin film sensors. The schematic of their sputtering system is shown in figure 2.18. The argon ion beam was generated in a separate ion beam chamber, and the uniform beam of argon ions was accelerated into the sputtering chamber and neutralized by means of electrons. The round sputter targets were arranged in an assembly of three targets, with only one target at a time in the sputtering position. The target was water cooled, and each target had a diameter of 125 mm.

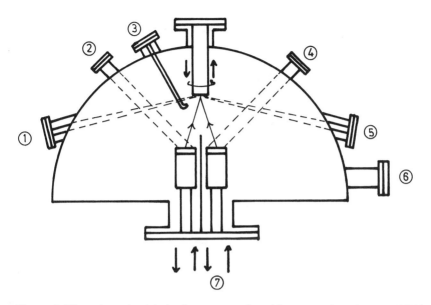

Figure 2.17   Schematic of the ion beam sputter deposition system featuring two UHV ion guns: 1, RHEEDS gun; 2, ion gun 1; 3, Hg source; 4, ion gun 2; 5, RHEEDS window; 6, to pump; 7, liquid N_2 cooling. (Data from Ref. 88c.)

Table 2.8   Deposition Parameters for Epitaxial CdTe and HgCdTe Films

Target	3 in diameter hot pressed CdTe for CdTe films, single-crystal pieces of composition $x = 0.2$ for $Hg_{1-x}Cd_xTe$ films
Substrate	CdTe (001) pieces, 1 cm^2
Substrate temperature	RT to 400°C
Substrate-to-target distance	7 cm
Ion current	30 mA
Deposition rate	1–3 μm/h

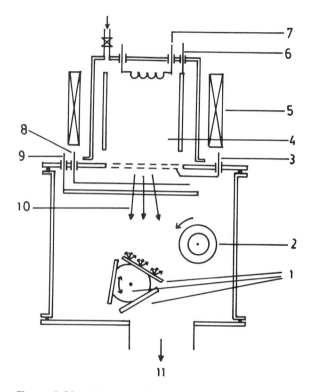

Figure 2.18 Schematic of the sputtering system for the deposition of thin films of platinum and manganine: 1, sputter targets; 2, substrate; 3, grid assembly; 4, ion beam chamber; 5, magnetic coil; 6, anode; 7, cathode; 8, neutralizing filament; 9, shutter; 10, Ar beam; 11, to cyropump. (Data from Ref. 89b.)

For ion beam sputtering, the ion beam impinges on the sputter target, causing the sputtering of the insulator and subsequently of the sensor film material; the film is deposited on the outer rim of the steel disk substrates rotating in front of the target.

The deposition details reported are as follows.

Chamber pressure	10^{-4} Pa
Energy of ion beams	1000 eV
Typical current density	1 mA/cm^2
Target:	
Al_2O_3	Deposition rate 100 Å/min
	Thickness, between 1-2 μm

Manganine/platinum   Deposition rate, several hundred
                                                angstroms per minute
                                                Thickness, 0.1–0.2 μm

---

Thin film sensors with high adhesive strength and sufficient durability have
been produced by this method.

Pellet et al. [89c] have reported epitaxial growth of YSZ (yttria-stabilized zirconia) films, with good physical and electrical properties, on Si <100> substrates
by ion beam sputter deposition of single-crystal $(ZrO_2)_{0.77}$ $(Y_2O_3)_{0.23}$ in an ultra
high vacuum system with in situ diagnostic equipment. The vacuum setup consisted of a duoplasmatron source and two differentially pumped vacuum chambers—an ion beam focusing chamber and a deposition chamber. A RHEEDS
system, an Auger spectrometer, and a residual gas analyzer were attached to the
deposition chamber.

To avoid charging of the target during sputtering, a neutralizer filament was im-
planted near the target. The experiments were carried out with xenon ions of 20
keV energy. The growth rate of the films was 0.08 nm/s, corresponding to an ion
beam current of 1.5 mA. The YSZ layers deposited in the temperature range of
700–900°C, under an oxygen partial pressure of about 10^{-4} Pa, exhibited a
monocrystalline cubic phase.

Ion beam sputtering is used as a technique to prepare HTS thin films. Ameen et
al. [89d] have recently reported experimental and computer simulation studies on
ion scattering and sputtering process in the ion beam sputtering of high Tc films.
Their work has indicated that sputtering by Kr^+ or Xe^+ ions preferable to the most
commonly used Ar^+ ions because the undesirable erosion of and inert gas incorporation in the growing films resulting from the scattering of neutralized ions from
the target are minimized for Kr^+ and Xe^+ ions.

Klien et al. [89e] have reported the deposition of oriented, superconducting
$YBa_2Cu_3O_7$ thin films on yttria-stabilized zirconia and $SrTiO_3$ substrates by ion
beam sputtering of a nonstoichiometric oxide target. The arrangement of their sputtering system is shown in Figure 2.19. The vertically oriented Ar ion source was
used for the sputtering and the target (uncooled) was kept inclined at 45° angle to
the vertical, with its center 10 cm from the face of the ion source. The resistance-
heated substrate holder was kept parallel to the surface of the target (center-to-cen-
ter distance, 7.5 cm). A substrate temperature of 670°C was maintained during
depositions. A flow of oxygen was supplied by flattened nozzles kept on either side
of the substrate. The chamber pressure was maintained at 4×10^{-4} torr. The chamber was backfilled with 40 torr oxygen during the postdeposition cool-down period. A horizontal Cu plate was kept 8 cm below the target, to reduce the possibility
of contamination from the sputtering of the chamber walls. A deposition time of
50–60 minutes was used to obtain films 150–500 nm thick. The films exhibited

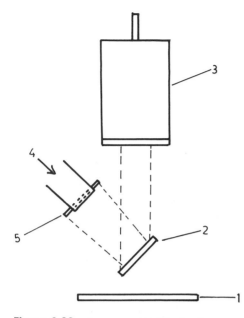

Figure 2.19 Arrangement of the ion beam sputtering system for the deposition of oriented, superconducting YBCO films: 1, Cu plate; 2, YBCO target; 3, Ar ion source; 4, oxygen; 5, heated substrate. (Data from Ref. 89e.)

zero resistance critical temperatures as high as 83.5 K on $SrTiO_3$ substrate without postannealing.

Several combinations of beam voltage and beam current were used, and the effect of beam power on the deposition rate and the lattice parameters of the deposited films were reported [89e].

2.8   AC SPUTTERING

Takeuchi et al. [90a] have reported the preparation of Bi-Sr-Ca-Cu-O (BSCCO) superconducting thin films by a simple ac sputtering system of their original design, reported earlier for the preparation of La-Sr-Cu-O (LSCO) thin films [90b]. The ac sputtering system used in their study is shown schematically in Figure 2.20. A pair of disk targets, which also served as the electrodes, were supported by the horizontally inserted copper rods at the center of the quartz cylinder reactor. The YSZ substrate was placed at the bottom of the cylinder. A focused IR lamp served to heat the substrate (up to 850°C).

Sputtering was initiated by the 50 Hz ac high voltage of 6.3 kV applied through a

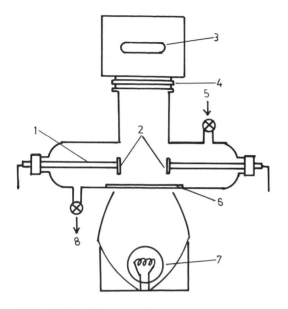

Figure 2.20   Schematic of the ac sputtering system: 1, copper rod; 2, target; 3, low pressure Hg lamp: 4, quartz window; 5, gas inlet; 6, substrate; 7, IR lamp; 8, to pump. (Data from Ref. 90a.)

transformer. The sputtering was carried out at a pressure of 100 mtorr using a 50:50 mixture of oxygen and argon.

Two processes were used for the preparation of the superconducting films. In one process, the substrates were heated, and at T_s higher than 700°C, the films crystallized into the superconducting $Bi_2Sr_2Ca_2Cu_3O_x$ phase. The films prepared at T_s between 750 and 815°C had a c-axis lattice parameter of 18 Å, which was characteristic of a T_c = 115 K phase. In the other process, the films were prepared without substrate heating and deposited films annealed in an oxygen atmosphere. UV irradiation from a low pressure mercury lamp, during the annealing step (680–700°C), was found to induce a T_c change in the film consisting predominantly of a $Bi_2Sr_2Ca_1Cu_2O_y$ phase (T_c = 80 K). They also found that T_c of the films with the 2212 phase was changed from 80 to 60 K when annealed with UV light irradiation at about 400°C. This T_c change has been attributed to the incorporation into the films of extra oxygen. The film prepared with UV light irradiation at the sputtering stage was found to be identical to that prepared without UV irradiation.

The main characteristic features of this ac sputtering method are ease of operation with simple and inexpensive equipment, and with the deposited film away from any direct exposure to plasma.

Hsieh and Yang [90c] have recently reported the preparation in situ of superconducting Bi-Pb-Ca-Sr-Cu-O thin films by an ac sputtering technique.

REFERENCES

1. W. R. Grove, *Phil. Trans. R. Soc. (London)*, 142:87 (1852).
2. L. I. Maissel, in *Physics of Thin Films*, Vol. 3 (G. Hass and R. E. Thun, Eds., Academic Press, New York, 1966, p. 61.
3. J. L. Vossen and J. J. Cuomo, in *Thin Film Processes* (J. L. Vossen and W. Kern, Eds.), Academic Press, New York, 1978, Ch. II-1.
4. A. von Hippel, *Ann. Phys.*, 81:1043 (1926).
5. J. Stark, *A. Elcktrochem.*, 14:752 (908); 15:509 (1909).
6. K. H. Kingdon and I. Langmuir, *Phys. Rev.*, 22:148 (1923).
7. F. Keywell, *Phys. Rev.*, 97:1611 (1955).
8. A. W. Wright, *Am. J. Sci.*, 13:49 (1877).
9. L. I. Maissel, *Phys. Thin Films*, 3L61 (1966).
10. K. L. Chopra, *Thin Film Phenomena*, McGraw-Hill, New YOrk, 1969, p. 23.
11. L. I. Maissel and R. Glang, Eds. *Handbook of Thin Film Technology*, MeGraw-Hill, New York, 1970.
12. G. N. Jackson, *Thin Solid Films*, 5:209 (1970).
13. J. L. Vossen, *J. Vac. Sci. Technol.*, 8:512 (1971).
14. P. D. Townsend, J. C. Kelley, and N. E. W. Hartley, *Ion Implantation and Sputtering and Their Applications*, Academic Press, New York, 1976, Ch. 6.
15. J. L. Vossen and W. Kern, Eds., *Thin Film Processes'*, Academic Press, New York, 1978, Part II-1.
16. R. Behrisch, Ed., *Topics in Applied Physics*, Vol. 47, *Sputtering by Particle Bombardment*, Springer Verlag, Berlin, 1981.
17. L. I. Maissel, in *Physics of Thin Films*, Vol. 3 (G. Hass and R.E. Thun, Eds.), Academic Press, New York, 1966.
18. L. I. Maissel and J H. Vaughn, *Vacuum*, 13:421 (1963).
19. L. I. Maissel, U.S. Patent 3,294,661 (December 1966).
20. J. D. Cobine, *Gaseous Conductors: Theory and Engineering Applications*, Dover, New York, 1958.
21. S. C. Brown, *Introduction to Electrical Discharges in Gases*, Wiley, New York, 1963.
22. H. Gawehn, *A. Angew. Phys.*, 14:458 (1962).
23. F. Vratny, *J. Electrochem. Soc.*, 114:505 (1967).
24. S. Maniv, *Vacuum*, 33:215 (1983).
25. J. Edgecumbe, L. G. Rosner, and D. E. Anderson, *J. Appl. Phys.*, 35:2198 (1964).
26. J. W. Nickerson and R. Moseson, *Res./Dev.*, 16(3):52 (1965).
27. R. C. Sun, T. C. Tisone, and P. D. Cruzan, *J. Appl. Phys.*, 44:1009 (1973).
28. R. C. Sun, T. C. Tisone, and P. D. Curzan, *J. Appl. Phys.*, 46:112 (1975).
29. W. W. Y. Lee and D. Oblas, *J. Appl. Phys.*, 46:1728 (1975).
30. R. O. Adams, C. W. Nordin, and K. D. Masterson, *Thin Solid Films*, 72:335 (1980).
31. J. W. Patten and R. W. Boss, *Thin Solid Films*, 83:17 (1981).
32. R. J. Sonkup, A.K. Kulkami, and D. M. Mosher, *J. Vac. Sci. Technol.*, 16(2):208 (1979).

33. D. M. Mosher and R. J. Sonkup, *Thin Solid Films*, 98:215 (1982).
34. P. Ziemann, K. Kochler, J. W. Comburn, and E. Kay, *J. Vac. Sci. Technol.*, B1:1 (1983).
35. (a) C. Cella and T. K. Vien, *Thin Solid Films*, 125:367 (1985). (b)P. Gallias, J. J. Hantzpergue, J. C. Remy, and D. Roptin, *Thin Solid Films*, 165:227 (1988). (c)S. Caune, Y. Mathey, and D. Pailhary, *Thin Solid Films*, 174:289 (1989).
36. H. C. Theuerer and J. J. Hauser, *J. Appl. Phys.*, 35:554 (1964).
37. H. C. Theuerer and J. J. Hauser, *Trans. Metall. Soc. AIME*, 233:588 (1965).
38. H. C. Theuerer, E. A. Nesbitt, and D. D. Bacon, *J. Appl. Phys.*, 40:2994 (1969).
39. H. C. Cooke, C. W. Covington, and J. F. Lisch, *Trans. Metall. Soc. AIME, 236:314 (1966)*.
40. M. Hong, E. M. Gyorgy, and D. D. Gacon, *Appl. Phys. Lett.*, 44:706 (1984).
41. D. D. Bacon, M. Hong, E. M. Gyorgy, P. K. Gollagher, S. Nakahara, and L. C. Feldman, *Appl. Phys. Lett.*, 48:730 (1956).
42. M. Hong, D. D. Bacon, R. B. Van Dover, E. M. Gyorgy, J. F. Dillion, and S. D. Albiston, *J. Appl. Phys.*, 57:3900 (1985).
43. R. L. Hines and R. Wallor, *J. Appl. Phys.*, 32:202 (1961).
44. G. S. Anderson, W. N. Mayer, and G. K. Wehner, *J. Appl. Phys.*, 33:2991 (1962).
45. P. D. Davidse and L M. Maissel, *Transactions of the 3rd International Vacuum Congress*, Stuttgart, 1965; *J. Appl. Phys.*, 37:574 (1966).
46. K. L. Chopra, *Thin Film Phenomena*, Krieger, New York, 1979, p. 39.
47. F. M. Penning, *Physica (Utrecht)*, 3:873 (1936); U. S. Patent 2,146,025 (1939).
48. E. Kay, *J. Appl. Phys.*, 34:760 (1963).
49. W. D. Gill and E. Kay, *Rev. Sci. Instrum.*, 36:277 (1965).
50. K. Wasa and S. Hayakawa, *Rev. Sci. Instrum.*, 40:693 (1969).
51. J. R. Mullay, *Res./Dev.*, 22(2):40 (1971).
52. P. J. Clarke, U.S. Patent, 3,616,450 (1971).
53. A. S. Penfold and J. A. Thornton, U.S. Patent, 3,884,793 (1975).
54. J. S. Chapin, *Res./Dev.*, 25(1):37 (1974).
55. I. G. Kesaer and V. V. Pashkova, *Sov. Phys. Tech. Phys.*, 4:254 (1959).
56. J. L. Vossen and W. Kern, Eds., *Thin Film Processes*, Academic Press, New York, 1978, pp. 3-173.
57. J. A. Thornton, *Thin Solid Films*, 80:1 (1981).
58. A. R. Nyaiesh, *Thin Solid Films*, 86:267 (1981).
59. J. A. Thornton, *Z. Metallkd.* 75(11):847 (1984).
60. M. Wright and T. Beardow, *J. Vac. Sci. Technol.*, A4:388 (1986).
61. T. Serikawa and A. Okamoto, *J. Vac. Sci. Technol.*, A3:1784 (1985).
62. S. Kobayashi, M. Sakata, K. Abe, T. Damei, O. Kasahara, H. Ohgishi, and K. Nakata, *Thin Solid Films*, 118:129 (1984).
63. S. A. Chang, M. B. Skolnik, and C. Altman, *J. Vac. Sci. Technol.*, A4:413 (1986).
64. (a) J. J. Cuomo and S. M. Rossnagel, *J. Vac. Sci. Technol.*, A4:393 (1986). (b) T. Hata, E. Noda, O. Morinoto, and T. Hada, *Appl. Phys. Lett.*, 37:633 (1980). (c) T. Hata, J. Kawahara, and K. Toriyama, *Jpn. J. Appl. Phys., Suppl.*, 22-1:1:505 (1983). (d) T. Hata, Y. Kamide, S. Nakagawa, and K. Hattori, *J. Appl. Phys.*, 59:3604 (1986. (e) M. Yoshimoto, A. Takano, H. Nagata, M. Kawasaki, and H. Koinume, *Rep. Res. Lab., Eng. Mater., Tokyo Inst. Technol.*, 14:71 (1989).

65. Y. Hoshi, M. Naoe, and S. Yamanaka, *Jpn. J. Appl. Phys.*, 16:1715 (1977).
66. M. Naoe, S. Yamanaka, and Y. Hoshi, *IEEE Trans. Magn.*, MAG-16:646 (1980).
67. Y. Niimura, S. Nakagawa, and M. Naoe, *IEEE Trans. Magn.*, MAG-22:1164 (1986).
68. (a) Y. Niimura and M. Naoe, *J. Vac. Sci. Technol.*, A5:109 (1987). (b) S. Ono, M. Nitta, and M. Naoe, *Cryst. Prop. Prep.*, 27:551 (1989). (c) H. Itto, M. Yamaguchi, and M. Naie, *J. Appl. Phys.*, 67:5307 (1990). (d) T. Hirala and M. Naoe, *J. Appl. Phys.*, 67:4047 (1990).
69. J. M. E. Harper, in *Thin Film Processes'* (J. L. Vossen and W. Kern, Eds.), Academic Press, New York, (1978), p. 117.
70. K. L. Copra and M. R. Randlett, *Rev. Sci. Instrume*, 38:1147 (1967).
71. C. Weissmantel, O. Fiedler, G. Hecht, and G. Reisse, *Thin Solid Films*, 13:359 (1972).
72. P. H. Schmidt, R. N. Castellano, H. Berz, A. S. Cooper, and E. G. Spencer, *J. Appl. Phys.*, 44:1833 (1973).
73. S. Yamanaka, M. Naoe, and S. Kawai, *Jpn. J. Appl. Phys.*, 16:1245 (1977).
74. R. N. Castellano, M. R. Notis, and G. W. Simmons, *Vacuum*, 27:109 (1977).
75. M. J. Mirtich, *J. Vac. Sci. Technol.*, 18:186 (1981).
76. J. Saraic, M. Kobayashi, Y. Fujii, and H. Matsunami, *Thin Solid Films*, 80:169 (1981).
77. C. Weissmantel, *Thin Solid Films*, 92:55 (1982).
78. (a) A. P. Semenov, M. V. Mokhosoev, and V. P. Bazarov, *Electron. Obrab. Mater. (USSR)*, 3:83 (1983); English translation in *Electrochem. Ind. Process Biol. (GB)*. (b) J. C. Angus, J. E. Stultz, P. J. Shiller, J. R. MacDonald, M. J. Mirtich, and S. Domitz. *Thin Solid Films*, 118:311 (1984). (c) J. W. Smits, S. B. Luitjens, and F. J. A. den Broeder, *J. Appl. Phys.*, 55:2260 (1984).
79. (a) J. W. Smits and F. J. A. den Broeder, *Thin Solid Films*, 127:1 (19850. (b) T. E. Veritimos and R. W. Tustison, *Thin Solid Films*, 151:27 (1987).
80. (a) D. Weller, W. Reim, and P. Schrijner, *IEEE Trans. Magn.*, MAG-24:2554 (1988). (b) S. M. Arnold and B. E. Cole, *Thin Solid Films*, 165:1 (1988). (c) F. Shoji, H. Tamguchi, O. Kusumoto, K. Oura, T. Hanawa, Y. Suzuki, and S. Ogawa, *Jpn. J. Appl. Phys., Reg. Pap., Short Notes*, 28:545 (1989).
81. (a) M. Ruth, J. Tuttle, J. Goral and R. Nouji, *J. Cryst. Growth*, 96:363 (1989). (b) C. Kim, S. B. Qadri, and P. Lubitz, *Mater. Sci. Eng. A. Struct. Mater. Prop. Microstruct. Process.*, A126:25 (1990). (c) J. Bergman, Y. Shapire, and H. Aharoni, *J. Appl. Phys.*, 67:3750 (1990). (d) O. Auciello, M. S. Ameen, T. Graettinger, S. H. Rou, C. Soble, and A. I. Kingon, *AIP Conf. Proc. (USA)*, 200:79 (1990). (e) K. Li, R. IIsiao, and C. Tang, *J. Appl. Phys.*, 68:3043 (1990).
82. M. Kitabatake and K. Wasa, *J. Appl. Phys.*, 58:1693 (1985).
83. F. Jansen, M. Machonkin, S. Kaplan, and S. Hark, *J. Vac. Sci. Technol.*, A3:605 (1985).
84. D. A. Gulino, *J. Vac. Sci. Technol.*, A4:509 (1984).
85. M. Takeuchi, K. Yanagida, H. Nagasaka, T. Tanabe, and H. Mase, *Thin Solid Films*, 144:281 (1986).
86. T. Toshima, A. Tago, and C. NIshimura, *IEEE Trans. Magn.*, MAG-22:1110 (1986).
87. C. Schwebel, F. Meyer, G. Gautherin, and C. Pellet, *J. Vac. Sci. Technol.*, B4:1153 (1986).
88. (a) J. Lo, C. Hwang, T. C. Haung, and R. Campbell, *J. Appl. Phys.*, 61:3520 (1987). (b) R. W. Tustison, T. Varitimos, J. Van Hook, and E. F. Schloeman, *Appl. Phys., Lett.*,

51:285 (1987). (c) S. V. Kirshnaswamy, J. H. Rieger, N. J. Doyle, and M. H. Francombe, *J. Vac. Sci. Technol.*, A5:2106 (1987).

89. (a) M. Nagakubo, T. Yamamoto, and M. Naoe, *J. Appl. Phys.*, 64:5751 (1988). (b) E. Kagerer and M. E. Koniger, *Thin Solid Films*, 182:333 (1989). (c) C. Pellet, C. Schwebel, and P. Hesto, *Thin Solid Films*, 175:23 (1989). (d) M. S. Ameen, O. Auciello, A. I. Kingon, A. R. Krauss, and M. A. Ray, *AIP Conf. Proc. (USA)*, 200:79 (1990). (e) J. D. Klein, Y. Yen, and S. L. Clauson, *J. Appl. Phys.*, 67:6389 (1990).

90. (a) K. Takeuchi, T. Yoshida, M. Kawasaki, S. Uchida, T. Hirayama, M. Yoshimoto, Y. Saito, S. Hayano, and H. Koniuma, *Rep. Res. Lab. Eng. Mater., Tokyo Inst. Technol.*, 14:77 (1989). (b) H. Koinuma, M. Kawasaki, M. Funabashi, T. Hasegawa, K. Kishio, K. Kitazawa, and K. Fuiki, *J. Appl. Phys.*, 62:1524 (1987). (c) M. H. Hsieh and H. C. Yang, Chin, *J. Phys.*, 28:287 (1990).

91. G. L. Christner, B. Bradford, L. E. Toth, R. Canter, E. D. Dahlberg, A. M. Goldman, and C. Y. Huang, *J. Appl. Phys.*, 50:5820 (1979).

92. A. I. Golovashkin, E. V. Pechen, *Sov. Phys. Solid State*, 21:725 (1979).

93. Y. Saito, H. Kinashi, and S. Suganomata, *Thin Solid Films*, 69:33 (1980).

94. S. Alexandrova, K. Kirov, and A. Szekeres, *Thin Solid Films*, 75:37 (1981).

95. J. Angilello, J. E. E. Baglin, F. Cardone, J. J. Dempsey, F. M. d'Heurle, E. A. Irene, R. MacInnes, C. S. Petersson, R. Savoy, A. P. Segmuller, and E. Tierney, *J. Electron. Mater.*, 10:59 (1981).

96. M. Naoe, S. Hasunuma, Y. Hoshi, and S. Yamanaka, *IEEE Trans. Magn.*, MAG-17:3184 (1981).

97. M. Levinson and M. Gurvitch, *J. Appl. Phys.*, 54:4683 (1983).

98. Y. H. Shing, Y. Chang, A. Mirshafii, L. Hayashi, S. S. Robens, J. Y. Joseto, and N. Tran, *J. Vac. Sci. Technol.*, A1:503 91983).

99. G. Lemperiere and J. M. Pitevin, *Thin Solid Films*, 111:339 (1984).

100. A. Azimkhan and J. A. Woollam, *Aloid State Electron. (GB)*, 27:385 (1984).

101. A. K. Dua, V. C. George, R. P. Agarwala, and R. Krishnan, *Thin Solid Films*, 121:35 (1984).

102. J. Battacharyya, S. Chaudhuri, D. De and A. K. Pal, *Thin Solid Films*, 128:231 (1985).

103. M. Hong, S. H. Liou, J. Kwo, and B. A. Davidson, *Appl. Phys. Lett.*, 51, 694 (1987).

104. Y. Matsui, M. Okuyama, N. Fujita, and Y. Hamakawa, *J. Appl. Phys.*, 52:5107 (1981).

105. A. J. Collins, C. J. Prior, and R. C. J. Hicks, *Thin Solid Films*, 86:165 (1981).

106. R. I. Christy, *Thin Solid Films*, 80:289 (1981).

107. A. Sathyamoorthy and W. Weisweiler, *Thin Solid Films*, 87:33 (1982).

108. P. L. Jones, D. Moore, and D. R. Cotton, *J. Cryst. Growth*, 59:183 (1982).

109. H. Volz, K. Koger, and H. Schmitt, *Ferroelectrics*, 51:87 (1983).

110. R. A. Roy, R. Messier, and S. V. Krishnaswamy, *Thin Solid Films*, 109:27 (1983).

111. J. R. Shappiro and J. J. Finnegan, *Thin Solid Films*, 107:81 (1983).

112. S. V. Krishnaswamy, A. S. Manocha, and J. R. Sxedon, *J. Vac. Sci. Technol.*, A1:510 (1983).

113. N. T. Tran, Y. C. Chang, D. A. Faragalli, S. S. Roberts, J. Y. Josefowicz, and Y. H. Shing, *J. Vac. Sci. Technol.*, A1:345 (1983).

114. S. Seki, T. Unagami, and B. Tsujiyama, *J. Vac. Sci. Technol.*, A1:1825 (1983).

115. T. L. Rose, E. M. Kelliher, A. N. Scoville, and S. E. Stone, *J. Appl. Phys.*, 55:3706 (1984).
116. I. Martil, G. Gonzalez-Diaz, and F. Sanchez-Quesada, *Thin Solid Films*, 114:327 (1984).
117. G. S. Sunatori and A. a. Benezin, *J. Appl. Phys.*, 55:3125 (1984).
118. F. S. Huang and Y. Y. Wang, *J. Vac. Sci. Technol.*, A2:461 (1984).
119. R. A. Roy and R. Meissier, *J. Vac. Sci. Technol.*, A2:312 (1984).
120. M. Gomi, T. Tanida, and M. Abe, *J. Appl. Phys.*, 57:3888 (1985).
121. J. Santamaria, F. Sanchez-Quesada, G. Gonzalez-Diaz, E. Iborra, and M. Rodrguez-Vidal, *Thin Solid Films*, 125:299 (1985).
122. S. Honda, M. Nawate, M. Ohkoshi, and T. Kusuda, *J. Appl. Phys.*, 57:3204 (1985).
123. A. Krzesinski, *Thin Solid Films*, 138:111 (1986).
124. J. K. G. Panitz, L. E. Pope, C. R. Hills, J. E. Lyons, and D. J. Staley, *Thin Solid Films*, 154:323 (1987).
125. C. T. Lee, Y. K. Su, and H. M. Wang, *Thin Solid Films*, 150:283 (1987).
126. T. Nagatomo, T. Kosaka, S. Omori, and O. Omoto, *Ferroelectrics*, 37:681 (1981).
127. S. Maniv, W. D. Westwood, F. R. Shepherd, P. J. Scandon, and H. Plattner, *J. Vac. Sci. Technol.*, 20:1 (1982).
128. J. W. Webb, *J. Appl. Phys.*, 53:9043 (1982).
129. S. B. Krupanidhi, N. Maffei, M. Sayer, and K. El-Assal, *J. Appl. Phys.*, 54:6601 (1983).
130. H. Nanto, T. Minami, S. Shooji, and S. Takata, *J. Appl. Phys.*, 55:1029 (1984).
131. M. Kitabatake, t. Mitsuya, and K. Wasa, *J. Appl. Phys.*, 56:1780 (1984).
132. A. Bensaoula, J. C. Wolfe, A. Ignatiev, F. O. Fong, and T. S. Leung, *J. Vac. Sci. Technol.*, A2:389 (1984).
133. R. Bichsel and F. Levy, *Thin Solid Films*, 124:75 (1985).
134. R. G. Goodchild, J. B. Webb, and D. F. Williams, *J. Appl. Phys.*, 57:2308 (1985).
135. M. Fernandez, J. P. Gonzalez, J. M. Albella, and J. M. Martinez-Duart, *J. Mater. Sci.*, 22:3703 (1987).
136. S. Suyama, A. Okamoto, T. Serikawa, and H. Tanigawa, *J. Appl. Phys.*, 62:2360 (1987).
137. D. Tzaneva, D. Dobrev, and M. Kanev, *Thin Solid Films*, 151:235 (1987).
138.. D. S. Burbidge, S. K. Dew, B. T. Sullivan, N. Fortier, R. R. Parsons, P. J. Mulhern, J. F. Carolan and A. Chaklader, *Solid State Commun.*, 64:749 (1987).
139. T. Shikama, Y. Sakai, M. Fukutomi, and M. Okada, *Thin Solid Films*, 156:287 (1988).
140. M. Hong, J. Kwo, C. H. Chen, A. R. Kortan, and D. D. Bacon, *AIP Conf. Proc. (USA)*, 182:107 (1988).
141. U. Poppe, J. Schubert, and W. Evers, *Physica, C.*, 153–155:776 (1988).
142. J. Schubert, U. Poppe, and W. Sybertz, *J. Less Common Met.*, 151:277 (1989).
143. F. Reniers, P. Delcambe, L. Binst, M. Jardiner-Offergeld, and F. Bouillon, *Thin Solid Films*, 170:41 (1989).
144. P. Vuoristo, J. Whalroos, T. Matyla, and P. Kettunen, *Thin Solid Films*, 166:255 (1988).
145. H. -P. D. Shieh, M. Hong, and S. Nakahara, *Thin Solid Films*, 181:101 (1989).
146. A. M. Klumb. C. R. Aita, and N. C. Tran, *J. Vac. Sci. Technol.*, A7:1697 (1989).

147. Z. Surowiak, A. M. Margolin, I. N. Zakharchenko, and S. V. Biryukov, *Thin Solid Films*, 176:227 (1989).
148. D. Giginoudi, N. Georgoulas, and A. Thanailakis, *J. Appl. Phys.*, 66:354 (1989).
149. M. Benmalek, P. Sainfort, and G. Regazzoni, *Thin Solid Films*, 175:25 (1989).
150. C. V. R. V. Kumar and A. Mansing, *J. Appl. Phys.*, 65:1270 (1989).
151. A. Valentini, A. Quirini, and L. Vasanelli, *Thin Solid Films*, 176:L167 (1989).
152. Y. Tokomitsu, A. Kawabuchi, H. Kitayama, T. Imura, Y. Osaka, and F. Nishiyama, *J. Appl. Phys.*, 66:896 (1989).
153. J. Gerblinger and H. Meizner, *J. Appl. Phys.*, 67:7453 (1990).
154. I. Maril, J. Santamaria, G. G. Diaz, and F. S. Quesada, *J. Appl. Phys.*, 68:189 (1990).
155. K. Onisawa, M. Fuyama, K. Temura, K. Taguchi, T. Nadayama, and Y. A. Ono, *J. Appl. Phys.*, 68:719 (1990).
156. H. Iwasaki, *J. Appl Phys.*, 67:5120 (1990).
157. T. Miyazaki, M. Mori, and S. Adachi, *Appl. Phys. Lett.*, 58:116 (1991).
158. K. Setsune, T. Kamada, H. Adachi, and K. Wasu, *J. Appl. Phys.*, 64:1318 (1988).
159. (a) S. N. Qiu, C. X. Qiu, and I. Shih, *Can. J. Phys.*, 67:435 (1989). (b) Y. Hakuraku, S. Higo, and T. Ogushi, *Appl. Phys. Lett.*, 55:1569 (1989).
160. M. Tosa, Y. Ikeda, and K. Yoshihara, *Thin Solid Films*, 177:107 (1989).
161. M. Scheib, H. Goebel, L. Hofman, D. Lengeler, H. Oechsner, and G. Zorn, *Thin Solid Films*, 174:5 (1989).
162. Y. Kobayashi, S. Terada, and K. Kubota, *Thin Solid Films*, 168:133 (1989).
163. J. Chang, M. Nakajima, K. Yamamoto, and A. Sayama, *Appl. Phys. Lett.*, 54:2349 (1989).
164. (a) Y. Hakuraku, Y. Aridome, D. M. Yagi, N. G. suresha, and T. Ogushi, *Jpn. J. Appl. Phys.*, 2(28):L819 (1989). (b) M. M. D. Ramos, J. B. Almeida, M. I., C. Ferreira, M. P. dos Santos, M. D., R. Cruz, and A. L. S. Gama, *Vacuum*, 39:735 (1989).
165. Y. Suzaki, S. Inoue, I. Hasegawa, K. Yoshii, and H. Kawabe, *Thin Solid Films*, 173:235 (1989).
166. P. Kus and S. Janos, *Mod. Phys. Lett.*, B3:37 (1989).
167. (a) K. Sreenivas, S. Mayer, and P. Garrett, *Thin Solid Films*, 172:251 (1989). (b) J. Willer, S. Pompl, and D. Ristow, *Thin Solid Films*, 188:157 (1990).
168. G. Subramanyam, F. Radpour, V. J. Kapoor, and G. H. Lemon, *J. Appl. Phys.*, 68:1157 (1990).
169. S. Y. Lee, T. S. Hahn, Y. H. Kim, and S. S. Choi, *J. Appl. Phys.*, 68:856 (1990).

3

Ion Beam and Ion-Assisted Deposition

Learning without thinking is labour lost; thinking without learning, perilous.
Confucius

Ion-related techniques of thin film preparation have been in use for more than 20 years [1-4], and a number of techniques have been developed, such as ion plating, ion beam sputtering, and ion beam deposition. Such modes of depositions make use of the inherent properties of the materials to achieve better adhesion, epitaxial growth at lower substrate temperatures, morphological changes, synthesis of compounds and other advantages by imparting kinetic energy and/or enhancing the chemical activity through ionization.

In ion beam deposition the desired film material is ionized, and the highly energetic ions of film material are extracted into a high vacuum region and decelerated before striking the substrate for direct deposition at low energy. The low energy ranges from a few to a couple of hundred electron volts. An ion-assisted process is a cross between evaporation and sputtering. Evaporation is fast, but the films have poor adhesion to the substrate, high porosity, poor uniformity of thickness, and other defects, whereas sputtering does not suffer from these drawbacks but is extremely slow. Ion-assisted deposition (IAD) is categorized as follows: (a) conventional ion plating (a combination of evaporation and the use of glow discharge) and its variants, (b) cathode arc deposition and hot hollow cathode (plasma electron beam) gun evaporation, where a significant percentage of the vapor of the source material is ionized, and (c) deposition, whether by sputtering or evaporation, where the substrate is bombarded directly with ions during film formation; these ions playing an important role in the growth and formation of the films (i.e., irradiation with an ion source during condensation). Sputtering—either glow discharge (plasma) sputtering and its variants or ion beam sputtering, where, unlike in other

101

ion-related techniques the ions or ion beams are directly involved in the process of ejection of the film material for deposition on the substrate—was discussed in Chapter 2.

Ions in general can transfer energy, momentum, and charge. A complex variety of processes occurs whenever energetic particles bombard the substrate surface and the growing films, and such energetic ion bombardment affects such components of the overall deposition processes as condensation, movement of the adatoms of the surface, incorporation of atoms at lattice defects, and nucleation. The ion surface interaction phenomena constitute the key factor in all the ion-assisted deposition techniques. The important ion surface interactions that occur [5,6] are as follows.

1.   Desorption or sputtering of the adsorbed impurities from the substrate surface by ion bombardment, a phenomenon often used for cleaning the substrate before deposition is started. This can be accompanied by forward recoil implantation of impurities into the substrate.

2.   Entrapment and penetration of coating atoms and support gas atoms into the immediate subsurface.

3.   Initial sputtering of the substrate and subsequently of the coating atoms. This reduces the film growth rate but may lead to atomic mixing.

4.   Displacement of the coating and substrate atoms and the generation of lattice defects. Atomic displacement can lead to significant intermixing of the substrate and film atoms, while enhanced defect densities can promote quick interdiffusion.

The ion impact desorption process is an important and useful factor both in the precleaning of the substrate and in the deposition by ion plating and similar allied techniques.

The particle energies involved in these processes are from thermal energies to about 1 keV. To achieve film growth, the rate of the coating atoms deposition must exceed the sputtering rate, and this generally requires that the total flux of depositing atoms greatly exceed that of the flux of energized support gas and coating atoms. It is also shown that to achieve atomic mobility in both the substrate and the coating which is necessary to ensure good adhesion and uniform deposition, the ratio of coating condensation rate to energized particle rate should not be too excessive [5].

3.1   ION PLATING

The ion plating technique as developed by Mattox [7] is a process in which the deposition source is a thermal evaporation source as in ordinary evaporation techniques, and a glow discharge is maintained at a pressure of $10^{-1}-10^{-2}$ torr between the source crucible or filament as the anode and the substrate as the cathode. In the simple ion plating system shown schematically in Figure 3.1, the substrate is made

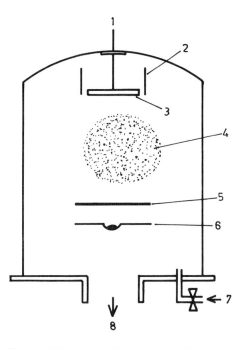

Figure 3.1 Schematic layout of the ion plate setup: 1, high tension negative; 2, ground shield; 3, substrate; 4, plasma; 5, shutter; 6, evaporation source; 7, argon inlet; 8, to pump.

the cathode of a diode dc discharge with a high voltage power supply (2–5 kV) as normally used in diode sputtering. The anode is attached to the deposition source—for example, a resistively heated crucible. A ground shield is placed around the cathode with spacing less than the width of the cathode dark space, to confine the ion bombardment to the substrate surface. Care should be taken to prevent the ground shield from shielding part of the substrate surface from ion bombardment. In a typical operation, the system is evacuated to an ultimate pressure of about 10-7 torr to reduce the level of the contaminating gases and the evaporant premelted. The inert gas (argon) is introduced and the discharge initiated when the operating pressure is 10–50 mtorr. The substrate is sputter-cleaned for a short period depending on the substrate material. The material is then allowed to evaporate slowly and the shutter is removed, allowing the material to deposit on the substrate. Here the vaporized material passes through the inert gas (say argon) glow discharge set up between the crucible and the substrate; as the vapor atoms pass through the glow discharge, some are ionized and accelerated toward the substrate (cathode) across the cathode dark space. Each ion will experience many collisions before it reaches the cathode, transferring its energy to neutral atoms and producing a number of energetic neutral atoms. According to Teer [8], the energetic neutral atoms (< 135 eV) are in the majority compared to high energy ions (≈ 300 eV) in the total flux for film formation. Lighter inert gases require higher pressures to maintain the discharge, while with heavier gas the discharge can be maintained at lower pressures. The sputter cleaning rate will be higher at a given voltage and current density with heavier gas.

The main benefits obtained from the ion plating process [9] are as follows.

1. The surface is sputter-cleaned before deposition of the film.
2. The high energy flux provided to the substrate surface results in a high surface temperature, enabling the enhancement of the diffusion and chemical reactions while keeping a low bulk substrate temperature.
3. The introduction of a high concentration of defects alters surface and interfacial structure and influences the nucleation and growth of the depositing films.

In ion plating the vapor can be produced by evaporation or sputtering [10] and has sometimes been applied to processes in which the substrate is subjected to purposeful ion bombardment both before and during film growth in a glow discharge environment.

There are a variety of available evaporation sources for providing the vapor to be deposited [11a]. Each has its advantages and disadvantages. Resistance-heated refractory metal boats or filaments, usually tungsten or molybdenum, are commonly preferred; but the range of materials is generally limited to elemental metals with low melting points. Flash evaporation has also been used successfully for the ion plating of alloys and compounds. By using electron beam heating, high melting point materials (refractory metals) have been deposited at high evaporation rates. A sputtering target is also used to provide the film material for ion plating; that is, the films are formed from a flux of atoms and ions that have been sputtered from a solid target. This is very similar to bias sputtering except that higher bias voltages are used. However, the high substrate potentials (2–5 kV) commonly used in ion plating cannot be applied, and because it is necessary to continuously decrease the substrate potential, at some point the deposition rate becomes greater than resputtered rate.

Ion plating using magnetron sputtering also has been reported. Using magnetron sputtering ion plating (MSIP), Wan et al. [11b] have studied the microstructure of Al films as a function of ion plating time. Wan and Kuo later reported [11c] a study of the relation between the distance of the target from the substrate and the microstructure of the Al films deposited. The schematic diagram of the ion plating system used is shown in Figure 3.2. The sputtering material Al (99.9%) was deposited onto Ni substrates using Ar as the sputtering gas. The distance between the target and the receiver sample could be changed from several millimeters to more than 350 mm. The Ni substrate was chemically cleaned in acetone before the film was deposited.

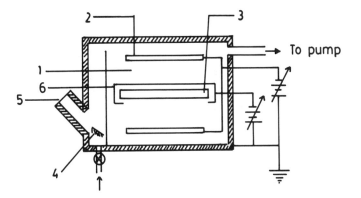

Figure 3.2 Schematic diagram of the sputtering ion plating system: 1, vacuum chamber; 2, sample; 3, target; 4, reflection mirror; 5, observation window; 6, auxiliary anode. (Data from Ref. 11c.)

The ion plating details reported are given below:

Power input to the target	650 V, 15 A (8.4 mA/cm^{-2})
Ion plating bias	-1500 V
Ar pressure	6.67×10^{-1} Pa
Distance	100 or 200 mm
Ion plating time	5 or 30 minutes
Magnetic field strength	400 G

Wan's study showed that sputtering distance affected the phase transformation and the microstructure of the films deposited.

The ion plating technique has been applied to the deposition of metals, alloys and compounds onto metals, insulators, and organic substances of varying sizes and shapes, including small screws and bearings. In many instances the technique has provided results superior to those from conventional deposition techniques in film–substrate systems where adhesion, corrosion protection, and electrical contact are important factors.

A bibliography of the preparation of thin films by ion plating is given in Table 3.1.

3.1.1 Triode Ion Plating

There are several methods by which the dc diode discharge can be enhanced and supported. A useful increase in ionization and lower permissible operating pressures can be achieved in the ion plating process by incorporating a positive

Table 3.1 Ion Plating

Material	Method of providing vapor	Discharge details	Remarks, if any	Ref.
Cu	Evaporated from Mo boat	2, 3, 4, and 5 kV at Ar pressure of 1, 2, 3, 4, and 5 × 10⁻² torr	Nickel substrates; deposition rate, 1 μm/min.	12
Cu and Au	Evaporated; 99.99% purity	3–5 kV; current density, 0.3–0.8 mA/cm² Ar pressure, 20 mtorr	304 Stainless steelsubstrates; sputter cleaned for 10 min before evaporation.	13
Al bronze: 14% Al, 4.5% Fe, 1% Ni, balance Cu	Resistively heated crucible, TiB₂ and BN	5 kV; current density, 0.1–0.25 mA/cm² Ar pressure, 10 mtorr	Mild steel and carbon tool steel substrates; distance between source and work piece 25 cm.; temperature not to exceed 550;C.	14
Al bronze: 14% Al, 4% Fe, balance Cu	Resistively heating crucible, TiB₂ and BN	4 kV; current density, 0.25 mA/cm² Ar pressure, 10 mtorr	Water-cooled specimens clamped to cathode.	15
Au and Pb	Thermal evaporation	3–5 kV; current density, 0.3–0.8 mA/cm⁻² Ar pressure, 20 mtorr	44°C stainless substrates, 6.4 cm diameter.	16
Al bronze: 14% Al, 4% Fe, balance Cu	Resistively heated crucible, TiB₂ and BN	4 kV; current densities, 0, 0.1, 1, and 2.5 mA/cm²; Ar pressure, 10 mtorr		17
Au, Al, In, Cu	Evaporated from Mo boats	Rf excited discharge; dc bias, −400 to 600 V; Ar pressure, 5 × 10⁻⁴ torr	Adhesion and vacuum sealing of metal films to PVDF sheet samples were studied.	18
Co and Co–Cr	Evaporated from EB gun	4kV; current density, 0.15 mA/cm² Ar pressure, 1.5 × 10⁻² torr	Mild steel plates 38 × 28 × 3 mm as substrates. Source-to-substrate distance 200 mm.	19

Material	Method	Parameters	Remarks	Ref.
Cu, Au, and Al	Evaporated from hot filament source	Rf discharge	Sputter-precleaned for 2 minutes at –500 V dc bias. Polyether sulfone (PES) and polyester used as substrates.	20
Au	Presistance heating; W boat	3–5 kV; current density, 0.3–.8 mA/cm²; 20 mtorr	Nickel and ion substrates; dc sputter-cleaned for about 10 minutes.	21
Au	Evaporated from resistively heated W		Copper, nickel, and iron substrates; dc sputter cleaned for about 10 minutes.	22
Ag	Resistively heated Ta boat	3 kV; current density, 0.2 mA/cm² Ar pressure, 2 × 10⁻² torr	Polished steel substrates; sputter-cleaned for 30 minutes.	23
Au	Evaporated from resistively heated W	3.5 kV; current density, 0.5 mA/cm²; Ar pressure 20 mtorr	Metallic glass in the form of a ribbon (foil 0.030–0.033 mm thick) used as substrates.	24a
Cu	Evaporated by electron bombardment	Rf discharge; Ar pressure, 2 × 10⁻³ torr	Polytetrafluoroethylene substrate, sputter-cleaned in argon-oxygen (or helium-oxygen) for good adhesion.	24b
Silver (99.99% purity)	Thermal evaporation from Mo boats	Negative potential of 3 kV, 85 substrate current density, 0.5 mA/cm²	Ion plating done in argon at a pressure of 20 mtorr. Care is taken that no surface heating occurs during evaporation. Films grown on air-cleaved rock salt substrates; deposition rate kept at 10 Å/s. Showed epitaxial growth from early stages. Films exhibited a very different behavior from vacuum-evaporated silver films.	

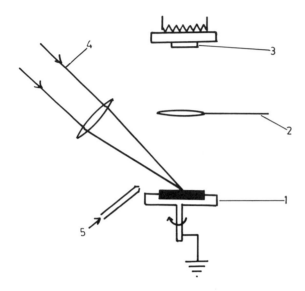

Figure 3.3   Schematic setup of the ion-assisted pulsed laser deposition system: 1, rotating target; 2, ring-shaped electrode; 3, heated substrate; 4, laser beam; 5, O_2 inlet. (Data from Ref. 25b.)

electrode between the substrate and the evaporation source to give a triode configuration. This arrangement is similar to that in triode sputtering although in triode sputtering it is the target, not the specimen, which is biased negatively.

Sauliner et al. [25a] have studied the influence of the potential V of the positively biased third electrode on the discharge current intensity I_D and on the ion energy distribution. Their observed increase in I_D with V has been qualitatively explained by considering the positive potential well created by the third electrode, which traps the electrons. As these electrons oscillate in this potential well, the effective electron path is longer, thus increasing the ionization. The increase in number of ions produces greater discharge currents for low pressures. It has also been shown that the biasing of the third electrode tends to decrease the dark space length, and the mean energy of the ions striking the substrate is increased, as well. Positive probe discharge support is useful under evaporation conditions when electrons emitted in the source region during evaporation can be accelerated to increase ionization.

Witanachchi et al. [25b] have used ion-assisted laser deposition to prepare HTS thin films, relying on the pulsed laser deposition system with a positive probe discharge support, shown schematically in Figure 3.3. The deposition was carried out in a vacuum chamber ($\sim 10^{-6}$ torr), where a rotating target (YBCO) was irradiated by an ArF excimer laser with a 193 nm output operated at 10 pps. The laser intensity on the target was approximately 3 J/cm^2.

The substrate on a substrate holder heated by a resistive heater was kept roughly 7.5 cm from the target, and the temperature of the substrate could be controlled from 425°C to room temperature. A ring-shaped electrode was placed in the middle between the substrate and the target and was held at +300 V. The ambient oxygen pressure was 10^{-4} torr. The dc discharge could be triggered by the first laser pulse and would sustain until the high voltage was switched off. A steady glow could be observed stretching from the target to the substrate, which was at a floating potential.

In this setup, the dc discharge served two purposes. The O_2^+ ions formed by electron impact between the ring electrode and the substrate were effective in enhancing and improving the film deposition by ion activation of the surface, whereas those formed between the target and electrode were repelled. Also the O_2^+ ions tend to enhance the oxygen content of the deposited film, thereby improving the superconducting properties.

Good control over ionization can be achieved in triode ion plating if a separate thermionic emitter is introduced independent of the vapor source. Baum [26] first reported the use of a thermionically assisted triode ion plating system. He claimed several advantages for this system.

Greater control over the discharge
Lower operating pressure
Use of lower bias voltage and improved stability
Reduced substrate heating

Kloos et al. [27] prepared thin, soft metallic films by thermionically assisted triode ion plating and compared the tribological behavior with that of films deposited by conventional diode ion plating. A schematic of their apparatus appears in Figure 3.4; the deposition parameters are given in Table 3.2. From the current-voltage characteristics, the investigators observed that the substrate current in the conventional diode ion plating with thermal evaporators is not significantly increased by the positive probe at low tensions. But when the hot filament is used in conjunction with the positive prove, the ionization markedly increases and high substrate current densities are achieved.

Mathews and Teer [28] used a thermionically assisted triode ion plating system with an electron beam gun evaporator; the schematic diagram of their setup is shown in Figure 3.5. The filament is a tungsten wire of 0.5 mm diameter and 12 cm long. The hot cathode can be operated in two basic states by selecting position s_1 or s_2 on the common point grounding switch. At position s_2, the filament and the positive probe circuit float independently of the rest of the system. Mathews and Teer have carried out extensive investigations on the influence of such system parameters as probe voltage, specimen bias, electron beam power, and chamber pressure

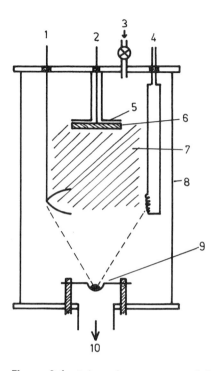

Figure 3.4 Schematic arrangement of thermionically assisted triode ion plating for the preparation of thin, soft metallic films: 1, positive probe; 2, high tension; 3, gas inlet; 4, hot filament; 5, dark space shielding; 6, substrate; 7, glow discharge; 8, vacuum chamber; 9, source; 10, to pump. (Data from Ref. 27.)

on the specimen and probe currents. They have shown that a positive electrode and a thermionic source that is electrically floating, so that the filament takes up a negative potential with respect to the grounded chamber constitute the best arrangement for the maximum specimen ion currents often required in ion plating. Ion efficiencies exceeding 3% and increases in specimen current density from less than 0.3 mA/cm^2 (with a single diode) to more than 3 mA/cm^2 were obtained with this thermionically assisted triode ion plating system. Thin film depositions could be made at pressures below 10^{-3} mtorr.

A triode ion plating apparatus with a rotating barrel, for coating of large quantities of small components, has been reported by Ahmed [29]. A thermionically assisted triode discharge was, used, with a tungsten wire serving as a hot filament source to emit electrons. Aluminum was evaporated from a resistance-heated boron nitride/titanium diboride boat, fed by a discontinuous slug feed system. Highly dense Al coatings at low argon discharge pressures (1 × 10^{-3} torr) have been ob-

Table 3.2 Deposition Parameters for Thermionically Assisted Triode Ion Plating

Cathode	
voltage	5 kV
current	250 mA
Positive probe	
voltage	510 V
current	36 A
Hot filament	
voltage	20 V
current	20 A
Pressure	7.5×10^{-3} torr
Substrate	Hardened ball-bearing steel
Metals deposited	Cu; 50% Cu/50% Pb; 50% Cu/50% In
	Ag;50% Ag/50% Pb; 50% Ag/50% In
	33% Cu/33% Ag/33% Pb
	33% Cu/33% Ag/33% In
	25% Cu/25% Ag/25% Pb/25% In

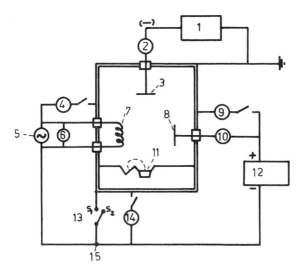

Figure 3.5 Schematic diagram of a thermionically assisted triode ion plating system with an electron beam gun evaporator: 1, HT supply; 2 and 10, to measure corresponding currents; 3, specimen; 4, 6, 9, and 14, to measure corresponding potentials; 5, filament supply; 7, filament; 8, positive probe; 11, vapor source (EB gun); 12, probe supply; 13, grounding switch; 15, common point. (Data from Ref. 28.)

tained by this method. This rotating barrel type of ion plating apparatus has been used to coat Al/Zn alloy onto small components such as steel fasteners [30]. Since this process is not confined to line-of-sight deposition, parts with complex shapes can be ion plated with fairly uniform coating.

The periodic pulsed gas process [31, 32] has been used for the preparation of multilayered coatings of aluminum/aluminum oxide on steel substrates by a thermionically assisted triode ion plating technique with pulsed oxygen [33a]. Films with high yield strengths and smooth surface finishes are produced owing to the suppression of columnar growth during film deposition. The details of the deposition parameters are given in Table 3.3.

Thin films of CdS doped with indium have been grown by an ionized deposition process by Kuroyanagi and Suda [33b]. Indium-doped CdS in the form of pellets was evaporated by an EB gun, ionized by electrons emitted from a filament in an ionizer, and accelerated by an accelerating voltage in preparation for deposition on a heated substrate (Figure 3.6).The deposition parameters are as follows.

Table 3.3 Deposition Parameters for the Periodic Pulsed Gas Process for Multilayered Coatings of Aluminum

Substrate	Polished flat steel plates, 37 mm × 37 mm × 3 mm
Chamber pressure before backfilling with Ar	0.75×10^{-5} torr
Sputter cleaning	Time 10 minutes Ar pressure, 0.75×10^{-2} torr negative bias, 2 kV
Filament voltage of triode assembly	10 V
Probe voltage	200 V
Discharge conditions	Pressure 0.75×10^{-3} torr substrate bias 2 kV (argon and oxygen flow rates adjusted to different values)
Oxygen pulsing	3 seconds duration at 3 second interval 30 seconds duration at 30 second interval
Method of evaporation of Al	Resistance heated BN–TiB$_2$ crucible using discontinuous slug feed system
Source-to-substrate distance	150 mm
Substrate temperature	≤200°C

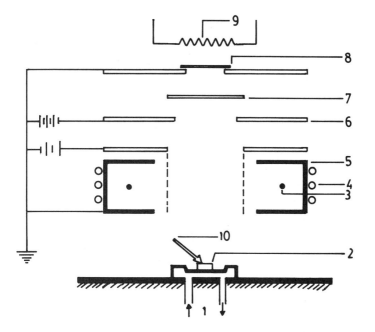

Figure 3.6 Schematic diagram of the setup for ionized deposition of thin films of indium-doped cadmium sulfide: 1; cooling water; 2, material (to be depositied); 3, filament; 4, cooling water; 5, ionizer; 6, accelerating electrode; 7, shutter; 8, substrate; 9, heater; 10, electron beam. (Data from Ref. 33b.)

Source material	indium-doped CdS pellets, 10 mm diameter, 5 mm thick
Electron current used for ionization	100 mA
Accelerating voltage	0–2 kV
Substrate	Glass
Substrate temperature	60–350°C
Pressure during deposition	< 1.0×10^{-5} torr
Rate of deposition	100–150 nm/min

Kuryoyanagi and Suda, who characterized these films by X-ray diffraction and compared them with films obtained by EB evaporation without ionization, found that the crystallinity of the films at temperatures as low as 60°C is improved by the ionization deposition. These films also exhibited a much stronger adhesion to glass substrates. Salmenoja et al. [33c] have studied the effect of a negatively biased, electron emitting hot filament on the ionization efficiency. They have derived an expression for ionization efficiency η:

$$\eta(\%) = 2.2 \times 10^{-3} \frac{j_c \, (MT)^{1/2}}{p}$$

where M (amu) is the molecular weight of the gas, j_c (mA/cm^2) is the cathode current density, and p(Pa) is the pressure. This equation has been used to evaluate the ionization efficiency. Two types of material—pure tungsten and thoriated tungsten containing 2% (by weight) thorium—were tested, and the difference in ionization efficiency were found to be negligible, although the thermionic electron emission of thoriated tungsten is greater than that of pure tungsten. Also at higher pressure, ionization efficiency with the thoriated tungsten filament was found to be even lower. The same authors used a tungsten filament to investigate the effects of process parameters on ionization efficiency in triode ion plating, and ionization efficiency was found to be strongly dependent on process parameters including cathode voltage, filament heating power, filament bias voltage, and pressure.

3.1.2 Hollow Cathode Discharge Ion Plating

In the early reported works, ion plating was mostly confined to the use of a dc abnormal glow discharge. In discharge devices with a cathode having a more or less cavity geometry (termed hollow cathode), such that a plasma region is enclosed or partially bound by a wall or walls kept at the cathode potential, the electrons within this region are reflected from the cathode wall or walls and will not escape as easily as they would otherwise have done. Such a discharge can therefore be sustained more easily, giving better performance. The major advantage of the hollow cathode discharge process is that the current density can be much higher than for a dc glow discharge under similar conditions of pressure and voltage [34]. Again in a hollow cathode discharge there is a higher percentage of ionized materials than in the abnormal glow discharge (characteristic of sputtering and ion plating), and so there can be more off-sputtering from the substrate, resulting in a more uniform deposition of materials by redistributing the coating during the process. Stowell and Chambers [35] prepared Cu and Ag coatings in a hollow cathode discharge and have investigated the hollow cathode effects on the structure of the films. The schematic diagram of their experimental setup is shown in Figure 3.7. The vacuum system was pumped down to about 10^{-6} torr before backfilling with argon for evaporation. The gas pressure was varied from 15 to 60 mtorr during the experiment. The authors showed that deposition efficiency can be varied by varying such parameters such as gas pressure, substrate voltage, and substrate separation.

Ahmed and Teer [36] ion plated aluminum films in a hollow cathode discharge onto stainless steel substrates and observed strong adherence of the films. The microstructure of the films was influenced by cathode separation, gas pressure, and

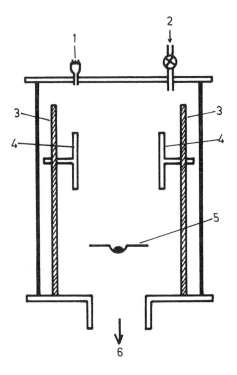

Figure 3.7 Schematic diagram of hollow cathode discharge ion plating: 1, thermocouple gage; 2, argon inlet; 3, insulated stand; 4, cathode; 5, boat; 6, to vacuum pump. (Data from Ref. 35.)

bias voltage. One disadvantage of ion plating in a hollow cathode discharge was the slight nonuniformity in thickness observed at small cathode separations and/or high gas pressures. The deposition details are given in Table 3.4.

3.2 CATHODIC ARC PLASMA DEPOSITION

Cathodic arc plasma deposition is a relatively new technique for thin film deposition and is in many respects similar to the ion plating processes. The advantage of cathodic arc evaporation for film deposition is due primarily to the high percentage of ions in the emitted flux (10–100%) and the high kinetic energies (40–100 eV) of these ions [37–40a]. Many of the benefits of ion-assisted deposition, such as enhanced adhesion, increased film packing density, and high reactivity for compound film formation [40b] are realized in this deposition process. The arc process has excellent throwing power for the non-line-of-sight film deposition of complex shapes. High deposition rates with excellent coating uniformity and low substrate

Table 3.4 Deposition Parameters for Aluminum Films in a Hollow Cathode Discharge Ion Plating

Negative bias voltage	0, 1, and 2 kV
Argon pressure	5, 10, and 15 × 10^{-3} torr
Cathode separation	10-80 mm
Substrate (serving as both substrates and cathodes)	Stainless steel, 25 mm × 25 mm
Distance between vapour source and substrate	Fixed: 150 mm above Movable: moved in a horizontal direction
Method of producing the vapour	Al evaporated from a resistance-heated BN-TiB₂ crucible

temperatures as well as easy-to-prepare stoichiometric compound or alloy films are some of the other salient features.

In cathodic arc deposition the material is evaporated by the action of vacuum arcs, the source material being the cathode in the arc circuit. Most of the basic processes of the arc occur in the region of cathode spots. The arc spot is typically a few micrometers in size and carries very high current densities [41].

The removal of the cathode material occurs by a thermal evaporation process that is due to the high current density and the resultant evaporant consists of electrons, ions, neutral vapor atoms, and microdroplets. Figure 3.8 is a simple schematic diagram of the emission process from the cathode spots based on the work of Daalder [40a]. Almost 100% of the material may be ionized within the cathode spot region [37], and these ions are ejected in a direction almost perpendicular to the cathode surface. The microdroplets are however thought to leave the cathode surface at small angles (≤ ≈ 30°) above the cathode surface. The electrons are accelerated toward the cloud of positive ions. Some of the ions are accelerated toward the cathode and may be responsible for the creation of new emission sites. The time of formation of such sites by ion impact has been estimated to be 1.2–4.5 ns for copper and 1.6–6.2 ns for molybdenum [42]. In addition to a high degree of ionization, the ions exist in multiple charge states [43,44]. Although the exact origin of the high ion energies is not well understood, a possible mechanism has been proposed by Plyutto et al. [45]. The high density of positive ions within the ion cloud region is sufficient to produce a hump in the electrical potential distribution. Such a hump may allow positive ions to escape the attraction of the negative cathode spots. A hump of about 50 V is sufficient to acquire high ion energies.

Cathodic arc deposition has been used to deposit a variety of metals, compound films, and other alloy films. Randhawa and Johnson [46] have given a review of the cathodic arc deposition technique and its various applications. A basic coating system for cathodic arc plasma deposition of thin films (Figure 3.9) consists of a vacuum chamber, the cathodic arc source, the arc power supply, substrate bias power

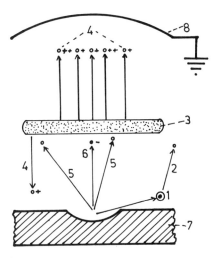

Figure 3.8 Schematic diagram of arc thermal evaporation process: 1, microdroplets; 2, neutral atoms; 3, positive ion cloud; 4, ion flux; 5, metal vapor; 6, electrons; 7, cathode; 8, anode. (Data from Ref. 40a.)

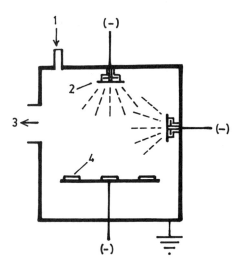

Figure 3.9 Schematic diagram of a basic coating system for cathodic arc plasma deposition: 1, gas inlet; 2, arc source; 3, to pump; 4, substrate. (Data from Ref. 46.)

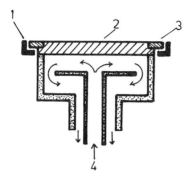

Figure 3.10 Schematic representation of a cathodic arc source: 1, anode; 2, cathode; 3, passive confinement ring; 4, cooling water. (Data from Ref. 46.)

supply, and gas inlet. The arc is a low voltage, high current discharge, and the arcs are sustained by voltages of 15–50 V, depending on the source material; currents in the range of 30–400 A are usually used.

The cathodic arc source consists of the source material serving as the cathode, an anode, an arc igniter, and some means of confining the arc spot to the surface of the cathode. A schematic diagram of a cathodic arc source is shown in Figure 3.10. Arc confinement can be accomplished by a passive confinement ring or by magnetic fields. With passive border confinement of the arc spot, the erosion of the cathode is found to be uniform. Thin films of titanium, copper, and chromium at high deposition rates and with excellent adhesion and density have been produced by this cathodic arc deposition technique.

Martin et al. [47], who used an arc evaporator system (Figure 3.11) to prepare thin films of titanium, characterized both the Ti vacuum arc and the structure of the deposited Ti films. The titanium cathode (purity > 99.5%) was 100 mm in diameter and water cooled. The arc was confined to the cathode surface by an alumina ring.

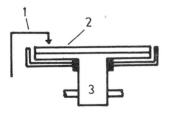

Figure 3.11 Titanium arc evaporator system for the preparation of thin films of titanium: 1, arc trigger; 2, Ti cathode; 3, water cooling. (Data from Ref. 47.)

The arc source was mounted inside a vacuum chamber 500 mm in diameter which acted as the anode. A tungsten trigger was used to initiate the discharge. Titanium films were deposited on polished plates of ASTM grade 316 stainless steel, 20 mm × 40 mm × 1 mm, over a range of substrate bias. The base pressure was around 1.5 × 10⁻⁵ torr; the deposition rate for a source-to-substrate distance of 160 mm and a 90 A arc was about 300 nm/min. The authors observed that the degree of negative bias applied to the substrate during deposition strongly influenced the crystal orientation of the films as revealed by X-ray diffraction. They also determined the energy of neutral and positive Ti ions in the region of the cathode spot of the Ti vacuum arc by measurement of the Doppler-broadened profiles of Ti (i) and Ti (ii) emission lines using Fabry–Perot interferometry. Also the energy analysis of Ti^+ and Ti_2^+ was done at a distance from the cathode using an electrostatic analyzer. The results were explained in terms of a simple model for the arc; the interested reader may refer to the paper [47] for more details.

As part of their study on the characteristics of titanium arc evaporation processes, Martin et al. [48a] deposited titanium film on glass and steel substrates. The arc evaporator used was as described elsewhere [47]. The influence of gas pressure and magnetic fields on photon and ion emission and cathode spot behavior was reported. The films were examined for microparticle content by SEM and for film texture by X-ray diffraction, as functions of the applied external and internal magnetic fields. The microparticle content in titanium metal films was reduced by approximately 40–50% with the external field, whereas the internal field had no effect.

Cathodic arc plasma deposition was used by Otsu et al. [48b] for the preparation of alloy thin films. Brass (60% cu, 40% Zn), nickel-silver (65% Cu, 25% Zn, 10% Ni and 65% Cu, 18% Zn, 17% Ni), and stainless steel (74% Fe, 18% Cr, 8% Ni) were deposited to study the chemical composition change from that of the target. A chemical analysis of the target and the thin films prepared revealed that both brass and nickel-silver alloy films differed in chemical composition from the target and that the chemical composition variation of these films could be minimized by controlling the substrate temperature. In the case of stainless steel, the chemical composition of the films was not different from that of the target. These investigators also examined the surface roughness of the as-deposited films.

In a recent review, Sanders [48c] has discussed the cathodic vacuum arc processes for film coatings. The different types of cathodic arc device, the control of the trajectories of the arcs, and the removal of the macroparticles are dealt with in this review, which cites relevant references and provides useful reading.

In a novel modification of the dc cathodic arc evaporation process for graphite reported recently by Rother et al. [48d], the arc spot position could be controlled, as indicated in Figure 3.12. The complex source plate assembly consisted of the cathode, cathode shield, an electromechanical trigger electrode, and a special magnetic system. A switchable magnetic field with components parallel to the erosion sur-

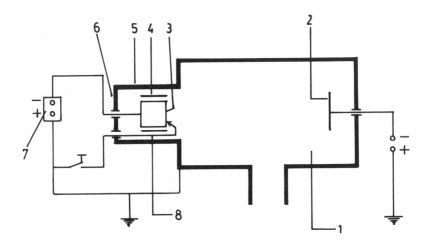

Figure 3.12 Schematic of the experimental arrangement for dc cathodic arc evaporation for graphite: 1, vacuum chamber; 2, substrate holder; 3, cathode; 4, cathode shield; 5, anode flange; 6, source plate; 7, arc power supply; 8, trigger system. (Data from Ref. 48d.)

face could be provided with this magnetic system. During cathode spot motion, stimulated by the magnetic fields, the electric potential of the cathode shield was analyzed by a computer-aided control unit, and the switching of the magnetic field was controlled via the electric potential of the shield. This setup improved considerably the uniformity of the cathode erosion face and the continuous discharge time; it is a very reliable method for the cathodic arc evaporation of porous cathode material such as graphite. The carbon films deposited onto silicon and cemented carbide substrates showed good adherence, and the investigators have also carried out some preliminary scratch and other tests on the coatings.

Shinno et al [48e] have recently reported the preparation of carbon–boron coatings in molybdenum by a vacuum arc deposition method; boron concentrations in the cathode ranged between 5 and 60 wt %. The cathode was made from a mixed powder of carbon and boron and pressed and sintered. The coatings prepared were characterized by X-ray diffraction, X-ray microanalysis, SEM, and Raman spectroscopy.

3.3 HOT HOLLOW CATHODE GUN EVAPORATION

Hot hollow cathode (HHC) gun evaporation is an arc generating device that can be used as a heat source by extracting electrons to form an electron beam; it serves to deposit metallurgical coatings of various kinds. The HHC gun (Figure 3.13) con-

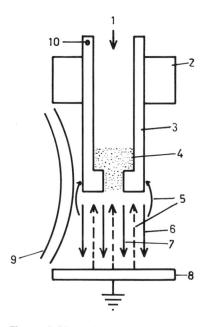

Figure 3.13 Schematic setup of a hot hollow cathode gun: 1, argon inlet; 2, water cooling; 3, hollow cathode; 4, plasma; 5, return ions to cathode; 6, electrons from refractory metal; 7, electrons from plasma; 8, anode; 9, magnetic field; 10, to main power supply. (Data from Ref. 53.)

sists of a hollow refractory tube, which acts as the cathode. An ionizable gas (usually argon) is introduced into the system through this cathode tube, and the argon flowing through the tube supports an arc discharge to the anode. The pressure drop across the orifice in the tube gives sufficient gas inside the cathode to sustain the plasma that generates the beam. A low voltage, high current dc power source is used, and the HHC gun operates in the arc region of the voltage current curve for a typical low pressure gas discharge.

When used to produce metallurgical coatings, the typical operating range of a hot hollow cathode gun has an arc current of 50–200 A, and the beam of electrons bombards the evaporant contained in the conducting crucible, which serves as the anode. The material is evaporated, is substantially ionized, and is deposited on the substrate, which is kept above the source. These ions have made possible a very high coating adhesion, and the technique can therefore be treated as an ion-assisted deposition method.

Morley and Smith [49] were the first to use HHC as an electron beam heating device for the preparation of vacuum coatings. Copper and quartz were evaporated from a water-cooled crucible about 76 mm in diameter onto substrates mounted about 50 cm above the source; at power levels of 15 kW (60 V, 250 A); evaporation rates of 4 g/min were obtained. Vacuum evaporation deposition using an HHC gun has since then been routinely conducted at Rockwell International, USA [50,51]. Larson and Draper [51] vapor deposited silver onto beryllium substrates. The HHC gun consisted of a shielded hollow tantalum tube and operated at 50–200 A and 15–20 V. The high operating pressure (chamber pressure of 10^{-6} torr increased to $3–20 \times 10^{-4}$ torr by the argon flow), together with the high beam current and low voltage, produced a rich source of argon and silver ions above the melt in the evaporation crucible. The substrates were sputter-cleaned before the actual deposition of the silver films. The relationship of the ion cleaning parameters to contaminants in the beryllium–silver interface and their effect on adhesion were investigated.

Wang et al. [52] deposited silver films onto steel substrates using a hollow cathode electron beam source; with the gun placed at an angle of 45° to the water-cooled copper hearth containing silver. the following are the deposition parameters.

Ar pressure	5×10^{-3} torr
Substrate bias	–50 V
Current density	0.8 mA/cm²
Hollow cathode power supply	30 V
Hollow cathode current	30 A
Substrate temperature	260°C

The investigators studied the composition profile of the interfaces with AES, combined with the ion sputter etching technique. The adhesion of the films was also examined to determine its relationship to the composition profile of the interfaces.

Kuo et al. [53], in a recent report, reviewed the applications of hot hollow cathode techniques in vacuum coating, and the interested reader may refer to this work for more details.

3.4 CONCURRENT ION BOMBARDMENT DEPOSITION

When a separate gas ion source is used to sputter clean the substrate prior to deposition and also to bombard the films during deposition in a controlled manner, thin films with high adhesion and through porosity can be prepared. Hirsch and Varga [54] have reported that germanium films, deposited under simultaneous argon irradiation, increased substantially the adherence of the films to glass and other substrate materials. Although the concurrent bombardment of condensed metal films was also reported briefly later [55,56], it was Hoffman and Gaerttner [57] who studied the property changes in the films by concurrent ion bombardment. Here thermally evaporated chromium films were simultaneously bombarded during condensation with inert ions (xenon and argon) from a separate ion source directed at the substrate. The inert gas ions and the evaporant fluxes could be controlled independently, and the resultant effects on the properties of the films were obtained by observation and measurement of the residual film stresses and optical reflectance of the deposited films. Representative examples of recent work on concurrent ion bombardment deposition are discussed below.

Thin amorphous hydrogenated silicon films have been prepared by the combined electron beam evaporation of silicon and ion bombardment [58]. The base pressure of the vacuum system was 0.75×10^{-6} torr, rising to 0.75×10^{-3} torr during the operation of the ion gun. The other deposition details are as follows.

Ion beam energy	1200 eV
Current density	30 $\mu A/cm^2$
Deposition rate	> 1 $\mu m/h$
Angle of incidence of the beam (mainly composed of a mixture of H_2^+ and H^+)	30° to be substrate surface
Substrate	High quality optical grade glass, 20 mm diameter
Substrate temperature	Room temperature

Nandra et al. [59] reported the deposition of thin films of gold onto copper substrates from a resistance-heated evaporation source and evaluated the effect of the irradiation of the films with low energy ions (< 10 keV) during deposition on such film properties as adherence, microstructure, and porosity. Figure 3.14 shows the schematic arrangement of their setup with the gas ion and evaporation sources. The details of the deposition system and the deposition conditions are given in Table 3.5. The investigators adapted the following procedure for the preparation and irradiation of the films. The chamber was first pumped to a vacuum of 0.75×10^{-6} torr. The resistance source was then switched on, with the shutter above the source, and the gold was outgassed above its melting point. The temperature of the source was then reduced to below the melting point of gold, the shutter was removed, and the substrate sputter-cleaned. Finally, the ion current density was reduced to the desired level and the temperature of the source raised, to permit the evaporation to take place, the thickness of the deposit being determined by the thickness monitor.

Argon ion bombardment before and/or during the thermal evaporation of gold onto copper was found to influence the adherence and through porosity of the films formed. Also sputter cleaning prior to deposition is a very effective means of producing highly adherent coatings, and the increase in the duration of sputter cleaning has shown a marked increase in through-porosity of the films.

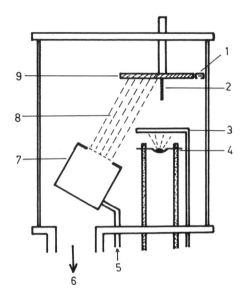

Figure 3.14 Schematic diagram of the gas ion and evaporation source for the deposition of thin films of gold onto copper substrates: 1, crystal monitor; 2, partition; 3, shutter; 4, evaporation source; 5, argon inlet; 6, to pump; 7, ion source; 8, ion beam; 9, substrate. (Data from Ref. 59.)

Table 3.5 Deposition Parameters for Concurrent Ion Bombardment to Produce Thin Films of Gold on Copper Substrates

Ion source	Kaufman type
Ion energy	0.8–6 keV
Current density	0.02–0.18mA/cm^2
Diameter of ion beam	25 mm
Distance from substrate to evaporation source	250 mm
Position of ion source	Inclined at an angle 30° from the normal to the substrate
Deposition rate	0.02–0.2 μm/min
Duration of sputter cleaning, using ion beam energy of 1.5 keV and current density of 1 mA/cm^2	30 seconds

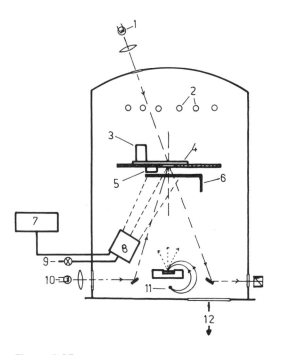

Figure 3.15 Schematic arrangement for the ion-assisted deposition of thin films of zirconium dioxides: 1, transmission light source; 2, radiant heater; 3, Faraday cup; 4, substrate; 5, rate monitor; 6, shutter; 7, ion control; 8, ion gun; 9, gas inlet; 10, reflection light source; 11, EB gun. 12, to pump. (Data from Ref. 60.)

In high precision optical coatings, ensuring the stability and reproducibility of the refractive index of the films has always been a problem. It can be seen that thin films usually exhibit columnar growth, and metallic films in general exhibit a preferred orientation in their crystal structure. When the voids between the columns in the columnar structure in thin films are exposed, they absorb water vapor from the atmosphere by capillary action and change the refractive index and stability of the coatings irreversibly. Thus it is very important that films be densely packed for best performance.

Ion-assisted deposition of dielectric ZrO_2 thin films has been reported by Martin et al. [60]. The films were prepared by electron beam evaporation of 99% pure ZrO_2, while low energy bombardment by oxygen and argon was used during the condensation of the films. The experimental arrangement for ion-assisted deposition is shown in Figure 3.15 and the deposition details are given below.

Ion energy	600–1200 eV Ar^+ or O^{+2}
Angle of incidence of ion beam on substrate surface	30°
Pressure during ion beam operation	0.75×10^{-4} torr
Substrates	Borosilicate crown glass, polished (for optical constant determination). Microscope slides (for structural and compositional analysis)
Source (EB)-to-substrate distance	420 mm
Rate of evaporation	0.8 nm/s

The substrate, which could be heated to 300°C, was mounted above the electron beam source on a slide so that the transmittance of the coated substrate could be referred to that of an uncoated area at any time. Upon investigating the influence of ion bombardment on the refractive index, film density, structure, and composition of these dielectric ZrO_2 films, the authors found that the film density has improved from 0.83 to unity, with a corresponding increase in the refractive index (from 1.84 to 2.19). Ion bombardment during condensation of evaporated films onto room temperature substrates resulted in the crystallization to the cubic phase; at elevated temperatures, the monoclinic phase is also present.

Gibson and Kennemore [61] have shown that ion bombardment during the deposition of MgF_2 can lead to significant improvement in the abrasion resistance and increased their adhesion to plastic substrates.

Ion-assisted deposition has been used to prepare very thin, semitransparent gold films on glass substrates for either solar control or transparent heat mirror applications [62]. The deposition details are as follows.

Ion energy	500 eV Ar^+
Ion current density, i_{ion}	$8 < i_{ion} < 33~\mu A/cm^2$
Ar pressure	10^{-4} torr
Substrate	Corning 7059 glass
Film preparation	Computer-controlled thermal evaporation
Deposition rate r_{Au}	$0.1 < r_{Au} < 0.2$ nm/s
Thickness, t	$4.5 < t < 18$ nm

It was found that these ion-assisted Au films have favorable properties for applications as window coatings. Improved transmittance of solar energy for low emittance coatings, reduced mass of metal required for solar control applications, and the ability to tune the near-infrared performance by varying the deposited mass are the main advantages of these ion-assisted gold films.

Panitz [63] used a dual-beam ion system with two different targets in the sput-

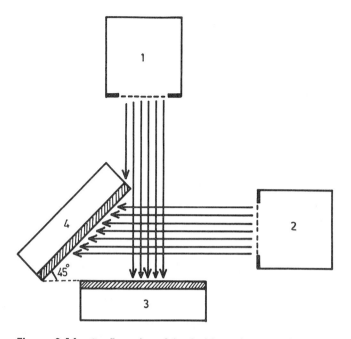

Figure 3.16 Configuration of the dual-beam ion sputtering system: 1, ion gun to bombard the substrate (substrate–gun distance, 30 cm); 2, ion gun for sputtering (taget–gun distance, 25 cm); 3, water-cooled platform with substrate; 4, water-cooled platform with target. (Data from Ref. 63.)

tering system to study the variation in the microstructure of nickel-based alloy coatings. A predominantly amorphous $Ni_{63.5}$ $Cr_{12.3}$ $Fe_{3.5}$ $Si_{7.9}$ $B_{12.8}$ foil and a crystalline $Ni_{55.3}$ $Cr_{16.9}$ $Si_{7.2}$ $B_{21.6}$ slab from a casting were used as the sputtering targets. The configuration of the sputtering system is shown schematically in Figure 3.16. One argon beam was used to sputter-deposit the material; and the other was used to bombard the film during deposition. The effects of microstructure on the nature of the films were investigated using TEM and electron diffraction. The effect of the composition of the source material also was studied. The coating composition was confirmed by AES.

Berg et al. [64] have demonstrated that selective deposition is possible by exposing the growing film to ion bombardment. Control of thin film orientation by glancing angle ion bombardment during growth has been reported by Yu et al. [65a].

Heat radiation from the evaporation source always degrades or destroys temperature-sensitive substrates during the deposition of thin films. The heat radiation in CW laser evaporation is considerably less than thermal and electron beam

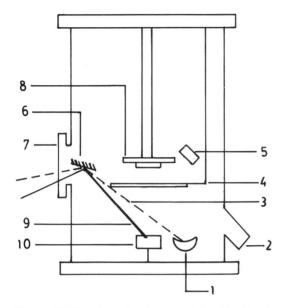

Figure 3.17 Schematic diagram of the thin film deposition system combining laser evaporation and ion bombardement for the deposition of calcium fluoride films at low temperatures: 1, Ge source; 2, ion gun; 3, pulsed laser beam; 4, movable shutter; 5, quartz thickness monitor; 6, mirror; 7, ZnSe window; 8, substrate holder; 9, CW laser beam; 10, CaF_2 source. (Data from Ref. 65b.)

evaporation, and Gluck et al. [65b] combined laser evaporation and ion bombardment to deposit high quality CaF_2 thin films at low temperatures (20°C). The schematic diagram if their thin film deposition system is shown in Figure 3.17.

The deposition details were reported as follows.

Substrates	Polished GaAs
Substrate temperature	20°C (water-cooled substrate holder)
Laser source	a CW CO_2 laser (operating at 10.6 μm)
Laser power	20 W
Rate of deposition	5-7 Å/s
Ion beam	Ar^+, O^{+2}, or CF^{+4} ions at an angle of 30° to the normal: low energies ranged from 100 to 700 eV; current density, 0.5-1 mA/cm^2

Ion-assisted deposition using laser evaporation produced smooth, durable coatings with bulk refractive indices, at low substrate temperatures. This technique was also applied [65b] to fabricate midwave IR dichroic filters using CaF_2 and Ge layers on cold substrates (20°C).

Erck and Fenske [65c] prepared silver films using an electron beam evaporator, with Ar ions from a 3 cm diameter broad beam ion source directed to the substrate during deposition. To reduce the substrate charging effect, the ion beam was neutralized. The power density of incident ion beam on the specimen was measured and the silver evaporation rate varied to attain different values of ion/atom arrival rate ratio. The deposition rate was monitored by a rate monitor. Polished Al_2O_3 substrates were sputter-cleaned before deposition. The angles of incidence of the vapor and the bombarding ions with respect to the plane of the substrate were 45° and 125° respectively. The deposition was carried out at ambient temperature under a 5×10^{-3} Pa pressure in the chamber. Erck and Fenske examined the effects of ion bombardment on film growth by studying the film morphology and the hardness of these evaporated films, which had been bombarded by argon ions during deposition.

Georgiev and Dobrev [65d] who studied the epitaxy of silver films deposited onto NaCl (100) substrates during Ar^+ ion bombardment, found that in addition to the Ag (100) ‖ NaCl (100) layer formation at lower substrate temperatures, high ion beam densities and temperature can produce oriented crystallites of close-packed directions other than <100>.

Mineta et al. [65e] have recently reported the preparation of cubic boron nitride films by CO_2 laser evaporation with simultaneous nitrogen ion bombardment. These authors used a high power CO_2 laser, where the focused beam irradiated onto the peripheral surface of a rotating ceramic ring target (sintered hexagonal boron nitride; h–BN) in the tangential direction. Figure 3.18 is the schematic of their setup, indicating the positions of the target, laser beam, substrate, and ion source. A continuous wave CO_2 laser beam was introduced into the vacuum chamber and focused by the focusing mirror. Since the vapor from the target flies in the direction perpendicular to the incident direction of the laser beam, contamination and damage of the optical system are avoided. The ceramic ring can be heated by a separate electric heater to several hundred degrees Celsius to avoid thermal fracturing of the ring caused by the rapid heating at the laser-focused point. The substrate can also be heated. The N_2 gas (99.999% purity), ionized in a Kaufman-type ion source and accelerated at a bias voltage of 0–2.0 kV, can be irradiated onto the substrates simultaneously with the laser evaporation process.

The deposition conditions reported are as follows.

Target (ring form)	Sintered h-BN (purity 99 and 95%)
Rotational speed of the ring	5–30 rpm

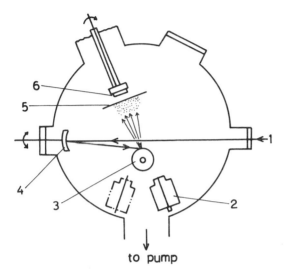

to pump

Figure 3.18   Schematic of the setup for the preparation of cubic boron nitride films by CO_2 laser evaporation with simultaneous nitrogen ion bombardment, showing the positions of target, laser beam, substrate, and ion source: 1, CO_2 laser beam; 2, ion source; 3, ceramic ring; 4, Cu concave mirror; 5, shutter, 6, substrate. (Data form Ref. 65e.)

Substrate	Mo, Si, Si_3N_4, Ti, WC–Co, TiN/WC–Co
Ultimate pressure	1×10^{-5} torr
Substrate temperature	500–600°C
Laser power	200–1000 W
Deposition time	30–90 minutes

By this method Mineta et al. [65e] prepared films rich in (c–BN) exhibiting very high hardness and a high wear resistance against Ni–Mo steel under dry sliding conditions.

The effects of low energy ion bombardment during sputtering on the microstructural, compositional, and electrical properties of ion beam sputtered YBCO thin films have been discussed by Doyle et al. [65f]. Ion bombardment of the growing film was carried out by an ion source of the Kaufman/ECR (electron cyclotron resonance) type with energy ranging up to 125 eV. The effects on the critical current and temperature also were reported.

Rossnagel and Cuomo [65g] have discussed in detail the effects of ion bombardment during deposition on such film properties as grain size and orientation, defect densities, electrical and optical properties, chemical stoichiometry, and surface morphology.

3.4.1 Unbalanced Magnetron Ion-Assisted Deposition

In conventional magnetrons, the discharge is confined by the magnetic field close to the surface of the cathode, and so the bombardment of growing films by energetic particles other than that of the depositing atoms is minimal.

Window and Savvides [66] have developed a new type of planar deposition source that gives a separate beam of ions (intensity can be varied independently of the deposition) along with the deposition flux. Such a source would more conveniently be used for ion beam assisted deposition, unlike the technology that uses separate sources for the deposition flux and the bombarding ion flux. The ion flux produced is a direct result of the unbalanced magnetic (UM) arrangement in the magnetron, and the authors have proposed the name UM gun for this source. These sources can provide large fluxes of ions whose energy can be varied upward from a few to several hundred electron volts—energy ranges not covered by conventional ion beam guns. The UM gun is a simple, reliable, low cost device that provides not only the flux of depositing atoms but also a flux of ions and a flux of electrons at the substrate. Moreover, in the case of insulated or insulating substrates, the presence of the electron flux neutralizes the ion flux and allows the substrate to reach significant self-bias voltages (>25 V) with respect to the plasma.

Savvides and Window [67] have reported using a UM gun capable of producing ion current densities up to 5 mA/cm^2 (corresponding to an ion flux of 3 × 10^{16} cm^{-2}/s) to prepare a variety of thin films of amorphous and crystalline materials by varying both the bombarding ion energy (2–1000 eV) and ion/atom arrival rate ratio over the range of 0.4–10. (The energy of the impinging ions and the arrival rate ratio of the ions to the condensing atoms are important parameters in ion-assisted deposition.) In the UM gun sputtering system (Figure 3.19), the magnet assembly consists of a permanent ring magnet fitted with a soft ion center pole, giving an effective sputtering area of about 20 cm^2. The substrate support is electrically isolated from the vacuum chamber and the fixtures and is placed off center with respect to the target. The substrate table can be rotated to accommodate up to eight individual specimens. A metal plate with an aperture masks the table and a shutter (not shown in Figure 3.19), which can be rotated, covers the aperture. The chamber fixtures are connected to the vacuum chamber, which is kept at the ground (anode) potential. The self-bias potentials of the insulating substrates and the table are normally in the range –25 to –10 V with respect to the ground. The authors have prepared diamondlike amorphous carbon films by the UM gun sputtering of a graphite target in Ar gas of 99.999% purity. These films possess outstanding diamondlike properties including extreme hardness, high electrical resistivity, high dielectric strength, optical transparency in the infrared, and an optical bandgap up to 0.74 eV. It is thought that the diamondlike properties are due to the tetrahedral bonding formed from the ion–surface interaction or to the interaction of ions with the bulk of the material during film growth.

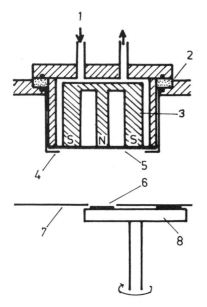

Figure 3.19   Schematic of the UM gun sputtering system for the preparation of thin films of amorphous and crystalline materials: 1, water cooling; 2, insulator; 3, magnet; 4, anode shield; 5, target; 6, substrate; 7, masking plate; 8, substrate table. (Data from Ref. 67.)

3.5   ION BEAM DEPOSITION

Ion beams are used in two basic configurations for the deposition of thin films. In the direct ion beam deposition (IBD), the ion beam consisting of the desired film material is extracted and is deposited at low energy ($\approx$ 100 eV) directly onto the substrate. The simple principle of ion beam deposition is shown schematically in Figure 3.20. In ion beam sputter deposition the ion beam at high energy (10^2–10^3 eV) is directed against the target of the material, which is sputtered and deposited onto the adjacent substrate, as discussed in Chapter 2.

In the direct ion beam deposition of film species, the energy of the depositing material is controlled directly, and the beam can be mass analyzed to produce a highly pure deposit. The main disadvantage of this technique is the limitation of the useful energy of the beam to avoid self-sputtering, giving deposition rates too low for the processing of large areas. Armour et al. [68] have discussed the different aspects of the kinetics of the film growth and the ion–surface interaction phenomena associated with low energy ion bombardment. From theoretical considerations, these authors have provided certain experimental conditions for an ion beam deposition system. The following are some of the requirements.

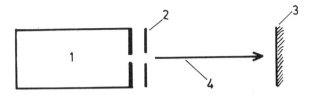

Figure 3.20 Schematic representation of the simple principle of ion beam deposition: 1, ion source; 2, ion extractor; 3, substrate; 4, ion beam.

1.   Ion beams should have a wide range of species in the energy range 20 eV to about 1 keV, with current densities of the order of 100 μA/cm². For uniform coating over useful areas, the beam area on the target should be about 20 × 5 mm and the beam should be stable for periods up to several hours, because of the low deposition rates.

2.   The beam must be mass analyzed, and the beam line must be designed to eliminate sputtered neutral particles.

3.   The energy spread of the ion beam from the source must be low compared with the minimum bombardment energy required.

4.   The residual gas pressure should be below about 10^{-9} torr when the deposition is made.

5.   The substrate mounting assembly must be movable.

6.   There should be facilities for in situ analysis.

Since 1961, when the first ion beam deposition method without mass analysis was proposed by Flynt [69] to fabricate microminiature electronic circuits, many have reported the deposition of thin films by the IBD method.

Table 3.6 summarizes some of the main work done in this field. The majority of the studies are carried out at the laboratory stage, and the interested reader can find details of the useful applications and the results achieved from these papers and the references cited therein. Examples of recent work on IBD are summarized in the following discussion.

Antilla et al. [82] prepared i-C coatings on WC-Co hard metal substrates using $12C^+$, CH_3^+, CH_4^+, and $C_2H_2^+$ beams. The gases led into the ion source were CO_2, CH_4, and C_2H_2, and the depositions were made in a vacuum of 7.5 × 10^{-7} torr. For $12C^+$ deposition, a deposition rate of 0.3 nm/s was obtained. The purity of the i-C films was checked by taking a Rutherford scattering spectrum from a self-supporting film. The hardness, abrasive wear resistance, and adhesion of the coatings prepared with C^+ ion beam were found to be superior to those prepared with other

Table 3.6 Ion Beam Deposition

Ion used	Energy (eV)	Ion current/density	Pressure during deposition (torr)	Substrate and substrate temperature	Remarks, if any	Ref.
Pb, Mg	24–500	10–15 A			Deposition rate, 0.3–1 nm/min	70
Ge	100	50–200 A	1×10^{-7}	Si (100), single crystal	300°C	71
Si	200		5×10^{-8}		740°C	
Pb, Mg	10 eV to 4 keV	Up to 20 μA	10^{-8} to 10^{-9}			72
Ag	50	4 A/cm^2		Boron-doped Si (111) slices RT	Base pressure in the chamber, 10^{-10} torr.	73
Ionized borazine ($B_3N_3H_6$) molecules	200–1000	100–200 $\mu A/cm^2$	1×10^{-4} 5×10^{-4}	Pyrex glass and stainless steel. Commercially obtained tool inserts	Thin films containing cubic phase boron nitride have been synthesized.	74
Ionized species from methane gas	200	10 mA	2×10^{-7}		Deposition rate, 0.5 Å/s; hydrogenated amorphous carbon films.	75
C	300	60 $\mu A/cm^2$	2×10^{-6}	n-type Si (100), glass plates deposited with Al	Deposition rate, 0.1 Å/s, 0.2 Å/s; films showed diamondlike characteristics.	76
	600	200 $\mu A/cm^2$				
Ar, C	100–150	1 mA/cm^2		Si and Si O_2	Deposition rate, 71 Å/min, CH_4-to-Ar ratio, 0.28, ideal for depositing films.	77

Material	Energy	Current	Pressure	Substrate	Comments	Ref.
Ionized borazine ($B_3N_3H_6$)	150–1800	0.2–1 mA/cm²	5×10^{-4}	Single crystal Si (polished), different cutting tools, glass microscope slides, fused silica slides	Deposition rate, 0.5–1 nm/s; etched for 15 minutes by a 1000 eV, 0.2 mA/cm² argon before deposition.	78
Pd	100–400	1–3 A	$2–4 \times 10^{-7}$	Single-crystal (111) Si	Continuous Pd films with resistivity of the same order as bulk; thickness as thin as a few nanometers.	79
C	90–250		5×10^{-5}	Si and fused silica	Growth rates measured for various beam energies and types of hydrocarbon gas used as the source of carbon.	80
Ge, Si	40–200 eV		10^{-9}	p-type Si (100); 27 and 412°C	Isotopic thin films ^{74}Ge and ^{30}Si. Deposits were uniform and continuous, with a sharp interface between IBD layers and substrate.	81

ions. The films prepared with hydrocarbon ion beams were brittle and had weak adhesion.

Direct ion beam deposition has been used by Appleton et al. [83] to deposit isotopic thin films and heterostructures and high quality epitaxial growth of ^{74}Ge on Ge (100) and ^{30}Si on Si (100) at temperatures as low as 400°C. A conventional ion implantation accelerator was used to produce isotopically pure beams at 35 keV. A specially constructed deceleration lens system and sample holder mounted in the chamber enabled the 35 keV incident ions to be slowed down for direct deposition onto the target. The energies of the deposited ions could be varied from 0 to 1000 eV. During deposition the chamber pressure was in the 10^{-9} torr range. The deposition rate was of the order of 1–5 nm/min. The samples were analyzed by a variety of techniques including ion scattering–ion channeling, cross-sectional transmission electron microscopy, X-ray diffraction analysis, and Auger electron spectroscopy. The results showed that uniform, continuous, isotopically pure thin films over large areas with atomically sharp interfaces could be obtained with the direct IBD system.

Wagal et al. [84] have reported the deposition of diamondlike carbon films onto clean unseeded substrates using a hybrid ion beam technique. Their experimental set up is shown in Figure 3.21. The output from a Nd-glass laser was focused onto the graphite target in the UHV chamber. The laser beam ejected a plume of carbon vapor ionizing as well the portion still being illuminated. These ions were drawn out of the plume and accelerated by the static fields between the grounded target and the negatively charged accelerating grind. The substrate was contained in a Faraday cage connected to the grid. A drift space was incorporated in the cage so that the path from the plume to the substrate totalled 15 cm.

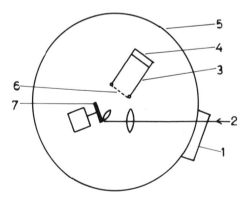

Figure 3.21   Schematic of the experimental setup for the deposition of diamondlike carbon films onto clean, unseeded substrates using a hybrid ion beam technique: 1, window; 2, laser; 3, Faraday cage; 4, silicon substrate; 5, UHV chamber; 6, accelerating grid; 7, graphite target. (Data from Ref. 84.)

Since highly pure graphite (99.999%) was used, only ions of carbon were collected. This avoided the use of the mass separation techniques necessary with molecular sources of C^+.

The laser was used both in a triggered mode (10 J; 10 ns) and in the Q-switched mode (1 J; 10 ns), and the acceleration potential ranged from 300 to 2000 V. The substrates were cut from Si (111) disks (75 cm in diameter).

The films of diamondlike carbon prepared exhibited a mirror-smooth surface finish, and areas of 10 cm² could be coated with uniform optical quality films at an average deposition rate of 20 μm/h. Again the growth of the films needed no seeding of the substrate with debris from abrasive treatment with diamond or any other particles.

REFERENCES

1. B. A. Probyn, *Br. J. Appl. Phys. (J. Phys. D)*, 2:457 (1968).
2. S. Aisenberg and R. Chabot, *J. Appl. Phys.*, 42:2593 (1971).
3. J. M. E. Harper, in *Thin Film Processes* (J. L. Vossen and W. Kern, Eds.), Academic press, New York, 1978.
4. T. Takagi, K. Matsubara, H. Takioka, and I. Yamada, Proceedings of an *International Conference on Ion plating and Allied Techniques, IPAT 79*, London, 1979, p. 174.
5. G. Carter and D. G. Armour, *Thin Solid Films, 80*:13 (1981).
6. D. G. Armour, P. Bailey, and G. Sharples, *Vacuum*, 36:769 (1986).
7. D. M. Mattox, *Electrochem. Technol.*, 2:295 (1964).
8. *D. G. Teer*, Proceedings of an International Conference on Ion Plating and Allied Techniques, IPAT 77, *Edinburg, 1977, p. 13.*
9. D. M. Mattox, *J. Electrochem. Soc.*, 12:1255 (1968).
10. D. M. Mattox, *J. Vac. Sci. Technol.*, 10:(1973).
11. (a) T. Spalvins, *J. Vac. Sci. Technol.*, 17:315 (1980). (b) L. J. Wan, B. Q. Chen, and K. H. Kuo, *J. Vac. Sci. Technol.*, A6:3160 (1988). (c) L. Wan and K. H. Kuo. *Vacuum*, 40:411 (1990).
12. D. G. Teer and B. L. Delcea, *Thin Solid Films, 54*:295 (1978).
13. T. Spalvins, *Thin Solid Films*, 64:143 (1979).
14. H. A. Sandquist, A. Mathews, and D. G. Teer, *Thin Solid Films*, 73:309 (1980).
15. E. Mini and H. Sandquist, *Thin Solid Films*, 80:55 (1981).
16. T. Spalvins and B. Buzek, *Thin Solid Films*, 84:267 (1981).
17. H. A. Sundquist and J. Myllyla, *Thin Solid Films*, 84:289 (1981).
18. R. E. Hurley and S. A. Wilkes, *Thin Solid Films*, 80:93 (1981).
19. R. D. Arnell and D. G. Teer, *Thin Solid Films, 84*:281 (1981).
20. R. E. Hurley and E. W. Williams, *Thin Solid Films*, 92:99 (1982).
21. K. Miyoshi, T. Spalvins, and D. H. Buckley, *Thin Solid Films*, 96:9 (1982).
22. K. Miyoshi, T. Spalvins, and D. H. Buckley, *Thin Solid Films*, 108:199 (1983).
23. L. W. Wang, *Thin Solid Films*, 105:319 (1983).
24. (a) K. Miyoshi, T. Spalvins, and D. H. Buckley, *Thin Solid Films*, 127:115 (1985). (b) A. Celerier and J. Machet, *Thin Solid Films*, 148: 323 (1987).

25. (a) P. Sauliner, A. Debhi, and J. Machet, *Vacuum*, 34:765 (1984). (b) S. Witanachchi, H. S. Kwok, X. W. Wang, and D. T. Sham, *Appl. Phys. Lett.*, 53:234 (1988).
26. G. A. Baum, Dow Chemical Co. Publ. FRP-686, Colorado, 1967.
27. K. H. Kloos, E. Brozeit, and H. M. Gabriel, *Thin Solid Films*, 80:307 (1981).
28. A. Mathews and D. G. Teer, *Thin Solid Films*, 80:41 (1981).
29. N. A. G. Ahmed, *Vacuum*, 34:807 (1984).
30. N. A. G. Ahmed and D. G. Teer, *J. Phys. E: Sci. Instrum.*, 17:411 (1984).
31. R. W. Springer and C. D. Hosford, *J. Vac. Sci. Technol.*, 20:462 (1982).
32. S. G. Noyes and H. Kim, *J. Vac. Sci. Technol.*, A3:1201 (1985).
33. (a) N. A. G. Ahmed, *Thin Solid Films*, 144:103 (1986). (b) A. Kuryoyanagi and T. Suda, *Thin Solid Films*, 176:247 (1989). (c) K. Salmenoja, J. M. Molarius, and S. S. Korhonen, *Thin Solid Films*, 155:143 (1987).
34. P. F. Little and A. Von Engel, *Proc. R. Soc. (London)*, 224;209 (1954).
35. W. R. Stowell and D. Chambers, *J. Vac. Sci. Technol.*, 11:653 (1974).
36. N. A. G. Ahmed and D. G. Teer, *Thin Solid Films*, 80:49 (1981).
37. C. W. Kimblin, *J. Appl. Phys.* 45:5253 (1974).
38. L. P. Harris, in *Arc Cathode Phenomena* (J. M. Lafferty, Ed.), Wiley-Interscience, New York, 1980.
39. I. I. Akseno, Yu. P. Antufev, V. G. Brew, V. G. Padalka, A. I. Popov, and V. M. Khoroshikh, *Sov. Phys. Tech. Phys. Tech. Phys.*, 26:184 (1981).
40. (a) J. E. Daalder, *Physica C*, 104:91 (1981). (b) P. J. Martin, *Vacuum*, 36:585 (1986).
41. J. E. Daalder, *J. Phys. D: Appl. Phys.*, 16:17 (1983).
42. J. Buttner, *J. Phys. D: Appl. Phys.*, 14:1265 (1981).
43. W. D. Davis and H. C. Miller, *J. Appl. Phys.*, 40:2212 (1969).
44. V. M. Lunev, V. G. Padalka, and V. M. Khoroshikh, *Sov. Phys. Tech. Phys.*, 22:858 (1977).
45. A. A. Plyutto, V. N. Pyshkov, and A. T. Koplin, *Sov. Phys. (J. Exp. Theor. Phys.)*, 20:328 91965).
46. H. Randhawa and P. C. Johnson, *Surface Coatings Technol.*, 31:303 (1987).
47. P. J. Martin, R. P. Netterfield, D. R. Mckenzie, I. S. Falconer, C. G. Pacey, P. Thomas, and W. G. Sainty, *J. Vac. Sci. Technol.*, A5:22 (1987).
48. (a) P. J. Martin, D. R. Mckenzie, R. P. Netterfield, P. Swift, S. W. Flipczuk, K. H. Muller, C. G. Pacey, and B. James, *Thin Solid Films*, 153:91 (1987). (b) M. Otsu, E. Ko, T. Yoshikawa, and K. Tsufi, *Thin Solid Films*, 181:351 (1989). (c) D. M. Sanders, *J. Vac. Sci., Technol.*, A7:2339 (1989). (d) B. Rother, J. Siegel, and J. Vetter, *Thin Solid Films*: 188: 293 (1990). (e) H. Shinno, T. Tanabe, M. Fujitsuka, and Y. Sakai, *Thin Solid Films*, 189:149 (1990).
49. J. R. Morley and H. R. Smith, *J. Vac. Sci. Technol.*, 9:1377 (1972).
50. D. G. Williams, *J. Vac. Sci. Technol.*, 11:374 (1974).
51. D. Larson and L. Draper, *Thin Solid Films*, 107:327 (1983).
52. L. W. Wang, F. Z. Wang, J. H. Zhang, L. H. Zhai, and P. G. Cao, *Thin Solid Films*, 105:319 (1983).
53. Y. S. Kuo, R. F. Bunshah, and D. Okrent, *J. Vac. Sci. Technol.*, A4:397 (1986).
54. E. H. Hirsch and I. K. Varga, *Thin Solid Films*, 52:445 (1978).
55. J. Dudonis, A. Jotautis, O. Meilius, A. Meskauskas, and L. Pranevicius, *Thin Solid Films*, 58:106 (1979).

56. J. Franks, P. R. Stuart, and R. W. Withers, *Thin Solid Films*, 58:128 (1979).
57. D. W. Hoffman and M. R. Gaerttner, *J. Vac. Sci. Technol.*, 17:425 (1980).
58. P. J. Martin, R. P. Netterfield, W. G. Sainty, and D. R. Mckenzie, *Thin Solid Films*, 100:141 (1983).
59. S. S. Nandra, F. G. Wilson, and C. D. DesForges, *Thin Solid Films*, 107:335 (1983).
60. P. J. Martin, R. P. Netterfield, and W. G. Sainty, *J. Appl. Phys.*, 55:235 (1984).
61. U. J. Gibson and C. M. Kennemore, Ill, *Thin Solid Films*, 124:27 (1985).
62. G. B. Smith, *Appl. Phys. Lett.*, 46:716 (1985).
63. J. K. G. Panitz, *J. Vac. Sci. Technol.*, A4:2949 (1986).
64. S. Berg, C. Neuder, B. Gelin, and M. Ostling, *J. Vac. Sci. Technol.*, A4:448 (1986).
65. (a). S. Yu, J. M. E. Harper, J. J. Cuomo, and D. A. Smith *J. Vac. Sci. Technol.*, A4:443 (1986). (b) N. S. Gluck, H. Sankur, and W. J. Glunning, *J. Vac. Sci. Technol.*, A7:2983 (1989). (c) R. A. Erck and G. R. Fenske, *Thin Solid Films*, 181:521 (1989). (d) N. Georgiev and D. Dobrev, *Thin Solid Films*, 189:81 (1990). (e) S. Mineta, M. Kohata, N. Yasunaga, and Y. Kikuta, *Thin Solid Films*, 189:125 (1990). (f) J. P. Doyle, R. A. Roy, J. J. Cuomo, S. J. Whitehair, L. Mahoney, and T. R. McGuire, *AIP Conf. Proc. (USA)*, 200:102 (1990). (g) S. M. Rossnagel and J. J. Cuomo, *Thin Solid Films*, 171:143 (1989).
66. B. Window and N. Savvides, *J. Vac. Sci. Technol.*, A4:196, 453 (1986).
67. N. Savvides and B. Window, *J. Vac. Sci. Technol.*, A4:504 (1986).
68. D. G. Armour, P. Bailey, and G. Sharples, *Vacuum*, 36:769 (1986).
69. W. E. Flynt, *Proceedings of the 3rd Symposium on Electron Beam Technology*, Boston, 1961, p. 168.
70. J. Amano and R. P. W. Lawson, *J. Vac. Sci. Technol.*, 15:118 (1978).
71. K. Miyaka and T. Tokuyama, *Thin Solid Films*, 92:123 (1982).
72. J. Amano, *Thin Solid Films*, 92:115 (1982).
73. J. J. Vrakking, L. J. Beckers, and G. E. Thomas, *Thin Solid Films*, 92:131 (1982).
74. S. Shanfield and R. Wolfson, *J. Vac. Sci. Technol.*, A1:323 (1983).
75. P. Oelhafen, J. F. Freeouf, J. M. E. Harper, and J. J. Cuomo, *Thin Solid Films*, 120:231 (1984).
76. T. Miyazawa, S. Miswa, S. Yoshida, and S. Gonda, *J. Appl. Phys.*, 55:188 (1984).
77. M. J. Mirtich, D. W. Swec, and J. C. Angus, *Thin Solid Films*, 131:245 (1985).
78. W. Halverson and D. T. Quinto, *J. Vac. Sci. Technol.*, A3:2141 (1985).
79. I Yamada, H. Inokawa, and T. Takagi, *Nucl. Instrum. Methods*, B6:439 (1985).
80. D. Nir and M. Mirtich, *J. Vac. Sci. Technol.*, A4:450 (1986).
81. N. Herbots, B. R. Appleton, T. S. Noggle, R. A. Zuhr, and S. J. Pennycook, *Nucl. Instrum. Methods*, B13:250 (1986).
82. A. Antilla, J. Koskinen, R. Lappalainen, and J. P. Hirvonen, *Appl. Phys. Lett.*, 50:132 (1987).
83. B. R. Appleton, S. J. Pennycook, R. A. Zuhr, N. Herbots, and T. S. Noggle, *Nucl. Instrum. Methods* B19/20: 975 (1987).
84. S. S. Wagal, E. M. Juengerman, and C. B. Collins, *Appl. Phys. Lett.*, 53:187 (1988).
85. R. Zimmerman, E. Broitman, and D. Latorre, *Thin Solid Films*, 165:L101 (1988).

4

Reactive Deposition Techniques

A half-truth won for ourselves is worth more than a whole truth learned from others.

Dr. S. Radhakrishnan

Compound thin films such as oxides, nitrides, and carbides are typically prepared by reactive evaporation, reactive sputtering, and so on—the metal is deposited in the presence of a reactive gas or component. But, to form certain oxides and nitrides, a high activation energy is needed for chemical reaction, and these compound films are not readily formed when evaporated in the presence of the reactive gas, as in ordinary reactive evaporation. The chemical reaction is enhanced through the presence of ionized atoms, and this plays an effective role in thin film formation. The presence of ions greatly influences the critical parameters of the condensation process even when only a small percentage of the atoms is ionized [1]. The kinetic energy of the ions also plays an important part in the formation of compound films [2]. The ions can be given additional kinetic energy by accelerating them in electric fields. The accelerated ions, when colliding with the neutral metal and gas atoms, effectively transfer part of their kinetic energies to the neutral particles, and this increased kinetic energy assists in the formation of good quality films as in reactive ion plating. This chapter presents the different reactive deposition techniques.

4.1 REACTIVE EVAPORATION

Reactive evaporation is a variant of Gunther's three-temperature method [3] and is based on the fact that continuous condensation of a given vapor at a given rate takes place only if the temperature of the substrate drops below a certain critical value. Differences in magnitude of the critical values (which are functions of interfacial energies) make it possible to condense a particular vapor or a combination of vapors (compound) preferentially on the substrate.

Reactive evaporation is a highly versatile technique for the preparation of compound semiconductor thin films from the individual components when one of the components is highly volatile. The advantages of this technique include the following.

1. The need to synthesize the compound from the elements prior to deposition, which is a tedious and sometimes expensive metallurgical process, is eliminated.
2. High temperatures (> 2000°C) are frequently needed to evaporate high melting point carbides, nitrides, and oxides as such and if resistive heating is used, the films will be contaminated as a result of evaporation of the source itself. This is avoided in reactive evaporation.
3. The decomposition of the compound upon heating in vacuum and the consequent lack of stoichiometry in the films are avoided.
4. High rates of deposition of the compounds are possible.
5. The lowest substrate temperature possible is dictated by the condensation temperature of the more volatile component (usually low for O_2, S_2, and Se_2), and as such this method can be used as an efficient technique for the preparation of amorphous films.
6. Dopants can be evaporated simultaneously, and a uniform dispersal of the dopant can be easily achieved.

This technique, which has been successfully used (as discussed later) for many technologically important compounds, has the following drawbacks.

1. The use of large volatile flux entails the loss of the volatile element.
2. When high deposition rates are required, use of large amounts of volatile flux leads to high volatile partial pressure (10^{-4}–10^{-2} torr), which reduces the mean free path and also scatters the nonvolatile beam away from the substrate surface. Again the high pressure in the vacuum system reduces the evaporation rate of the nonvolatile component.
3. Because of the high volatile elemental pressure, some unreacted volatile element is likely to be entrapped in the growing film, changing the film properties, especially at low substrate temperatures.

For purposes of reactive evaporation, the rate at which metal atoms arrive at a substrate is best expressed in terms of the deposition rate as observed from the same source at the same temperature and substrate-to-source distance, but in the absence of the reactive element flux [4].

$$\frac{dN_m}{A_r dt} = \frac{N_a \varrho_m d'}{M_m} \text{ atoms} \cdot \text{cm}^{-2}\text{s}^{-1} \tag{4.1}$$

where ρ_m is the density of the metal film (gcm^{-3}), M_m is the molar mass of the metal (gmol^{-1}), d' is the pure metal condensation rate (cms^{-1}), and A_r is the receiving surface area (cm^2).

The impingement rate of the reactive element molecules is given by

$$\frac{dN_{R_2}}{A_r dt} = 3.513 \times 10^{22} \, (^M R_2 T)^{-1/2} \, {}^P R_2 \text{ molecules} \cdot \text{cm}^{-2}\text{s}^{-1} \tag{4.2}$$

where $^M R_2$ is the molar mass of the reactive element, T is the vapor temperature, and $^P R_2$ is the partial pressure (torr).

Equations 4.1 and 4.2 can be used to calculate the number of atoms or molecules impinging on the substrate for given deposition rates or partial pressures.

Reactive evaporation has been used to prepare stoichiometric films of many compounds. For example, Tschulena [5] has reported the preparation of BeO films with good reproducibility by evaporating pure beryllium from a resistance-heated tantalum boat in an oxygen atmosphere at reduced pressure. Reactive evaporation has been successfully used to prepare superconducting Nb3Ge films by introducing germane gas (GeH4) into the vacuum evaporation system in which a niobium billet was evaporated from a electron beam heater source [6]. Germane gas was fed at a controlled flow rate from a nozzle in the vicinity of the niobium billet. For the preparation of thin films of sulfide compounds by the reactive evaporation technique, H2S has been used as the reactive gas. Reactively evaporated copper molybdenum sulfide films were prepared on sapphire substrates by bleeding H2S into the vacuum system with copper and molybdenum evaporated from separate sources [7a]. Bue H2S is a poisonous gas, and its preparation and purification to the level needed in thin film work is not easy. The controlled evaporation of sulfur itself for the creation of a sulfur atmosphere is used in the author's laboratory for preparing sulfide compounds. Preparation of tin sulfide films, for example, is discussed below in detail.

Thin films of tin sulfide were prepared by evaporating tin (99.999% pure) and sulfur (recrystallized three times from solution) from two different sources, keeping the substrate at an elevated temperature in a conventional vacuum system at a pressure of about 10^{-6} torr (7b). A glass crucible placed in a conical basket of molybdenum wire was used as the sulfur source (S$_1$). The temperature of the source, hence the flux, could easily be controlled by adjusting the current through the molybdenum wire. A helical filament made of molybdenum wire (S$_2$) was used to evaporate tin. The helical filament was covered with stainless steel heat shields to minimize substrate heating due to radiation from the high temperature source; thus

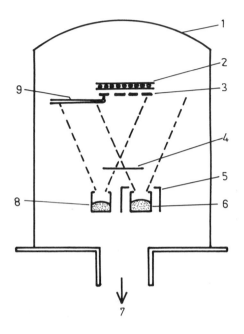

Figure 4.1 Schematic diagram of the setup for the evaporation of tin and sulfur: 1, bell jar; 2, substrate heater; 3, substrates; 4, shutter; 5, heat shield; 6, metal source (S_2); 7, to pump; 8, sulfur source (S_1). (Data from Ref. 7b.)

the temperature could be controlled to within 2°C. A schematic diagram of the setup is shown in Figure 4.1.

When the substrate (glass or quartz slides) had attained the required temperature and stabilized, with a shutter placed over the sulfur and tin sources, the current through S_1 was switched on and adjusted such that a sulfur partial pressure of 5×10^{-5} to 9×10^{-5} torr was maintained in the chamber. Then the current through the tin source was switched on and increased to a preset value, which gave a deposition rate of 0.1–0.2 nm/s of the metal film. The shutter was then withdrawn and the deposition of the compound films allowed to take place. At the substrate temperature and sulfur partial pressure used in this experiment, the supersaturation of sulfur atoms is not sufficient to form a film by itself. Only the metal or compound can form at this temperature. When the shutter is withdrawn, the tin atoms reaching the substrate react with sulfur atoms present on the substrate, and the compound film is deposited.

It has been found that a stiochiometric interval exists for the reactive evaporation of tin sulfide from its individual components with the following parameters (already mentioned):

Metal film deposition rate 0.1 - 0.2 nm · s⁻¹

Metal film deposition rate $0.1 - 0.2 \, \text{nm} \cdot \text{s}^{-1}$
Sulfur partial pressure 5×10^{-5} to 9×10^{-5} torr
Substrate temperature $22-62°C$

The corresponding metal atom flux and chalcogen flux can be calculated using equations 4.1 and 4.2, as follows:

Metal atom flux 4×10^{14} to 1.5×10^{15} atoms cm⁻²s⁻¹
Chalcogen flux 1.5×10^{16} to 2.8×10^{16} molecules cm⁻² s⁻¹

Good stoichiometric films were thus obtained; these films were golden yellow and extremely transparent.

Examples of recent work reported on reactive evaporation are compiled in Table 4.1.

4.2 ACTIVATED REACTIVE EVAPORATION

Many metals will not readily form thin nitride or oxide phases when reactively evaporated in nitrogen or oxygen atmosphere because of the high activation energy required for chemical reaction (1-3 eV) to form the nitride or oxide phase. Another disadvantage of ordinary reactive evaporation is the low deposition rate. In the activated reactive evaporation (ARE) technique developed by Bunshah and Raghuram [16], these difficulties have been overcome. In their setup (Figure 4.2), the secondary electrons from the molten pool of the metal are attracted by a low voltage electrode (biased positively around 100 V), and these electrons ionize the reactive gas, creating a thick plasma. Because of the presence of the plasma, the chemical reaction rate is greatly increased: Bunshah and Raghuram were able to deposit TiC at a rate of 12 μm/min at a source distance of 15 cm using this technique. It may be mentioned that using ordinary reactive evaporation, titanium and the reactive gas acetylene do not react to form the compound.

High rates of deposition of several oxides, nitrides, and so on have been reported [17]. For more details on ARE developments and applications, the interested reader is referred to Bunshah's paper [18]. In a modification of ARE reported by Randhawa et al. [19], the electron gun deposition source is dispensed with and the metal is evaporated from a resistively heated boat. A magnetic field together with a thoriated tungsten electron emitter generate the plasma. Films of SnS$_2$, CuS, and In$_2$S$_3$ have been prepared by ARE [20a], using the setup described earlier (Figure 4.1) but modified by the addition of an electron emitter, a low voltage anode, and a longitudinal magnetic field (Figure 4.3). The longitudinal DC magnetic field was produced using a magnet (especially constructed for the purpose) placed outside the bell jar. With this setup, it was possible to maintain the glow discharge down to 10^{-4} sulfur partial pressure. These films exhibited more optical transmission than films prepared by ordinary reactive evaporation for the same thickness.

Jacobson et al. [20b] have used the ARE technique to prepare carbonitride [Ti

Table 4.1 Reactive Evaportion Deposition

Film	Components 1	Components 2	Volatile vapor/reactive gas pressure (torr)	Substrate and substrate temperature (°C)	Rate of deposition	Remarks, if any	Ref.
Sb_2S_3	Sb (99.899% pure)	S_2 (99.98% pure)		Optically flat glass plates 30–50		Resistive heating p-type films; temperature dependence of conductivity was studied.	8
CuS	Cu	S_2	10^{-4}	Optically flat glass/quartz plates 22–42	2–5 Å/s	Three-times-crystallized sulfur and two-times-electrolyzed copper; resistive heating; amorphous n-type films.	9a
CuS	Cu	S_2	10^{-4}	Optically flat glass/quartz plates 52–172	2–5 Å/s	Resistive heating; shows that CuS is a semiconductor; p-type crystalline films.	9b
TiO_2	TiO	O_2	15×10^{-5}	Schott BK7 and machine-drawn glass 40–300	10–15 Å/s	Evaporated using EB gun.	10
Bi_2Te_3	Bi (99.999% pure)	Te (99.999% pure)		Optically flat glass/quartz plates 257–272	3–4 Å/s	Resistive heating for evaporation; n-types films. Conductivity, Hall effect, and thermoelectric power were investigated.	11
In_2O_3	In (99.999% pure)	O_2	1×10^{-3}	Heated substrates $\approx 145°C$	7 Å/s	For structural studies, films were deposited on freshly cleaved NaCl crystals.	12

SnTe	Sn (99.999% pure) Te (99.999% pure)		Optically flat glass slides 295–326	9 Å/s	Resistive heating; polycrystalline films; p-type Hall coefficient, dc conductivity, and Hall mobility were measured.	13a
PbS	Pb (99.999% pure) S$_2$ (99.999% pure)	10^{-4}	Optically flat glass quartz plates 27–377	7–15 Å/s	Resistive heating. Films prepared at RT substrates are amorphous, Increase in temperature makes the film polycrystalline. Films were p-type.	13b
Bi$_2$O$_3$	Bi (99.999% pure) O$_2$	10^{-3}	Optically flat glass/ quartz plates 27–302	3–20 Å/s	Films prepared at substrate temperature of 27°C were amorphous.	24
SnTe:Ge	Sn and Ge (99.999% pure) Te		Optically flat glass slides 312 ± 3		Resistive heating. Films obtained were p-type. Films with 0.8 wt % Ge had the maximum mobility (1150 cm^2 V^{-1}s^{-1} at RT) and minimum carrier concentration (3 × 10^{19} cm^{-3} at RT)	14a
In$_2$S$_3$	In (99.99% pure) S$_2$	10^{-4}	Optically flat glass slides RT to 327	200–300 Å/min	Resistive heating. In$_2$S$_3$ films were obtained.	14b
In$_2$O$_3$	In O$_2$	8×10^{-5} to 16×10^{-4}	Sodalime glass substrates 200	20–24 Å/min	Resistively heated tungsten source. Films were characterized electrically and optically, and all were found to be n-type.	15

Table 4.1 (Continued)

Film	Components 1	Components 2	Volatile vapor/reactive gas pressure (torr)	Substrate and substrate temperature (°C)	Rate of deposition	Remarks, if any	Ref.
$YBa_2Cu_3O_{7-x}$	Y, Cu, and Ba	Ozone		MgO (100) 590–650	0.8 Å/s	Deposition rate is for a total thickness of 1000 Å. Y and Cu evaporated from two EB-heated sources and Ba evaporated from a Knudsen cell. As-deposited films exhibited zero resistance at 86 K. Critical current density, 6×10^{-5} A/cm².	110
TiO_2	TiO	Oxygen	1.9×10^{-4}	Polished fused quartz 25–250	10.0 nm/min	Starting material TiO was evaporated from a 6 kW electron beam gun. Influence of substrate temperature and postdeposition heat treatment in air on the optical properties of the films was studied. As-deposited films were oxygen deficient.	111a
Fe_3O_4	High purity iron	Oxygen	Varied	Polished (001) face of sapphire single crystals	5 Å/s	Iron was thermally evaporated from an alumina crucible. Films of various ion oxides (e.g., FeO, Fe_3O_4, α-Fe_2O_3)	111b

					were formed. Formation range of iron oxides as a function of oxygen partial pressure (PO_2) and substrate temperature (T_s) were studied. Films were characterized by X-ray diffraction and conversion electron Mossbauer spectroscopy. (111) oriented epitaxial magnetite films formed at T_s 523–623 K and PO_2 1.0–5.0 × 10⁻⁴ torr.	112
SiO_2	SiO	Oxygen	10⁻¹	Si (100) RT	Laser evaporation using a pulsed ArF excimer laser (193 nm); laser beam focused onto the target at normal incidence. Target was rotated during laser irradiation. Role of oxygen during laser processing on the stoichiometry and properties of the oxide film discussed.	

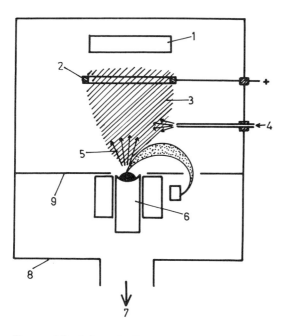

Figure 4.2 Schematic diagram of the setup for activated reactive evaporation: 1, substrate; 2, electrode; 3, plasma; 4, reactive gas inlet; 5, coating flux; 6, EB evaporator; 7, to vacuum pump; 8, vacuum chamber; 9, pressure barrier. (Data from Ref. 16.)

(C,N)] films on high speed steel substrates by evaporating titanium in C_2H_2–N_2 plasma. The effects of deposition parameters such as rate of evaporation, substrate temperature, and partial pressures of C_2H_2 and N_2 on the structural and mechanical properties of these films were studied.

A simple, inexpensive, and well-controlled deposition technique for the preparation of cubic boron nitride films by ARE of H_3BO_3 in an NH_3 plasma is reported by Chopra et al. [21]. A schematic diagram of their setup is shown in Figure 4.4. The plasma is created by electrons emitted from a resistively heated filament (A) and the anode (B). H_3BO_3 evaporated from a molybdenum boat passes through the plasma and is decomposed to react with NH_3 at a pressure of about 3×10^{-5} torr. Ammonia is introduced through the inlet D. The external magnetic field H (~60 G) enhances the plasma and also sustains the discharge. The film is deposited on the heated (500°C) substrate S. The investigators obtained a typical deposition rate of 150 nm/min, and the films were found to be highly adherent.

Another modification to the setup of Randhawa et al. [19] has been reported by George et al. [22] for preparing Bi_2O_3 films. Any type of vapor source can be used,

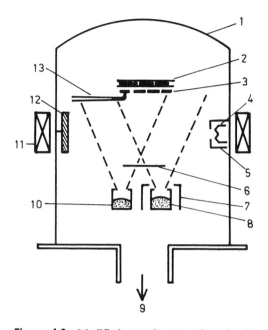

Figure 4.3 Modified setup, incorporating a longitudinal dc magnetic field, for activated reactive evaporation: 1, bell jar; 2, substrate heater; 3, substrates; 4, electron emitter; 5, shield; 6, shutter; 7, heat shield; 8, metal source; 9, to pump; 10, sulfur source; 11, field coil; 12, anode; 13, thermocouple.

and for the first time, the need for a magnetic field is dispensed with. Two heated electron emitters were used to obtain a high emission current with a negatively charged reflector to direct the electrons toward the anode kept on the opposite side, as shown in Figure 4.5. The electron emitters were two heated 0.5 mm tungsten wires having an effective length of 3 cm; the anode was a thick aluminum block $(8.5 \times 8.5 \times 1 \text{ cm}^3)$, which is biased 100 V positive with respect to the cathode. The reactive gas (oxygen) was admitted through a needle valve. Ordinary microscope slides placed in contact with a resistively heated substrate holder served as substrates.

After the substrates had attained the required temperature, oxygen was admitted into the chamber to a pressure greater than 10^{-3} torr. The cathode current was slowly increased to 20 A and the anode supply turned on, initiating the discharge. The oxygen pressure inside the chamber was reduced 5×10^{-4} torr by a slight adjustment of the needle valve, and a steady discharge current of 1.6–2 A was obtained. Bismuth (99.999% pure) was then evaporated from a molybdenum boat to form Bi_2O_3 films. In this setup, the plasma was sustained without the magnetic

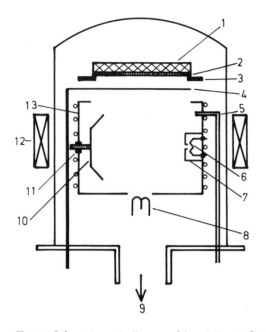

Figure 4.4 Schematic diagram of the ARE setup for the preparation of cubic BN films: 1, substrate heater; 2, substrate (S); 3, substrate holder; 4, shutter; 5, gas inlet (D); 6, electron emitting filament (A); 7, filament cover; 8, evaporation source; 9, to pump; 10, anode (B); 11, ceramic insulators; 12, external magnet; 13, water-cooled discharge chamber. (Data from Ref. 21.)

field because of the increased number of electrons. The grounded reflector also acts as a shield, preventing tungsten atoms from the cathode reaching the substrate.

With ordinary reactive evaporation, bismuth will not form the oxide phase easily, and the deposition rate reported [23] is only 0.55–0.22 Å/s. With the method above, good quality Bi_2O_3 films at a deposition rate of about 50 Å/s were obtained. X-ray diffraction studies of Bi_2O_3 films prepared by this method have also been reported [24], and these films show excellent optical transmission and very good adhesion to the substrate. This setup has also been used for preparing Bi_2O_3 films as a dielectric for a Bi_2O_3–Au–Bi_2O_3 structure for use as heat mirrors [25a].

More recently, ARE has been used for the deposition of SiN films using nitrogen as a reactive gas instead of ammonia, with the objective of obtaining films with a low hydrogen content. Films with near-stoichiometric composition were obtained at a deposition rate of approximately 10 Å/s [25b].

Bando et al. [25c] have succeeded in preparing for the first time single-crystal films of YBCO with (001) orientation on $SrTiO_3$ (100) at temperatures below

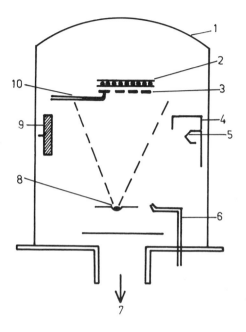

Figure 4.5   Modified setup for ARE for the preparation of Bi_2O_3 films: 1, glass bell jar; 2, substrate heater; 3, substrates; 4, electron reflector and shield; 5, tungsten filament; 6, O_2 inlet tube; 7, to vacuum pump; 8, melt source; 9, anode; 10, thermocouple.

600°C by ARE, using the deposition system reported by them earlier [25d] and shown in Figure 4.6. Here the three metals were coevaporated from separate evaporation sources. A relatively high pressure (10^{-1}–10^{-2} torr) oxygen atmosphere was obtained by introducing oxygen gas near the substrate. Yttrium and barium were evaporated from electron beam heated sources and the oxygen plasma introduced by rf oscillation (13.56 MHz, 100 W) in the background oxygen pressure of 10^{-5} torr. The evaporation rates of the metals measured by an oscillating quartz sensor were adjusted to yield the 1:2:3 stoichiometry of metals in YBCO. The substrate temperatures were around 600°C, and the typical overall film growth rates were 3.5 Å/s. After deposition, the films were oxidized in situ below the growth temperatures by introducing oxygen gas in the chamber (<200 torr).

Films 1000 Å thick exhibited high T_c ($R = 0$) of 90 K and J_c (at 77 K) of 4×10^{-6} A/cm^2.

ARE has also been used to prepare YBCO superconducting films in situ [25e] without postdeposition annealing.

In situ preparation of good quality ferroelectric BaTiO$_3$ thin films by ARE has

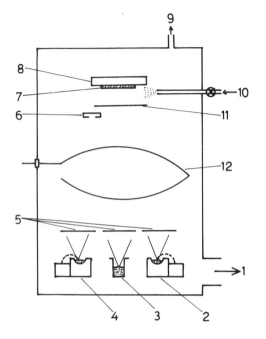

Figure 4.6 Schematic diagram of the ARE deposition system for the preparation of single-crystal films of YBCO: 1, to vacuum; 2, Ba source; 3, Cu source; 4, Y source; 5, shutter; 6, thickness monitor; 7, substrate; 8, substrate holder; 9, vacuum gauge; 10, O_2 inlet; 11, shutter; 12, rf coil. (Data from Ref. 25c.)

been recently reported by Iijima et al. [25f]. Barium and titanium were individually evaporated from electron beam heated sources and their rate of evaporation adjusted to give the stoichiometry of these metals in $BaTiO_3$. The oxygen plasma was generated by an rf power supply. The $BaTiO_3$ films were grown epitaxially on single-crystal $SrTiO_3$ and on Pt epitaxial film substrates. The temperature of the substrate was around 600°C, and the as-grown films exhibited ferroelectric behavior as confirmed by the typical D-E hysteresis loop.

Terashima et al. [25g] prepared YBCO superconducting films for the first time by the reactive plasma evaporation method [25h], using specially designed rf plasma systems [25i]. In their deposition setup (Figure 4.7), a mixture of $BaCo_3$, Y_2O_3, and CuO powders (size, 1 μm) was fed into the rf plasma flame with the Ar carrier gas coevaporated completely. (The rf plasma torch was made of double silica tubes.) High temperature vapor mixtures generated in the plasma flame are codeposited onto the (100) MgO single-crystal substrate placed in the tail of the plasma flame.

The experimental conditions reported are as follows.

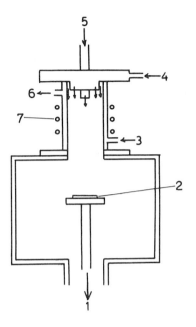

Figure 4.7 Schematic diagram of the deposition setup for the preparation of YBCO films by reactive plasma evaporation: 1, to exhaust; 2, substrate; 3, water inlet; 4, plasma gas and sheath gas; 5, powder + carrier gas; 6, water outlet; 7, rf coil. (Data from ef. 25 g.)

Rf power	45 kW (4 MHz)
Total flow rate of plasma	55 L/min (Ar)
sheath and carrier gases	30 L/min (O_2)
Pressure	atmospheric
Powder feeding rate	0.3 g/min
Substrate temperature	600–650°C

The deposition rate was much more than 10 μm/h, and no annealing was done after deposition. The films were characterized by X-ray diffraction, SEM, and resistivity measurements (four-probe). The as-deposited films without postannealing showed a superconducting transition temperature T_{cm} (midpoint—defined as 50% drop of resistivity) of 94 K. The crystal structure of the films was identified to be the orthorhombic oxygen–deficient perovskite phase. Some of the prepared films showed the preferred orientation of (001).

4.3   REACTIVE ARC EVAPORATION

Reactive deposition using the arc evaporation process is now commonly used for the deposition of carbides, nitrides, and so on. The ionization efficiency (number of ions arriving at the substrate per unit area as a percentage of the total number of atoms arriving per unit area) is significantly higher than that of other similar deposition process such as ion plating, ARE, where the ionization efficiency is only in the range 2–8%. The ionization efficiency in the cathodic arc deposition process is high as 30–50%, and this high degree of ionization in the plasma and the high energies of the ions increase the reaction efficiency and result in coatings with enhanced adhesion and density.

The basic process of vacuum arc evaporation was discussed in Section 3.2, and the preparation of compound films (oxides, nitrides, etc.) involves the arc evaporation of the metal in the respective reactive gas. Some of the advantages of this deposition process are as follows.

1.   Stoichiometric films with enhanced adhesion and high density can be obtained for over a wide range of reactive gas pressures and evaporation rates.
2.   High deposition rates are possible, with good uniform coating.
3.   Low substrate temperatures can be used.
4.   Preparation of reacted compound films is easy.

Little has been reported on the relatively new process of reactive vacuum arc evaporation of film coatings, and the following discussion summarizes some recent reports.

Shinno et al. [26] have reported the deposition of films of titanium nitride and titanium carbide using a vacuum arc source of the ion gun type developed by them for the purpose of reactive deposition. Films can be coated at a much higher deposition rate, and their source (Figure 4.8) has the following salient features.

1.   The cathode is a cylindrical rod with a small hole along the central axis that will serve as a gas channel. The arc spots will be concentrated around the gas nozzle on the cathode surface.

2.   The cylindrical cathode is cooled by a water jacket on the lateral side.

3.   The anode is a cylindrical shell made of stainless steel or molybdenum brazed with copper tubes for water cooling.

4.   A titanium shield surrounds the cathode, with a small gap to confine the arc within a certain area on the cathode.

5.   To enhance chemical reactivity, the reactive gas is preionized in a quartz tube by rf discharge before flowing into the vacuum chamber through the channel along the central axis. Preionization also assists arc discharge by providing gas ions in the cathode surface.

A titanium cathode and reactive gases such as nitrogen and acetylene were used for preparing films of titanium nitride and titanium carbide. The vacuum was main-

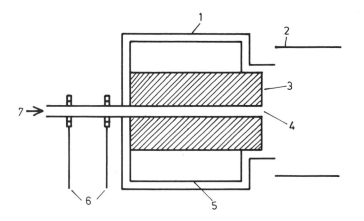

Figure 4.8   Schematic diagram of the vacuum arc source of the ion gun type for the deposition of TiN and TiC film: 1, shield; 2, anode; 3, cathode; 4, gas nozzle; 5, water jacket; 6, rf discharge; 7, gas inlet. (Data from Ref. 26.)

tained by an oil diffusion pump system, and the ultimate vacuum was about 1×10^{-6} torr. The reactive gas flow was regulated by a leak valve.

Three cases of depositions are reported: (a) arc discharge maintained in a uniform nitrogen gas. (b) gas fed through the nozzle on the cathode surface, and (c) gas preionized by the rf discharge and fed through the nozzle on the cathode surface.

The investigators found that the deposition rate increased when the gas was fed through the nozzle and also with the preionization of the gas by the discharge. The deposited nitride films were found to be deficient in nitrogen, whereas the deposited carbide films contained much excess carbon. Stoichiometric titanium nitride films could not be obtained even for an ambient nitrogen pressure as high as 3.75×10^{-2} torr. This shows that the reactivity is insufficient to produce stoichiometric films of titanium nitride.

The same group [27] have reported the preparation of titanium nitride, titanium carbide, and carbon films on molybdenum substrates and have given the deposition conditions for the different coatings (Table 4.2).

Randhawa and Johnson [28] have presented a review of the cathodic vacuum arc deposition process emphasizing the role of the plasma environment for the reactive deposition of carbides, nitrides, carbonitrides, and oxides. Various applications of this technique have also been discussed. The arc system used also was discussed previously (see Section 3.2). Other reports from this group have been published in the literature. The surface morphology, crystal structure, microhardness, and composition of zirconium nitride films prepared by this technique have been published, as well [29]. The authors' deposition parameters were as follows.

Table 4.2 Deposition Conditions for TiN and TiC Coatings

Condition	TiN	TiC
Cathode–substrate distance, mm	65–180	100
Reactive gas flow rate, cm^3/min	10–30 (N$_2$)	
Reactive gas pressure, torr	38×10^{-4}– 22×10^{-3} (N$_2$)	38×10^{-4}– 15×10^{-3} (C$_2$H$_2$)
Arc current, A	60–120	50–90
Deposition duration, min	5–10	10
Bias	0 to –200 V	
Deposition rate, μm/min	≈ 0.8	≈ 0.5
Substrate temperature, °C	223–527	377–527

Substrate	High speed steel and stainless steel
Substrate temperature	300–450°C
Reactive gas	Nitrogen
Total chamber pressure	2×10^{-3}–10^{-2} torr
Deposition rate	200–300 nm/min
Film thickness	0.1–2.0 μm
	3.0–5.0 μm

Johnson and Randhawa have shown from wear test results that ZrN is marginally superior to TiN in conventional metal cutting applications and betters TiN by a factor of 2 for cutting titanium alloys. They have also found from reflectivity measurements that doped ZrN films are very similar to gold films and offer a good substitute for gold coatings for various decorative applications.

Also TiC and TiC$_x$N$_{1-x}$ films have been prepared by reactive arc deposition as reported by Randhawa [30]. The carbide films were prepared using titanium and CH$_4$ gas; for the TiCN films, titanium and a mixture of CH$_4$ and N$_2$ gases were used. The deposition conditions are given in Table 4.3.

For TiC$_x$N$_{1-x}$ the compositions were varied from 100 vol % CH$_4$ to 100 vol % N$_2$ in the following proportions (vol %) of CH$_4$ to N$_2$: 80:20, 70:30, 50:50, 30:70; the composition of the gas mixture was subjected to relative flow rate controls. The films were analyzed for adhesion, microhardness, composition, and crystal structure. Randhawa found that TiC films deposited above 600°C were virtually free of

Table 4.3 Deposition Conditions for TiC and TiC$_x$ N$_{1-x}$

Film	Substrates	Substrate temperature (°C)	Total pressure in chamber (μm)	Source-to-substrate distance (cm)	Deposition rate (Å/min)
TiC	High speed steel and stainless steel	350,500, 600, and 700	2-10	25	3000
TiC$_x$N$_{1-x}$	High speed steel and stainless steel	400, 500, and 600	2-10	25	3000

hydrogen, whereas the film deposited at 500°C and below contained significant amounts of hydrogen. The presence of hydrogen markedly affected the wear characteristics of the films, which were found to be unsuitable for cutting applications at moderate temperatures. In cutting tests, TiC$_x$N$_{1-x}$ (x $\approx$ 0.15-0.30) were found to be superior to TiN films.

Freller and Haessler [31] used the experimental setup shown in Figure 4.9 to prepare Ti$_x$Al$_{1-x}$N coating by reactive arc evaporation and magnetron sputtering. Aluminum concentrations between 10 and 50 at. % in the source material were used for the film deposition. The deposition conditions were as follows.

Working pressure	7.5 × 10^{-4} to 4 × 10^{-2} torr
Reactive gas pressure	7.5 × 10^{-4} to 4 × 10^{-2} torr
Deposition temperature	<300°C
Bias voltage	100V(0-250 V)
Specific deposition rate	0.1 μm min^{-1} kW^{-1}

The magnetron process produced coatings in an extremely narrow region of nitrogen partial pressure, whereas the arc process allows deposition in a broader pressure region of the reactive gas. The results of Freller and Haessler also show that in deposition by arc evaporation, the aluminum content in the deposited film is less than that of the source material, whereas films deposited by magnetron sputtering show a small increase in aluminum content compared with that of the target. The

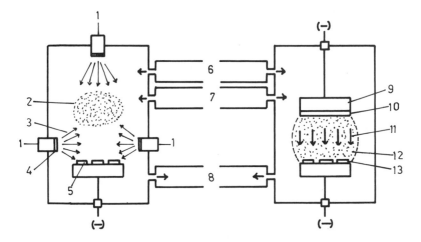

Figure 4.9 Schematic diagram of the experimental setup for reactive arc evaporation and magnetron sputtering: 1, Arc evaporator; 2, plasma; 3, coating material flux; 4, source material; 5, substrates; 6, neutral gas; 7, reactive gas; 8, vacuum pump; 9, magnetron cathode; 10, source material; 11, material flux; 12, plasma; 13, substrate. (Data from Ref. 31.)

aluminum concentration also depends on the bias voltage in the cathodic arc deposition, and this result offers the possibility of changing the atomic concentration in the growing films by varying the bias voltage.

Several others have also reported the preparation of nitrides (e.g., TiN, ZrN, HfN) by the reactive vacuum arc evaporation technique [32–34a].

Schemmel et al. [34b] have reported a new process for achieving a high rate deposition of Al_2O_3 by a modified cathodic arc. The modified cathodic arc apparatus they used (Figure 4.10) was described previously by Randhawa [34c]. The films were deposited in a vacuum system with a typical base pressure of 1×10^{-6} torr. Working pressures in the range 1 to 8×10^{-3} torr were adjusted with a combination of 99.9% pure argon and oxygen, the ratio being adjusted to obtain the best rate while maintaining film quality; 99.9% pure aluminum was used as the target.

The plasma flux generated from an arc source passes through one-fourth of an electromagnetic torus. The axial magnetic field confines the electrons in the plasma to helical paths along the axis of the torus, while the radial electric field forces the aluminum ions to the central region along the torus. The neutral particles and macroparticles, not affected by the fields, condense on the walls of the torus, and the material flux at the exit post is very highly ionized [34d].

Alumina films were deposited on quartz and stainless steel substrates. The

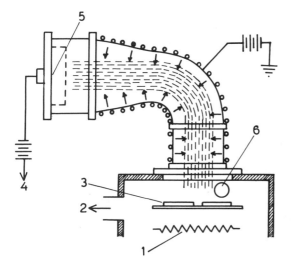

Figure 4.10 Schematic diagram of the modified cathodic arc apparatus for the rapid deposition of Al_2O_3: 1, substrate heater; 2, vacuum port; 3, substrate; 4, anode; 5, cathode; 6, exit port. (Data from Ref. 34b.)

substrates were heated from the back side by a quartz lamp heater, and the deposition temperature varied from ambient to 130°C.

It was found that the gas inlet location influenced the film properties, and high optical quality films with high deposition rates were prepared by introducing argon at the target end of the torus and oxygen at the exit port.

The depositions made without intentional heating of the substrate (t < 75°C) required 35% oxygen, whereas at 130°C, an oxygen content of 24% was used to ensure clear films at the maximum deposition rate of 800 Å/min. For all depositions, the arc current was kept at 110 A. The other process parameters (longitudinal magnetic field inside the torus, average field strength in the vicinity of the cathode and exit plane, bias voltage on the walls, etc.) were optimized, and the deposited films exhibited correct stoichiometry, good adhesion, high density, and good optical properties.

4.4 REACTIVE SPUTTERING

Sputtering of metal targets in the presence of a reactive gas when the sputtered material from the target and gas react to form a compound (e.g., oxides or nitrides) is called reactive sputtering. When the compound target that chemically decomposes during inert gas sputtering results in a film deficient in one or more constituents of the target, the reactive species is added to make up for the deficient constituent and is also treated as a case of reactive sputtering.

In a typical reactive sputtering system, the reactive gas reacts with the target and a compound is formed on the surface of the target. This is termed target poisoning. When it happens, the sputtering rate drops significantly because the sputtering rate of the compound (here the compound is exposed to ion bombardment, not the target material) is only about 10-20% of the metal sputtering rate [35]. Figure (4.11) shows the common salient feature of reactive sputtering, the plot of the reactive gas partial pressure V_s (the reactive gas flow at constant power) exhibiting a hysteresis curve. Here when the reactive gas is first turned on, the reactive gas pressure initially stays at the background level because at low flow rates complete reaction of the gas with the sputtered material takes place, and even when the flow rate is increased, the reactive gas pressure changes very little until at A (Figure 4.11) the pressure rises sharply. At this point the chamber contains sufficient reactive gas, which reacts with the target surface itself, to form the gas–metal compound. The

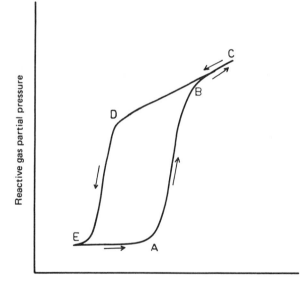

Figure 4.11   Hysteresis curve for reactive sputtering: reactive gas partial pressure versus reactive gas flow.

surface layer then reduces the metal sputtering yield; consequently less reactive gas is consumed, and the reactive gas pressure rises sharply to B.

Here it should also be noted that the compound sputters at a much lower rate than the metal. When more reactant is added, there is a proportional increase in the partial pressure (C in Figure 4.11), when the reactive gas flow decreases, the pressure falls proportionally until at D, the supply of gas is too low to maintain the compound surface layer on the target and the pressure drops more rapidly to E (the sputtering rate reverts to that of the metal). Thus the reactive gas partial pressure and the reactive gas flow exhibit a hysteresis loop.

The effect of target poisoning on reactive sputter deposition depends on the particular combination of metal and reactive gas and on the properties of the compound surface layer that is formed.

Hohnke et al. [36a] have presented an analysis of the reactive dc sputter deposition of compound films (metal oxides and nitrides) and have developed a model for reactive sputter deposition. The model establishes that the ratio W/G, where W is the sputtering power and G the reactive gas flow rate, is a fundamental parameter of reactive sputtering. This ratio is independent of the reactive gas pressure and, within certain approximations, depends only on the sputtering yield of the metallic target. The reactive gas–metal systems TiN, Al_2O_3, TiO_2 were examined and reported on, and proposed model [36a] is in good agreement with experimental results. Because of its simplicity, the model is easily applied to the estimation of sputtering conditions for the stoichiometric deposition of compounds.

Other models of reactive sputtering have been published recently [36b,c] and the interested reader is referred to these papers.

Reactive sputtering, the low temperature plasma vapor deposition (PVD) process that offers the best reproducibility, has been used for the preparation of a variety of compound films (e.g., Si_3N_4, SiO_2, Ti_2O_5, Al_2O_3, ZnO, Cd_2SnO_4, TiN, HfN) for applications from decorative coating to cutting tools and microelectronic components, depending on the compound films. Quite a number of reports on reactive sputtering have been published in the literature, and for the early works, the reader is referred to the extensive bibliography presented by Vossen and Cuomo [37]. A summary of the more recent work is given in Table 4.4. In the following discussions examples of particular interest, with special features, are described.

Tool steels can be nitride coated without changing their properties or distorting their shapes by high rate reactive sputtering. Sproul [38], who reported reactively sputtered TiN, ZrN, and HfN coatings using a commercial sputtering system, found that it was necessary to operate at the knee of the hysteresis loop (i.e., at point A (in Figure 4.11), where the reactive gas partial pressure suddenly jumps up. Instead of the manual control of the reactive gas flow, using a patented [39] automatic feed control device, it was possible to operate at the knee. This allowed Sproul to obtain very high deposition rates [40], and for all three nitrides, the hardness of the coating was greater than that of the bulk material.

Table 4.4 Reactive Sputtering

Deposited film	Target	Reactive gas	Sputtering power/density	Deposition rate	Pressure (torr)	Substrate and substrate temperature (°C)	Type of system	Remarks, if any	Ref.
ZnO	Zn (99.99% pure)	High purity oxygen	500–1000 W	1–10 μm/h	7×10^{-3}	Metallized silica, lattice matched sapphire, glass, and silicon 350–500	Rf planar magnetron	Target-to-substrate spacing, 4–7 cm.	49
TiC	Ti	Ar (99.9997% pure) + methane (99.9998% pure)	1000 W	250 Å/min	5×10^{-3}		Rf diode		50
a-Si:H	Boron-doped cast polycrystalline Si	Ar + H₂ (99.999% pure)		0.05 μm/min			Dc planar magnetron		51
Al₂O₃	Al	Ar + O₂	12 W/cm²	36 Å/s	15×10^{-3}		Dc planar	Reaction at substrate separated from reaction at target surface to get high deposition rates. Separate control of target and	52
SiO₂	Si	Ar + O₂	9.5 W/cm²	24 Å/s	15×10^{-3}		Dc planar		
SnO₂	Sn	Ar + O₂	7.5 W/cm²	61 Å/s	17×10^{-3}		Dc planar		
Ta₂O₅	Ta	Ar + O₂	13 W/cm²	52 Å/s	15×10^{-3}		Dc planar		
TiOₓ	Ti	Ar + O₂	12 W/cm²	24 Å/s	15×10^{-3}		Dc planar		

Material	Target	Sputtering gas	Power	Deposition rate	Pressure	Substrate	System	Remarks	Ref.
TiC	Ti	Argon (99.998% pure) + 0.7% methane (99.8% pure)	0.75–2 kW	5.1×10^{-11} ms^{-1}	10^{-3}–10^{-4}	325–525	Dc diode	substrate processes. Oxygen mass flow is varied. Target-to-substrate distance, 3×10^{-2} m	53
In$_2$O$_3$:Sn	Alloy of In and Sn (90 wt % In + 10 wt % Sn)	Ar + O$_2$		3–18nm/min		Quartz plates or si wafers Not heated	Dc magnetron		54
TiN	Ti (>99.95% pure)	N (99.9992%) + Ar (99.9997% pure)	1.9 kW	2000 Å/min	5×10^{-3}	High speed steel and stainless steel long strips	Dc magnetron	6 in. target; target-to-substrate distance, 6 cm.	55
BN	12.7 cm diameter 1C grade BN disk	Ar/Ni mixture (99.999% pure)	300 W	5 nm/min	7×10^{-3}	Water-cooled glass, Si (100), and (Sapphire) Al$_2$O$_3$ (0001)	Rf diode	Anode–cathode spacing 9 cm. Effect of sputtering gas composition on optical and chemical film behavior was studied.	56

Table 4.4 (Continued)

Deposited film	Target	Reactive gas	Sputtering power/density	Deposition rate	Pressure (torr)	Substrate and substrate temperature (°C)	Type of system	Remarks, if any	Ref.
ZnO	Zn	Ar + O₂			7.5 × 10⁻²	Polished glass and Si (100) wafer	Dc Sputtering, conventional diode and magnetron sputtering	Conventional system substrate: (a) in front (35 mm from cathode) and (b) adjacent to the cathode outside the discharge (10 mm from edge of cathode).	57
WC	W sheet, 1 mm	Ar + acetylene	4.5 W/cm²	600 Å/min	2 × 10⁻²	Stainless substrates, 5 × 5 × 1 mm 200–500	Rf magnetron	Tungsten sheet bonded to the front surface of a 6 in. planar magnetron cathode holder by a conducting paste. Sputter-cleaned for 10 minutes in argon before sputter deposition. A single phase monocarbide is formed at high temperatures.	58

166

Material	Target	Gas	Power	Rate	Pressure	Substrate	Source	Remarks	Ref.
NbN	Nb (99.99% pure)	Ne-N$_2$, Kr-N$_2$ (Inert gas purity> 99.999%)				Unheated sapphire	S-gun	Target-to-substrate distance 4 in. Temperature of substrate during deposition 90°C; using Ne or Kr in place of Ar gives no advantage.	59
AlN	Al disk	Ar-N$_2$	500–1000 W	0.1–1.2 μm/h	(2–8) 10^{-3}	Si (100) SiO$_2$, Al/Si, GaAs 100–450	Rf planar magnetron	Gas composition, Ar/N$_2$, 0.25:75, 1:1 0.75:0.25. Target bias, −100 to −200 V. Target-substrate spacing, 68 mm.	60
TiN	Ti (99.9% pure)	Ar + N$_2$	1.5–2 kW	0.12–0.63 nm/s		Number of steel substrates (Ni-Cr, low carbon, and tool steels) dur-Al, and glass 200	Rf	Mechanical properties and structural characteristics studied.	61
LiNbO$_3$	Hot pressed LiNbO$_3$ (4 in. dia. 99.99% pure)	Ar 10% + O$_2$			5 × 10^{-3}	Corning glass 380	Rf	Sample composition analyzed using electron diffraction, electron microscopy and Auger electron spectroscopy.	62

Table 4.4 (Continued)

Deposited film	Target	Reactive gas	Sputtering power/ density	Deposition rate	Pressure (torr)	Substrate and substrate temperature (°C)	Type of system	Remarks, if any	Ref.
Cd_2SnO_4	66% Cd/ 34% Sn (99.999% pure)	Ar/O_2 mixture			4.5×10^{-2}	Corning glass 7059 377	Dc	Measurements of the electron concentration, mobility and thermoelectric power as functions of temperature showed that the films were degenerate semiconductors.	63
Si:H	5 in. poly-crystal-line Si	H_2/Ar		Undoped, 30 Å/min	40×10^{-3}	325	Rf	Phosphorus and boron doping done by premixing the argon with 2000 ppm of PH_3 or B_2H_6.	64
SnO_2 (Sb doped)	Sn–Sb complex target (99.999% metal purities)	$Ar + O_2$ 20–100 vol % O_2	50 W	20–210 Å/min	$(5{-}40) \times 10^{-3}$	Sodalime glass and quartz	Rf	Distance between substrate and target, 45 mm.	65
In_2O_3:Sn	In (80%) Sn (20%)	$Ar + O_2$	100–200 W	0.8 nm/s	3×10^{-3} to 4.5×10^{-3}	Glass, Si or CaF_2	Dc magnetron	$Ar/O_2 = 3$ with 182 W gives excellent optical properties. Size of target	66

168

Material	Target	Gas	Power	Deposition rate	Pressure	Substrate	Mode	Remarks	Ref.
ITO	In/Sn alloy	O_2			40×10^{-3}		Dc diode	50×12.5 cm².	67
BeO	Be	O_2 + Ar 0.1 ratio	400 W	< 2.5 nm/min	7.5×10^{-4}	Optically flat quartz glass and pyrographite plates	Rf	Optical and electrical properties were invesgated.	68
SiO_2	Si (111) single crystal slices	Ar/O_2 mixture			3.8×10^{-3}	Silica	Dc magnetron	High rate reactive deposition reported.	69
Aluminum oxyfluorine	Al (99.5 wt %) pure	O_2 and CF_4					Dc planar magnetron	Various flow rates and partial pressures of CF_4. Cathode substrate distance, 4 cm; also used rf planar magnetron. Produced low refractive index ($n \approx$ 1.40), transparent ($K \leq 10^{-3}$) dielectric material. Extremely high deposition rates were obtained.	70
SiO_2	Si in platelets form $1 \times 5 \times 10$ cm	O_2	20 kW		23×10^{-5}	Polyester web	Biplanar magnetron	Target surface sputter-cleaned in pure argon for 5 minutes before introducing oxygen. High rate, large-area	71

Table 4.4 (Continued)

Deposited film	Target	Reactive gas	Sputtering power/ density	Deposition rate	Pressure (torr)	Substrate and substrate temperature (°C)	Type of system	Remarks, if any	Ref.
								sputtering for industrial applications. Equally applicable to AlN, ZnO, Al$_2$O$_3$, ZrO$_2$, and ITO.	
Si:H	Semi-conductor grade polycrystalline Si disk (10 cm dia.)	H$_2$ + Ar mixture	215 ± 5 W		$P_H/P_A =$ 0.1, 0.2, 0.3, 0.45, and 0.6	Corning 7059 glass 200	Rf	Dark conductivity and photoconductivity measured as functions of temperature. Distance between target and substrate, 5 cm.	72
ZnSe	Zn	H$_2$Se/Ar mixture; H$_2$Se (99.999% pure)				Glass and conducting SnO$_2$ coated glass ≤ 500	Magnetron sputtering s-gun	Separate gas manifolds for Ar and H$_2$Se. Ar directed downward sputter targets and H$_2$Se towards rotating substrate. Highly conducting films were obtained.	73
TiN	Ti	Ar (99.999% pure) + N$_2$	10 kW			High speed steel drills	Magnetron sputtering	Nitrogen partial pressure was regulated by the automatic feedback	74

Material	Target	Reactive gas	Power density	Rate	Pressure	Substrate	Method	Remarks	Ref
Zirconia alumina	Composite target of Zr and Al	Ar + O$_2$ (99.999% pure)		3 Å/s	3×10^{-3}	Ultrasmooth graphite, glass, single-crystal sapphire	Magnetron sputtering	control system. High rate of sputtering with excellent properties was reported.	75
(Ti, Al)N	Ti, 50 at. % Al alloy		dc: 7–8 W/cm^2 rf: 1–15 W/cm^2	rf: 10–50 nm/min dc: 60–130 nm/min	1.5×10^{-5} (dc) 1.5×10^{-6} (rf)	Polished flat high speed steel	Dc and rf magnetron	Oxygen was directed at the substrate with an annular supply that surround the substrate.	76
ITO	90 wt % In 10 wt % Sn, 5 × 18 in.^{2} rectangular	Ar + O$_2$			3×10^{-3}	Sodalime silicate glass 25,125,200	Dc magnetron	Film morphology (SEM) and composition (AES) were studied and correlated to mechanical properties.	77
Higher nitrides of Hf, Zr and Ti	Hf, Zr, and Ti metal	N$_2$			30 mtorr	Sapphire, MgO, Si, fused quartz, C, and W 196–750	Rf	Deposition was carried out by moving the substrate over the target at constant speed.. Electrical and optical properties of these films are highly dependent on substrate temperature and target voltage.	78

171

Table 4.4 (Continued)

Deposited film	Target	Reactive gas	Sputtering power/density	Deposition rate	Pressure (torr)	Substrate and substrate temperature (°C)	Type of system	Remarks, if any	Ref.
InSb	Sb	Ar + metal organic vapor (trimethylindium)	25 W	1.2 Å/s		Si (100) 200	Magnetron	Argon saturated with TMI vapor was introduced near the substrate through a tube with holes. Vapor was injected radially inward.	79
AlN	Al (99.999% pure)	Ar + N_2			15×10^{-3}	Corning 7059 glass	Dc reactive sputtering	Influence of target erosion on the flow discharge, deposition rate, and film conductivity were investigated.	80
$Ti_{0.5}Al_{0.5}N$	TiAl alloy (>99.7% pure)	Ar (99.999% pure) + N_2 (99.9995% pure)		0.2 μm/min	8.2 mtorr	Polished austenitic stainless steel plates 400–500	Dc magnetron double cathode	Substrate-to-target separation 6 cm. Nitrogen partial pressure, 2.9 mtorr.	81a
BN	BN	Ar + N_2	650 W	50 Å/min	15×10^{-3}	C, Ta, thin carbon films, Si single crystals RT and 350	Rf diode	Target, cerac hexagonal BN; 42 wt % B, 53.5 wt % N, 0.05 wt % C, 1.5 wt % Ca, 0.2	81b

Material	Target	Gas	Deposition conditions	Substrate	Method	Comments	Ref.
Ti-Si-N	Two component Ti-Si	Ar/N$_2$	5×10^{-4}	Crystalline glass 27, 200, and 300	Triode sputtering	Substrate temperature was increased from RT to about 100°C during deposition because of ion bombardment. Surface area of the silicon plate was 31% of the total surface area. Under the sputtering conditions (target voltage –500 V) a mixture of Si$_3$N$_4$ and TiN was obtained.	113
Fe-B-Si	Mild steel	Ar, B$_2$H$_6$, and SiH$_4$	5.8, 1.0, and 2.0 × 10^{-3} torr partial pressures of Ar, B$_2$H$_6$, and SiH$_4$, respectively		Planar magnetron	Amorphous magnetic thin films (76% Fe: 16.5% B: 3% Si: 3% C) have been prepared Amount of boron or silicon depends on partial pressures of reactive gases.	114
a-Si:H	High purity Si slab	H$_2$ and Ar	P_{H2}, 0.05 × 10^{-3} to 1.2 × 10^{-3} P_{Ar}, 1×10^{-3}; 20–220 Å/min	Double polished crystalline Si and	Dc planar magnetron	Films deposited in a UHV chamber with base pressure of about 10^{-8} torr.	115

The silicon plate was 31% of the total surface area. Under the sputtering conditions (target voltage –500 V) a mixture of Si$_3$N$_4$ and TiN was obtained.

wt % Cl, 0.1 wt % B$_2$O$_3$.

Table 4.4 (Continued)

Deposited film	Target	Reactive gas	Sputtering power/density	Deposition rate	Pressure (torr)	Substrate and substrate temperature (°C)	Type of system	Remarks, if any	Ref.
SnO$_x$	99.999% Sn	Ar/O$_2$	V_{dc} = 2 kV I_{dc} = 8 mA	30 Å/min	P_{Ar}, 2 × 10^{-4} P_{O2}, 1 × 10^{-4}	Corning glass 100–350 Oxidized Si < 100	Dc magnetron	Optical and electrical properties of the films were obtained. Films were characterized by conversion electron Mossbauer spectroscopy, X-ray diffraction, nuclear resonance scattering analysis, and sheet resistance measurements. Amorphous highly resistive and defective tin oxide films with α–Sn precipitates were obtained.	116
TiB	5 in. titanium	Diborane/Ar mixture			10 × 10^{-3}	Si and vitreous C	Rf diode	Reactive sputtering process was studied. Effect of a hysteresis region, normally found in reactive sputtering, was absent.	117

Material	Target	Gas	Power	Pressure	Deposition rate	Substrate	Method	Remarks	Ref.
NiCr–O	Sintered targets; Ni:Cr ratios 50:50 and 40:60	Ar/O₂	100 and 800 W	3×10^{-3}	0.17–1.2 nm/s	Thermally oxidized Si wafers and vitreous C	Dc magnetron	During deposition, the substrates were clamped on a heated metal block (300–775 K). A qualitative description of the incorporation of oxygen is given.	118
Bi-Sr-Ca-Cu-O	5 cm diameter, 0.5 cm thick stoichiometric Bi₂Sr₂CaCu₂Oₓ disk	Ar and O₂ oxygen fraction varied from 0 to 0.5	5 W/cm²	Ar and O₂ pressure varying from 1×10^{-3} to 1×10^{-1} (O₂ fraction varying from 0 to 0.5)		(100) MgO, water cooled	Rf planar magnetron	Effect of deposition conditions on stoichiometry and superconducting properties was studied. Films had T_c (onset) of 115 K, T_c ($R = 0$), 87 K. Critical current density was 5×10^5 A/cm² at 4.2 K.	119
V₂O₅	V disk, 99.9%	Ar and O₂	100–500 W	1×10^{-2}	2.4 and 1.0 nm/s in metallic and oxidized target mode, respectively	Uncoated glass or SnO₂:F coated glass	Dc magnetron	No substrate heating. Substrate temperature rose from 22°C to less than 60°C during deposition. Films were amorphous when deposited at RT.	120
TiN	Ti (99.9% pure)	Ar–10% N₂		5×10^{-3}	0.1 μm/min	Surface-hardened structural	Dc planar magnetron	Deposition was shown to decarborize the hardened	121a

Table 4.4 (Continued)

Deposited film	Target	Reactive gas	Sputtering power/density	Deposition rate	Pressure (torr)	Substrate and substrate temperature (°C)	Type of system	Remarks, if any	Ref.
						steel		surface even at low substrate deposition temperature (<200°C).	
WO_3	W	Pure O_2	2–10 W/cm²			Corning glass, Corning glass coated with ITO	Dc magnetron	Influence of substrate temperature on the crystallanity of the films was studied. Refractive indices, optical bandgap, and dielectric properties were determined.	121b
RBa_2 Cu_3O_{7-x} (R = Y, Er, and Nd)	Metallic	Xe/O_2				YSZ, MgO, ZrO_2-buffered sapphire and Ag-buffered sapphire	Four-gun dc magnetron	Post annealed in flowing oxygen. Erbium-based films were produced with good characteristics.	122a
$YBa_2Cu_3O_7$	Y, Ba, and Cu metal targets	$Ar + O_2$			10^{-4}	YSZ 300	Magnetron	High quality films were prepared by simultaneous sputtering Targets were	122b

Material	Target	Gas	Power	Rate	Pressure (Torr)	Substrate	Method	Comments	Ref.
SnO_2,	Sn	O_2 + Ar	6 W	60 Å/min	0.07	Glass slides	Magnetron	yttrium and copper, with dc power supplies, and barium, using a 13.56 MHz rf power supply. Superconductivity onsets of 92 K with transition widths of 1.5 K (10–90%) have been obtained with YSZ substrates. Optical properties electrical resistivity and crystallinity of of the films were	123a
Al_2O_3,	Al	O_2 + Ar	92 W	80 Å/min	0.08				
ZnO,	Zn	O_2 + Ar	31 W	170 Å/min	0.06				
AlN	Al	N_2 + Ar	113 W	285 Å/min	0.04				
AlN	Al	Ar + N_2	100 W	0.5–1.5 K/s for all films	15×10^{-3}	Sapphire C (123b) plate with the c-axis parallel to growth direction	Rf magnetron	Substrate temperatures were 500, 500, 400, and 500°C for AlN, GaN, InN, and $Al_xIn_{1-x}N$, respectively. Films were characterized by X-ray diffraction, Raman scattering, optical absorption, and electrical measurements and were of high quality.	123b
GaN,	Ga	Ar + N_2	50 W		20×10^{-3}				
InN,	In	Ar + N_2	30 W		20×10^{-3}				
Al_xIn_{1-x}	Composite target of Al sheet with several slits put on In plates	Ar + N_2	100 W		15×10^{-3}				

Table 4.4 (Continued)

Deposited film	Target	Reactive gas	Sputtering power/ density	Deposition rate	Pressure (torr)	Substrate and substrate temperature (°C)	Type of system	Remarks, if any	Ref.
YBa2Cu3-O7-x	Sintered planar YBCO target	Ar + O2 Ar/O2 10:1		30 Å/min	2×10^{-1}	(100), (110) SrTiO3 and (100) YSZ 800	Modified planar dc magnetron	Perfect epitaxial growth has been achieved with in situ annealing at 400°C for 1 hour. Films exhibited zero resistances at 87–90 K with transition width of about 2 K.	123c
Iron Nitride	Iron (99.97% pure)	N2 (99.995% pure) + Ar (99.995% pure)		2.2–4.3 Å/s	4×10^{-4}	Si (111) wafers	Planar magnetron	Total gas pressure kept constant by controlling the gas flow rate. Compositional and magnetic properties under various nitrogen partial pressures were investigated.	124
Ta2O5	Ta disk (99.99% pure)	Oxygen		1.2 μm/h		Fused quartz 450	Planar magnetron dc diode	X-Axis oriented films were deposited on quartz. Material constants were determined.	125

Material	Target / Gas	Power	Rate	Pressure	Substrate / Temp	Method	Description	Ref.
TiN	Ti (99.99% pure) N$_2$ (99.999% pure)	2.2 kW, 13.56 MHz	200–300 Å/min	5×10^{-3}	Si with thin oxide layer, low carbon steel	Rf magnetron	Deposited TiN films were heat treated at 300, 400, and 550°C for 4 hours in air and nitrogen. As-deposited and heat-treated films were analyzed using SEM, TEM, X-ray diffraction, secondary ion mass spectroscopy, and X-ray photo-electron spectroscopy.	126
LiNbO$_3$	99.999% pure LiNbO$_3$ compensated with Li$_2$O (99.9% pure) (90:10) O$_2$/Ar ratio, 50:50 to 0:100	Rf power, 50–250 W	400–500 Å/h	1×10^{-3} 20×10^{-3}	Si (111) 550–600	Rf	For most depositions the power kept at 100 W. Oxygen to-argon ratio of 40:60 gave the highly oriented films. Under optimum conditions the films were optically transparent and adhered well to the substrate.	127

Stoichiometric as well as nonstoichiometric polycrystalline films of TiN have been prepared by a variety of different techniques. But the first reported single-crystal TiN films were deposited on cleaved (111) oriented MgO wafers using dc reactive magnetron sputtering [41]. The schematic diagram of the dc diode magnetron sputter system used by Johnson et al. [41] is shown in Figure 4.12. Sputtering was carried out in either mixed N₂/Ar or pure N₂ discharges.

The deposition parameters reported are as follows.

Target	15 cm Ti dish (99.9% pure)
Current	1.4 A
Target voltage	415–430 V
Substrate	Cleaved (111) MgO wafers
	(0.5 × 0.5 × 0.1 cm)
Target-to-substrate separation	13 cm
Sputtering gas	Mixed N₂/Ar or 99.9999% pure N₂
Total pressure	3.5 mtorr
Deposition rate	~1 μm/h
Film thickness	1–2 μm

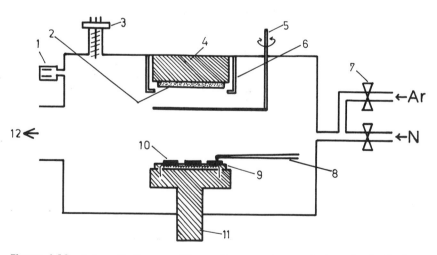

Figure 4.12 Schematic diagram of the reactive magnetron sputtering setup for the deposition of single-crystal TiN films onto cleaved, oriented MgO wafers: 1, capacitance manometer; 2, Ti target; 3, Ion gauge; 4, magnetron target holder; 5, shutter; 6, ground shields; 7, leak valves; 8, thermocouple; 9, substrate heater; 10, substrate; 11, height adjustable substrate table; 12, to pump. (Data from Ref. 41.)

After deposition, the substrates were allowed to cool to below 50°C before dry nitrogen was admitted into the system.

The preparation and study of phosphosilicate glass (PSG) films are of interest because of their importance of silicon devices. Highly doped (with phosphorus) PSG films are difficult to obtain by conventional sputtering because highly phosphorus-doped targets cannot be easily prepared. Serikawa and Okamoto [42] have reported a method of preparing highly doped PSG films by reactive sputtering of a nondoped silicon target in an oxygen–argon mixture using an rf planar magnetron sputtering system (Figure 4.13). A nondoped silicon target was bonded to the water-cooled electrode. The substrate holder with silicon was mounted approximately 80 mm in front of the target. The crucible containing solid red phosphorus (5.0 g) was placed 40 mm in front of and 30 mm below the target and was heated by a platinum wire. The deposition parameters are given in Table 4.5.

PSG films with phosphorus concentrations up to 3×10^{21} cm^{-3} were success-

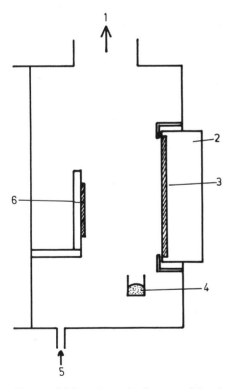

Figure 4.13 Schematic diagram of the planar magnetron sputtering apparatus for the preparation of highly doped PSG films: 1, to cryopump; 2, planar magnetron cathode; 3, silicon target; 4, red phosphorous; 5, gas (O$_2$ + Ar) inlet; 6, silicon substrate. (Data from Ref. 42.)

Table 4.5 Deposition Parameters for Reactive Sputtering Using the Setup in Figure 4.13

Target	Silicon (99.995% pure)
Substrate	p type (100) oriented Si ($\approx$ 10 mm diameter)
Base pressure	7.5×10^{-8} torr
Sputter gas	Oxygen/Argon mixture
Sputter pressure	5×10^{-3} to 15×10^{-3} torr
Sputter power	3.0 kW
Substrate temperature	Room temperature
Crucible temperature	RT to 260°C (measured by a chromel–alumel thermocouple)
Film thickness	0.5 μm

fully prepared at low substrate temperature. The investigators had experimentally shown [42] that the phosphorus concentration in the films can be controlled by crucible temperature and sputtering gas pressure.

Polyphospides are phosphorus-rich inorganic semiconductor compounds with unusual crystal structures. Schachter et al. [43] were the first to report the sputter deposition of amorphous polyphosphide thin films from a compound target. They showed that the growth rate of this semiconductor compound (KP_{15}) is determined by the alkali metal flux and that the stoichiometry is determined by the phosphorus flux. A new technique has been used to provide excess phosphorus by the generation of reactive phosphorus (P) species from P_4 molecules injected into the system via Ar carrier from an independent P_4 delivery system. The phosphorus delivery system is safer then PH_3, a gas used in an earlier reactive sputtering system [44]. The rf diode sputtering system (Figure 4.14), with high purity Ar as the sputtering gas, was evacuated to a base pressure of 10^{-7} torr prior to deposition. Details of the sputtering parameters were as follows.

Target	5 cm diameter disk (pressed polycrystalline charge of KP_{15})
Substrate	Alkali-free Corning 7059 glass
Target-to-substrate distance	5 cm
Total pressure (Ar + P) in chamber	25 mtorr
Partial pressure of phosphorus	$\leq$ 1% of total pressure
Deposition rate	5–200 Å/min

The high quality amorphous KP_{15} films obtained were dense and hard, with shiny surfaces and good adhesion to the substrate. The microstructure showed the intermediate range order characteristic of these materials.

Parsons et al. [45] reported the deposition of amorphous Si, Sn:H alloy films by reactive magnetron sputtering from separate silicon and tin targets. In their experimental setup (Figure 4.15), the deposition was done in a VHV compatible chamber in which the substrate was introduced through a load lock with a magnetic transfer device. The target materials were silicon (99.9999% pure) and tin (99.999% pure). Argon (99.9999% pure) and hydrogen (99.999% pure) were used with the operating pressure in the millitorr range. The substrate was mounted about 18 cm from each of the targets. A tubular geometry configuration for ground shield was used for the silicon target, while a conical geometry ground shield structure was used for tin. In their previous reports [46,47a] on amorphous Si, Ge:H alloys, these investigators used the same tubular geometry configuration for both Si and Ge. Figure 4.16 shows the conical ground field structure for the Sn target.

In the dual-magnetron system, the alloy composition is controlled by the power fed to each of the sputtering targets; and a discharge at a target with conical ground shield geometry could be sustained with less power input than was needed for tubular geometry. Specifically, 5 W would sustain a discharge with a conical ground

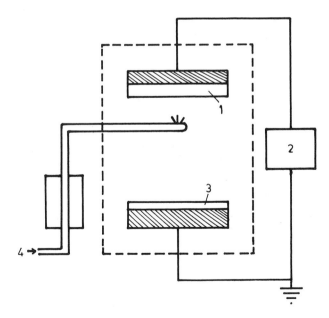

Figure 4.14 Schematic diagram of the setup of the rf diode sputtering system with P_4 injection: 1, target, KP_{15}; 2, rf power; 3, substrate; 4, Ar + P_4 injection. (Data from Ref. 43.)

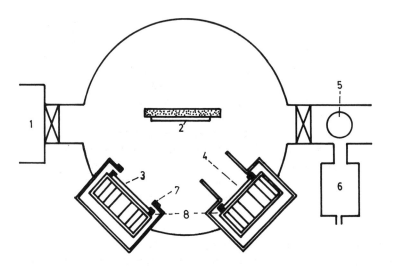

Figure 4.15 Schematic diagram of a dual magnetron sputtering system for the deposition of Si, Sn:H alloy films: 1, cryopump; 2, substrate; 3, Si target; 4, Sn target; 5, load lock; 6, Turbo pump; 7, Si washer ring; 8, Si Clips. (Data from Ref. 45.)

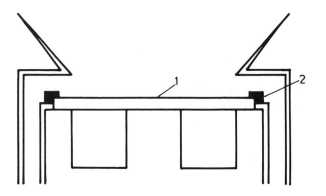

Figure 4.16 Conical ground field structure for Sn target (see Figure 4.15): 1, Sn target; 2, Si clips. (Data from Ref. 45.)

shield configuration, which enabled investigators to obtain alloy films with Sn concentrations in the range of 0.5 at. % and higher. Since the sputtering rate of Sn is much higher than that of Si, with a fixed target power of 200 W for Si, films with an alloy range of 0.5-26 at. % Sn could be prepared. With a tubular ground shield target for Sn, the alloy films had an average Sn concentration only in excess of 40 at. % (200 W power to the Si target). Crystalline Si and fused silica were used as substrates, and the films were simultaneously deposited onto these substrates. The chemical bonding of the constituent atoms was studied by IR absorption spectroscopy, Auger electron spectroscopy (AES), and X-ray photoelectron spectroscopy (XPS). Also the optical and electrical properties of the films were investigated.

Spencer et al. [47b] have reported the use of unbalanced magnetrons [see ref. 66 of Chapter 3] for reactive sputtering to prepare thin films of indium oxide and titanium oxide. Their results show that the plasma bombardment activates metal-gas reactions on the substrate, increasing the utilization of the reactive gas, with the result that a smaller amount of reaction products should form on the target and the deposition rate should consequently be higher.

The film properties also improved. In particular, the refractive index of titanium dioxide was shown to be increased by plasma bombardment.

A modified technique for the preparation of Al_2O_3 films by direct current reactive magnetron sputtering has been reported by Pang et al. [47c]. To overcome such effects as reduction in the sputtering rate due to the formation of the nonconducting layer on the target surface, target surface arcing, and changes in film quality with slight changes in the deposition parameters (commonly seen in reactive sputtering, as pointed out earlier), a gas separation was established by creating a physical barrier between the inlet of the reactive gas on the target surface.

The schematic of the setup (Figure 4.17) used in dc reactive sputtering of Al_2O_3 illustrates the principle of the gas separation. The main elements consist of the metallic cathode shield all around, except for an aperture directly opposite the aluminum target. There is a pair of inlets for the sputtering gas argon within the shields; outside, in the vicinity of the substrate surface, are inlets for the reactive gas. The metallic shield physically separates the reactive gas from the target surface and also provides a gettering area for excess reactive gas. If the mean free path of the reactive gas at the operating pressure is less than the separation effected by the physical barrier, sputtering will take place in a relatively inert surroundings. Several other parameters (inert gas flow, reactive gas flow, target power, substrate bias, process pressure, etc.) have been optimized to produce the best results, and a statistical modeling approach has been adopted to find the most suitable set of operating conditions.

Using commercial sputtering machines, several tests have been performed for Al_2O_3 films in the rf mode (without gas separation) and in the rf magnetron and dc reactive magnetron (with gas separation) modes. Deposition rate and performance of the modified technique are compared with other systems and presented in the

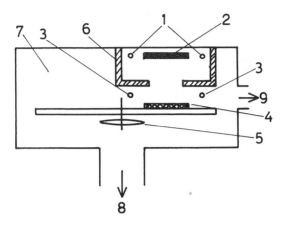

Figure 4.17 Schematic diagram of the setup for the reactive sputtering of Al_2O_3: 1, sputtering gas (argon) inlet; 2, aluminum target; 3, oxygen inlet; 4, substrate; 5, substrate holder; 6, aluminum shielding; 7, vacuum chamber; 8, to pumps; 9, to partial pressure controller. (Data from Ref. 47c.)

paper [47c]. The results of the study of such film properties as hardness, optical absorption, and electrical conductivity of Al_2O_3 are also compared. The authors have also investigated the deposition of other nonconducting materials using the gas separation technique described above.

The growth of high quality thin films of superconducting oxide $YBa_2Cu_3O_{7-x}$ has been reported by Char et al. [48a], using reactive magnetron cosputtering from three metal targets, Y, Ba, and Cu, in an Ar/O_2 atmosphere. The films were deposited onto heated substrates. Three magnetron sputtering guns were set up in a triangular arrangement, such that they pointed toward the center of the substrate holder, which kept two rows of 10 substrates (6 mm × 6 mm) 10 cm above the targets.

The following procedure was used to prepare the films. Argon gas of 2×10^{-3} torr pressure was introduced into the system (base pressure 1×10^{-6} torr) at the Ba and Y sources to keep the targets from oxidizing. Oxygen was injected toward the substrate through a thin slit to produce a homogeneous gas distribution at the growing film surface. This way a high oxygen impingement rate at the substrate surface was obtained while a relatively low partial pressure of oxygen (1×10^{-4} torr in the chamber) was maintained. With this arrangement of gas flow, the targets, especially Ba and Y, do not become oxidized, and a uniform and stable deposition rate

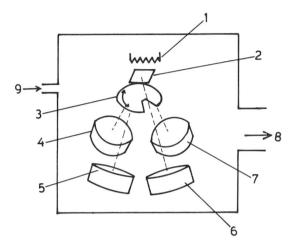

Figure 4.18 Schematic diagram of the multitarget dc magnetron sputtering system for reactive, sequential deposition: 1, heater, 2, substrate; 3, shutter; 4, SrCu target; 5, CaCu target; 6, SrCu target; 7, Bi target; 8, to vacuum; 9, sputtering gas inlet. (Data from Ref. 48b.)

is obtained. This arrangement of the gun and substrate provides a composition spread such that one row has a little more copper than the other row, and the Y/Ba ratio varies across the row of substrates. The metal composition was obtained by electron microprobe analysis. The composition spread from the first specimen to the tenth sample on the copper-rich row was from $Y_{13} Ba_{37} Cu_{50}$ to $Y_{22} B_{24} Cu_{54}$; on the other row the composition spread was found to be from $Y_{13} Ba_{41} Cu_{46}$ to $Y_{22} Ba_{28} Cu_{50}$, giving an overall composition variation of about 1 at. % in a single substrate.

Barium was sputtered by rf power, while dc sputtering was used for Y and Cu. The samples from the chamber were not superconducting, and only after high temperature annealing with O_2 gas flowing did the films exhibit superconductivity.

In a continuation of their work on a superconducting bismuth system, Setsune et al. [48b] reported the preparation of high T_c superconducting bismuth layered perovskite by reactive sequential deposition of Bi, SrCu, and CaCu, using dc magnetron multitarget sputtering in an Ar/O_2 (5:1) atmosphere (3.5 Pa). The schematic of their sputtering system is shown in Figure 4.18. Four metal targets Bi (99.999%), two SrCu (99%), and CaCu (99%) were tilted against the substrate normal. The computer-controlled sputtering rates and turns of these targets were adjusted by the dc power input and the shutter between the target and the substrate. The substrates were MgO (100), T_s 600–680°C, and thin films with controlled c-axis lattice spacing have been successfully prepared. The as-grown films need postannealing to impart the desired superconductivity.

4.5   REACTIVE ION PLATING

Reactive ion plating is a useful and important development of the basic principle of ion plating in which the residual atmosphere in the vacuum system contains a reactive gas—for example, oxygen or nitrogen. This process is similar to reactive evaporation, in which the metal atoms and reactive gases react, aided by the plasma, to form a compound. The plasma cannot be supported at low pressure in simple diode ion plating, and Kobayashi and Doi [82] introduced an auxiliary electrode biased to a low positive voltage to initiate and sustain the plasma at a pressure of about 10^{-3} torr. This process, which the authors called reactive ionplating (RIP), actually is the same as that of the original ARE discussed in Section 4.2, with a negative bias as reported by Bunsha, who used the term "biased activated reactive evaporation" (BARE) [83].

Preparation of thin films of In2O3, TiN, and TaN by rf reactive ion plating had been reported earlier by Murayama [84], and the schematic diagram of his setup is shown in Figure 4.19. The rf coil electrode was placed between the evaporator (anode) and the substrate holder (cathode). The oscillating field was applied between the coil and the evaporator.

In rf reactive ion plating, a portion of the vaporized material is ionized by the gas discharge and will be accelerated toward the cathode. Formation of oxides or nitrides is enhanced by the excitation of the rf oscillating field. An optimum deposition rate must be selected for each reaction, because oxidation or nitriding is controlled by the evaporation rate.

Indium, tin, and tantalum were evaporated in an oxygen/nitrogen atmosphere at a pressure of the order of 10^{-4} torr, and the film prepared by rf ion plating was found [84] to be uniform and pinhole-free. The In2O3 films showed excellent electrical properties and optical properties, the TiN and TaN films exhibited stable electrical properties, and good epitaxy was found in all the films formed.

Since 1975, reactive ion plating has been used for the preparation of compound films, oxides, nitrides, and so on by the dc and rf discharge methods. Fukutomi et al. [85a] have reported the use of reactive ion plating by dc and rf discharge in a single setup (Figure 4.20) for the preparation of titanium carbide films on molybdenum. For the dc discharge method, a probe made of molybdenum (3 cm × 5 cm × 0.3 cm) was placed above the evaporation source opposite the EB gun and was biased positively with respect to the ground. The lower end of the probe was roughly 3 cm from the molten pool. When the EB power input was raised above a certain minimum value, an arc-type discharge was initiated between the molten pool and

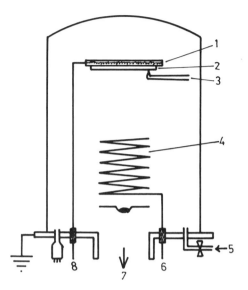

Figure 4.19   Schematic diagram of the rf reactive ion plating system for the preparation of thin films of In_2O_3, TiN, and TaN: 1, cathode base; 2, substrate; 3, thermocouple; 4, rf coil; 5, gas inlet; 6, to rf power supply; 7, to vacuum pump; 8, to dc power supply. (Data from Ref. 84.)

this probe and could be sustained at probe voltages of 50–60 V and at currents up to 20 A. The probe current was sensitive to the EB power input, and so the evaporation rate of the source material could be controlled by monitoring the probe current and regulating the EB power. For the rf discharge method of ion plating, an rf coil electrode made of stainless steel tube of six turns, 10 cm in diameter, was placed directly above the evaporation source. The temperature of the specimen was measured with a chromel–alumel thermocouple.

For the comparative study of the structure and stoichiometry of titanium carbide films, Fukutomi et al. used the following deposition parameters.

Reactant gas	C_2H_2
Pressure	$(8–80) \times 10^{-4}$ torr
EB gun power	0.6–1.2 kW
Coating temperature	450–900°C

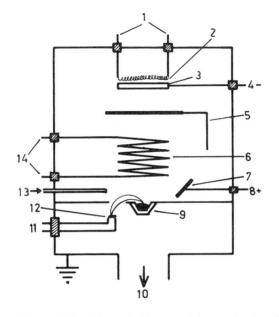

Figure 4.20 Schematic diagram of the reactive ion plating setup for the preparation of TiC films on molybdenum: 1, ac power supply; 2, substrate heater; 3, substrate; 4, to dc power supply; 5, shutter; 6, rf coil; 7, probe; 8, to probe dc power supply; 9, water-cooled copper crucible; 10, to vacuum pump; 11, EB power supply; 12, EB gun; 13, gas inlet; 14, rf power supply. (Data from Ref. 85a.)

Source-to-substrate distance	20 cm
Substrate bias	–200 V
Film thickness	10–15 μm

They found that in the dc discharge, efficient ionization of the vapor species was achieved, and stoichiometric TiC depositions could easily be obtained in a wide range of pressure ratios (C_2H_2 gas to titanium vapor). The rf discharge method calls for careful optimization of the deposition conditions to prepare stoichiometric films.

Edlinger et al. [85b] have deposited single-crystal films of Nb_2O_5 on various substrates by reactive low voltage ion plating using a patented ion plating system [85c]. In this process a low voltage, high current argon arc is established between the hot filament plasma source and the crucible of a specially designed electron beam gun. Reactive gas is admitted directly into the chamber. The substrate holder is electrically insulated and immersed in the plasma, leading to a negative self-bias of the substrates with respect to the plasma. As a result, the positive ions of the depositing material, the reactive gas, and the argon impinge on the substrate with a kinetic energy corresponding to the sheath potential, resulting in dense, hard, adherent films on unheated substrates.

Here for the deposition of Nb_2O_5 films the starting material was granulated Nb_2O_5 (99.9%), and for all coating runs the base pressure was from 4×10^{-6} to 8×10^{-6} mbar. With argon admitted, the pressure came to 2×10^{-4} mbar in the process chamber. The arc current was set to 70 A, with the arc voltage becoming 50–70 V.

To study the properties of the films, partial oxygen pressure, substrate temperature, and rate of deposition were varied.

Wang and Oki [85d] have prepared Cr-N films by reactive ion plating using the setup schematically shown in Figure 4.21. The chromium metal (99.99% pure) was evaporated from a resistance heating system, with water cooling. A filament of tantalum wire (0.8 mm diameter) was set at 2 cm above the evaporator, and a negative bias voltage of 300 V was applied between this ionization filament and evaporator. The ionization current (i.e., the current between the filament and the evaporator)

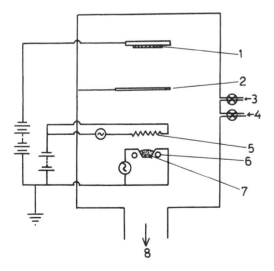

Figure 4.21 Schematic diagram of the reactive ion plating setup for the preparation of Cr-N films: 1, substrate; 2, shutter; 3, Ar inlet; 4, N_2 inlet; 5, filament; 6, cooling water; 7, metal; 8, to pump. (Data from Ref. 85d.)

Table 4.6 Reactive Ion Plating

Film	Reactive gas	Pressure (torr)	Substrate and substrate temperature (°C)	Bias voltage (V)	Remarks, if any	Ref.
In_2O_3	O_2, CF_4		Microscope glass slides		Metal (In) was evaporated in an O_2/CF_4 rf discharge. For fluorine doping of In_2O_3 CF_4 introduced through a stainless steel tube positioned about 5 cm below the substrate holder (rf electrode)	86
In-Sn oxide and Cd_2Sn oxide	O_2 + Ar mixture		Glass RT		Planar magnetron sputtering of In–10% Sn and Cd_2Sn targets respectively.	87
TiC	Acetylene	$(8-70) \times 10^{-4}$	Mo		Electron beam evaporation of Ti; rf discharge power was 230 W (13.56 MHz).	88
ITO	O_2	$(4.5-10) \times 10^{-3}$	Ordinary glass plates 300–350		Indium and tin were evaporated separately from boats; cathode supporting the substrates was biased by an rf generator (13.56 MHz); growth rate was 4–6 μm/h.	89
TiN	N_2		Modular cast iron		Electron beam gun was used for evaporation; deposition rate, 20 Å/s.	90
SiC	Acetylene	3.8×10^{-3} to 7.5×10^{-3}	Sintered Mo 227–1027		Rf discharge; silicon evaporated by electron beam.	91
HfN	Ar + N_2		Stainless steel 400	200–300	Metallic hafnium was evaporated by electron beam, ion etched before deposition. Microstructural characteristics were studied.	92

Material	Reactive gas	Pressure	Substrate/Temperature	Bias	Description	Ref.
Ti–N (N content, 17–50 at. % N$_2$)	N$_2$	4.5×10^{-3}	Austenitic stainless steel sheet 357–527		In two series of experiments, depsoition rate was varied from 50 to 200 nm/min. Commercial grade Ti (purity 99.2 wt %).	93
TiN	N$_2$		Sheets of stainless steel and high speed steel 300–500	–30 to –40	Hollow cathode discharge ion plating. Nitrogen gas was introduced into the vacuum chamber a few minutes after the beginning of Ti evaporation. Deposition rate was 0.04–0.05 μm/min.	94
TiN	N$_2$		Phosphorus bronz; single-crystal NaCl, and p-type Si (100) ≤ 700		Titanium source was heated by electron beam. Influence of plasma conditions during deposition on the preferred orientation and resistivity was studied.	95
Cr–C	C$_6$H$_6$	0.75×10^{-4}	NaCl cleavage planes, Si wafers (111) and (100) orientations and Al foils ≤ 400 K	≤ –3000	Chromium of purity higher than 99.9 wt % was evaporated from a directly heated tungsten crucible. Evaporation rate was monitored by quartz thickness monitor. Composition of the Cr–C films was varied by regulating the heating power of the evaporator.	128
Ni–TiC	C$_2$H$_2$	0.8×10^{-3}	Stainless steel 673 K	Biased negative (2 kV) with respect metal pool	Ni–Ti alloy (different compositions) was ingot melted and evaporated by EB gun. Deposition rate was 0.5 μm/min. Deposited films were characterized by X-ray diffraction, electron microprobe analysis, and TEM. Films obtained were fine-grained mixture of Ni and TiC.	129

was adjusted by controlling the ionization filament current. Cold rolled steel plates (25 mm $\times$ 25 mm $\times$ 1 mm), polished and then ultrasonically cleaned in an acetone bath, were used as substrates. These plates were kept 11 cm from the source, with the substrate temperature (measured on the back of the substrate) was controlled at about 260 $\pm$ 30°C.

After the initial evacuation to a pressure around 6.6 $\times$ 10^{-3} Pa, the chamber was backfilled with argon gas to a pressure of 1.3 Pa. A 1 kV dc bias voltage was applied between the evaporator and the substrate, and the system was pumped down to about 6.6 $\times$ 10^{-3} Pa after ion bombardment. The chromium metal was evaporated with N$_2$ (99.8% pure) introduced into the chamber. The deposition rate obtained was about 0.1 μm/min.

The depositions were carried out by these authors under different conditions of reactive nitrogen gas pressure, substrate bias voltage, and ion current density, and the influence of these preparation conditions on the crystal orientation, morphology, and microhardness of the Cr-N films obtained was investigated.

Various techniques (hollow discharge, magnetron sputtering, etc.) have been used to improve ionization, in reactive ion plating, and a summary of recent reports is presented in Table 4.6.

4.6   REACTIVE ION BEAM SPUTTERING

In the reactive deposition methods described in the preceding sections for the preparation of compound thin films, the metal species is deposited in the presence of a background gas (e.g., oxygen, nitrogen). Hence only external variables such as gas pressure, flow ratios, and target power are controlled. Reactive ion beam sputter (RIBS) deposition, which is a modification of general ion beam sputter deposition (material is sputtered onto a substrate by a collimated ion beam striking the target as described in Section 2.7), permits greater control over many deposition parameters (metal atom arrival rate, direction of arrival and also ion energy, angle of incidence of the beam, total pressure in the chamber, etc.), which can be varied independently over a wide range. Again thin film formation occurs in a field-free region, isolated from the plasma generation and the primary sputtering process; hence plasma interactions at the target and growing film are avoided. Moreover, sputtering under high vacuum conditions is possible.

Reactive IBS differs from other reactive deposition methods in fact that the reactive species can be introduced either in the ion beam or in the gaseous phase. With reactive gas introduced in the ion beam, the reactive sputtering and compound film formation are accomplished by:

1.   Sputtering the compound target/solid component of the binary compound with the reactive ion beam or a mixture of inert and reactive ions.

2.   Sputtering the solid component of a binary compound with an inert ion beam; a reactive ion beam of the second component, generated from a separate

source, is directed onto the growing film on the substrate. (An example of a basic dual-ion beam deposition is given in Chapter 3; see Ref. 63.)

Reactive ion beam sputtering with reactive species in the gaseous phase is accomplished when the deposition proceeds with the inert gas bombarding the target in the presence of the reactive gas background pressure (reactive gas usually is admitted in the vicinity of the substrate).

The three different processes for introducing the reactive species are shown schematically in Figure 4.22.

Reports of thin film materials prepared by RIBS generally include a-Si:H (a hydride), as well as several oxides (ITO, SiO_2, Al_2O_3) and nitrides (Si_3N_4, AlN, NbN). Deposition of Si_3N_4 has been extensively studied by many authors and is applicable to other nitrides. Since molecular nitrogen does not react with silicon [96], the introduction of the reactive ionic species in the sputtering beam (Figure 4.22a,b) is used for the preparation of stoichiometric Si_3N_4 films. Also since molecular hydrogen does not react directly with Si [97], and the hydrogen intake from the residual gas atmosphere of hydrogen for direct IBS with argon is low [98], reactive ion beam sputtering in an atmosphere of H_2 for the preparation of amorphous hydrogenated (silicon) alloys is not commonly used. RIBS of the metal target in the presence of oxygen background has been extensively used for the preparation of oxides, including vanadium oxide [99], and yttrium oxide [100].

A summary of reactive ion beam sputtering is given in Table 4.7.

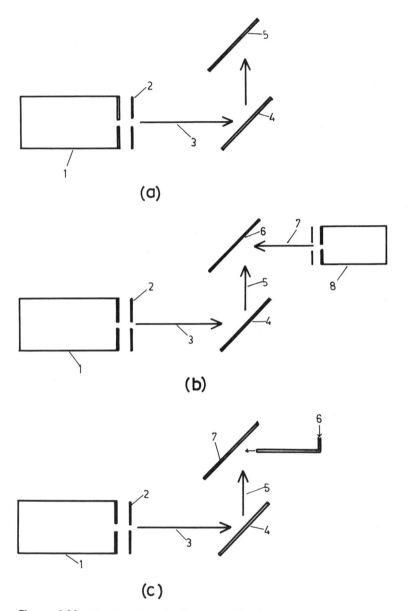

Figure 4.22 Simple schematic diagrams of the three ways of introducing the reactive species in reactive ion beam sputtering: 1, Target ion source; 2, ion extractor; 4, target. Additional components as follows: (a) 3, reactive or inert + reactive gas ion beam; 5, Substrate; (b) 3, inert gas ion beam; 5, sputtered material; 6, substrate; 7, reactive gas ion beam; 8, substrate ion source; and (c) 3, inert gas ion beam; 5, sputtered material; 6, reactive gas; 7, substrate.

Table 4.7 Reactive Ion Beam Sputtering

Film	Target	Ion beam energy (keV)/or acceleration voltage (kV)	Ion source and reactive ion	Pressure during deposition	Deposition rate	Substrate and substrate temperature (°C)	Remarks, if any	Ref.
ITO	Varied from 100% SnO_2 to 100% In_2O_3	1.5 keV	Ar, Xe, or Kr, O_2 reactive gas	$(1-7) \times 10^{-5}$ torr; O_2 balanced to 10^{-4} torr with Ar, Xe, or Kr	80 Å/min	Single-crystal InP and GaAs 27–277	Optimum composition, 8% SnO_2 + 92% In_2O_3, ITO/p-type InP and ITO/p-type GaAs solar cells.	101
PZT	PZT hot pressed disks	2 kV	Ar ion, O_2 atmosphere	2×10^{-4} torr	35 Å/min	NiCr-Au film on Pyrex or Invar. 750	Film was characterized for composition, crystal structure, dielectric, and ferroelectric properties.	102
a-Si:H	Si	10 keV	Ion beam of Ar or 50/50 Ar and H_2	With pure Ar, working pressure of 7.5×10^{-6} torr H_2 is maintained	10–15 nm/min	Doped Si and Al precoated quartz	Hydrogen incorporation was much higher for films prepared with mixed beams than with residual gas.	98

Table 4.7 (Continued)

Film	Target	Ion beam energy (keV)/or acceleration voltage (kV)	Ion source and reactive ion	Pressure during deposition	Deposition rate	Substrate and substrate temperature (°C)	Remarks, if any	Ref.
a-Si:H	High purity (>99.999%) undoped Si	500 eV	Ion beam of Ar and H_2	10^{-4} torr	0.6 μm/h 25–600	Heatable substrates	Gives high resistivity films ($\rho \geq 10^9$ Ωcm); low density of defect states in the bandgap. Deposition rate is for 3:1 H_2/Ar mix.	103
a-Si:H	Crystalline Si wafer	1–5 kV	Ar ion beam, H_2 atmosphere	3.7×10^{-5} torr	1.1 Å/s at 4 mA and 1.8 Å/s at 6 mA (acceleration voltage, 3.5 kV)	Corning 7059 glass plates and high resistivity Si wafers 130–350	Dopant in the target was boron or phosphorus. Hydrogen content was found to be a major factor governing film properties.	104
AlN$_x$O$_y$	Al	5 kV	N_2 ion beam, O_2 reactive gas	10^{-5}–10^{-4} torr	15 Å/min	Signle-crystal GaAs 300	For the fabrication of metal–insulator–semiconductor (MIS) based devices.	105
SiN$_x$O$_y$ AlN$_x$O$_y$	Si Al		N_2 ion beam, O_2 reac-	10^{-5}–10^{-4} torr	10–15 Å/min	Si substrate 300	Pure molecular oxygen introduced in the vicinity of the substrate to control y/x ratio. For the fabrication of MIS	106

SiO_xN_y	Polycrystalline Si	10 keV	tive gas Ions of N_2/O_2 gas mixtures	10^{-5} torr			structures. Angle of incidence to the target, 45°. Dependence of film properties on oxygen concentration in N_2/O_2 mixtures was studied.	107
a-Si:H	Si wafer	800 eV	Ion beam from a 50/50 mixture of high purity Ar and H_2 (99.999% pure) or with Ar beam only	1.5×10^{-5} torr	0.3 μm/h	300	Ion beam bombarded the target at an angle of 45°. Substrate can be rotated and heated with radiation. Chamber was backfilled with hydrogen when sputtered with argon beam.	108
Y_2O_3	Y_2O_3	2 kV	Ar ion beam, O_2 reactive gas			Diamond-turned Al	Presence of oxygen to enhance film stoichiometry was studied. Y_2O_3 was investigated as a protective coating that meets optical and environmental requirements.	100
AlN	Al	100–500 eV (substrate source)	Ar and N_2				Argon ion beam sputtering with simultaneous bombardment by N_2 ions was studied. N/Al composition ratio in the film increases linearly with N/Al arrival ratio up to 1.0. Above this, excess nitrogen is rejected from the film.	109

Table 4.7 (Continued)

Film	Target	Ion beam energy (keV)/or acceleration voltage (kV)	Ion source and reactive ion	Pressure during deposition	Deposition rate	Substrate and substrate temperature (°C)	Remarks, if any	Ref.
YBCO	Metallic Y and Cu		O_2 as reactive gas	Reactive gas pressure, 10^{-4} torr		(100) Sr TiO_3	Barium flux was obtained from the resistively heated source containing either Ba or BaF_2. Two ion sources were directed independently on metallic Y and Cu. Composition of the films obtained was near 1:2:3 stoichiometry. Superconductivity onset was at 93 K with zero resistance at 90 K.	130
$Ni_{53}Ti_{47}$	$Ni_{49}Ti_{51}$	Highest energy of the ions, 1.5 keV	N_2 for reactive deposition	5×10^{-3} Pa	0.1 nm/s	Stainless steel	Substrate was attached to a water-cooled rotating holder. Sputtering with 1.5 keV, N^+ ions gave amorphous $Ni_{53}Ti_{47}$ films with a high content of nitrogen and containing small (5 nm) TiN precipitates.	131
WN_x	W (99.9%)		Ar and N_2	2×10^{-4} torr	16.3 Å/min	n-type GaAs, (100) orientation	Argon ion beam was used for sputtering and nitrogen ion for reactive deposition. Nitrogen concentration in the films was found to be almost proportional to the nitrogen ion dose. Deposited films were characterized by X-ray photoelectron spectroscopy and X-ray diffraction.	132

REFERENCES

1. F. Kitagawa and K. Takaharshi, *Thin Solid Films*, 77: 273 (1981).
2. T. Takagi, *Thin Solid Films*, 92: 1 (1982).
3. K. G. Gunther, in *The Use of Thin Films in Physical Investigations* (J. C. Anderson, Ed.), Academic Press, New York, 1966, p. 213.
4. R. Glang, in *Handbook of Thin Film Technology* (L. I. Maissel and R. Glang, Eds.), McGraw-Hill, New York, 1970, p. 1–81.
5. G. Tschulena, *Thin Solid Films*, 39: 175 (1976).
6. K. C. Chi and R. F. Bunshah, *Thin Solid Films*, 72: 285 (1980).
7. (a) K. C. Chi, R. O. Dillion, R. F. Bunshah, S. Alterovitz, and J. A. Woolam, *Thin Solid Films*, 54: 259 (1978). (b) J. George and K. S. Joseph, *J. Phys. D: Appl. Phys.*, 15: 1109 (1982).
8. J. George and M. K. Radhakrishnan, *Solid State Commun.*, 33: 987 (1980).
9. (a) J. George and K. S. Joseph, *Solid State Commun*, 48: 601 (1983). (b) J. George and K. S. Joseph, *J. Phys. Chem. Solids*, 45: 341 (1984).
10. A. J. Perry and H. K. Pulker, *Thin Solid Films*, 124: 323 (1985).
11. J. George and B. Pradeep, *Solid State Commun.*, 56: 117 (1985).
12. A. Gupta, P. Gupta, and V. K. Srivastava, *Thin Solid Films*, 123: 325 (1985).
13. (a) J. George and T. I. Palson, *Thin Solid Films*, 127: 233 (1985). (b) J. George and T. I. Palson, *Solid State Commun.*, 58: 605 (1986).
14. (a) J. George, T. I. Palson, and K. S. Joseph, *Solid State Commun.*, 64: 161 (1987). (b) J. George, K. S. Joseph, B. Pradeep, and T. I. Palson, *Phys. Stat. Sol. (a)*, 106: 123 (1988).
15. S. Naseem, I. A. Rauf, K. Hussian, and N. A. Malik, *Thin Solid Films*, 156: 161 (1988).
16. R. F. Bunshah and A. C. Raghuram, *J. Vac. Sci. Technol.*, 9: 1389 (1972).
17. R. F. Bunshah, in *Science and Technology of Surface Coatings* (B. N. Chapman and J. C. Anderson, Eds.), Academic Press, London, 1974, p. 361.
18. R. F. Bunshah, *Thin Solid Films*, 80: 255 (1981).
19. H. S. Randhawa, M. D. Mathews, and R. F. Bunshah, *Thin Solid Films*, 83: 267 (1981).
20. (a) K. S. Joseph, Ph.D. thesis, University of Cochin, 1983. (b) B. E. Jacobson, C. V. Deshpandey, H. J. Doerr, A. A. Karim, and R. F. Bunshah, *Thin Solid Films*, 118: 285 (1984).
21. K. L. Chopra, V. Aggarwal, V. D. Vankar, C. V. Deshpandey, and R. F. Bunshah, *Thin Solid Films*, 126: 307 (1985).
22. J. George, B. Pradeep, and K. S. Joseph, *Rev. Sci. Instrum.*, 57: 2355 (1986).
23. A. Milch, *Thin Solid Films*, 17: 231 (1973).
24. J. George, B. Pradeep, and K. S. Joseph, *Thin Solid Films*, 148: 181 (1987).
25. (a) J. George, B. Pradeep, and K. S. Joseph, *Phys. Stat. Sol. (a)*, 100: 513 (1987). (b) J. S. Yoon, H. J. Doerr, C. V. Deshpandey, and R. F. Bunshah, *Thin Solid Films*, 181: 603 (1989). (c) Y. Bando, T. Terashima, K. Iijima, K. Yamamoto, K. Hirata, T. Takada, K. Kamjijake, and H. Terauchi, *Thin Solid Films*, 181: 147 (1989). (d) T. Terashima, K. Iijima, K. Yamamoto, J. Takada, K. Hirata, H. Mazaki, and Y. Bando, *J. Cryst. Growth*, 95: 617 (1989). (e) S. Prakash, D. M. Umarjee, H. J. Doerr, C. V. Deshpandey, and R. F. Bunshah, *Appl. Phys. Lett.*, 55: 504 (1989). (f) K. Iijima, T. Tershima, K. Yamamoto, K. Hirata, and Y. Bando, *Appl. Phys. Lett.*, 56: 527 (1990). (g) K. Terashima, K. Eguchi, T. Yoshida, and K. Akashi, *Appl. Phys. Lett.*, 52: 1274 (1988). (h) T. Yoshida,

K. Akashi, A. Kawasaka, and K. Nakagawa, *J. Mater. Sci.*, 14: 1624 (1979). (i) S. Takeuchi, T. Okasa, T. Yoshida, and K. Akashi, *Proceedings of the 8th International Symposium on Plasma Chemistry* (K. Akashi and A. Kinbava, Eds.), 1987, p. 1928.

26. H. Shinno, M. Fukutomi, M. Fujitsuka, and M. Okada, *J. Nuclear Mater.*, 133 and 134: 749 (1985).

27. H. Shinno, M. Fukutomi, Y. Sakai, M. Fujitsuka, Y. Yamauchi, T. Shikama, and M. Okada, Private communication.

28. H. Randhawa and P. C. Johnson, *Surface Coatings Technol.*, 31: 303 (1987).

29. P. C. Johnson and H. Randhawa, *Surface Coatings Technol.*, 33: 53 (1987).

30. H. Randhawa, *Thin Solid Films*, 153: 209 (1987).

31. H. Freller and H. Haessler, *Thin Solid Films*, 153: 67 (1987).

32. O. A. Johnson, J. H. Dontja, and R. L. D. Senner, *Thin Solid Films*, 153: 75 (1987).

33. E. Erturk and H. J. Heuvel, *Thin Solid Films*, 153: 135 (1987).

34. (a) P. J. Martin, D. R. Mckenzie, R. P. Netterfield, P. Swift, S. W. Filipczuk, K. H. Muller, C. G. Pacey, and B. James, *Thin Solid Films*, 153: 91 (1987). (b) T. D. Schemmel, R. L. Cunningham, and H. Randhawa, *Thin Solid Films*, 181: 597 (1989). (c) H. Randhawa, *J. Vac. Sci. Technol.*, A7: 234 (1989). (d) I. I. Aksenov, V. G. Padalka, N. S. Repalov, and V. M. Khoroshikh, *Sov. J. Plasma Phys.*, 6: 173 (1980).

35. T. Abe and T. Yamashina, *Thin Solid Films*, 30: 19 (1975).

36. (a) D. K. Hohnke, D. J. Schmatz, and M. D. Hurley, *Thin Solid Films*, 118: 301 (1984). (b) S. Berg, T. Larsson, C. Nender, and H.-O. Blom, *J. Appl. Phys.*, 63: 887 (1988). (c) T. Larsson, S. Berg, and H.-O. Blom, *Thin Solid Films*, 172: 241 (1989).

37. J. L. Vossen and J. J. Cuomo, in *Thin Film Processes* (J. L. Vossen and W. Kern, Eds.), Academic Press, New York, 1978.

38. W. D. Sproul, *J. Vac. Sci. Technol.*, A3: 580 (1985).

39. W. D. Sproul and J. R. Tomashek, U. S. Patent, 4,428,811 (1984).

40. W. D. Sproul, *Thin Solid Films*, 107: 141 (1983).

41. B. D. Johnson, J. E. Sundgren, and L. E. Greene, *J. Vac. Sci. Technol.*, A3: 303 (1985).

42. T. Serikawa and A. Okamoto, *J. Vac. Sci. Technol.*, A3: 1788 (1985).

43. R. Schachter, M. Viscogliosi, J. Baumann, and P. M. Raccah, *J. Appl. Phys.*, 58: 332 (1985).

44. J. A. Thornton and A. D. Jonath, *Conf. Rec. IEEE Photovoltaic Spec. Conf.*, 12: 549 (1976).

45. G. N. Parsons, J. W. Cook, Jr., G. Lucovsky, S. Y. Lin, and M. J. Mantini, *J. Vac. Sci. Technol.*, A4: 470 (1986).

46. R. A. Rudder, J. W. Cook, Jr., and G. Lucovsky, *Appl. Phys. Lett.*, 45: 887 (1984).

47. (a) R. A. Rudder, J. W. Cook, Jr., and G. Lucovsky, *J. Vac. Sci. Technol.*, A3: 567 (1985). (b) A. G. Spencer, K. Oka, R. P. Howson, and R. W. Lewin, *Vacuum*, 38: 857 (1988). (c) T. M. Pang, M. Scherer, B. Heinz, C. Williams, and G. N. Chaput, *J. Vac. Sci. Technol.*, A7: 1254 (1989).

48. (a) K. Char, A. D. Kent, A. Kapitulnik, M. R. Beasley, and T. H. Geballe, *Appl. Phys. Lett.*, 51: 1370 (1987). (b) K. Setsune, M. Kitabatake, T. Matsushima, Y. Ichikawa, H. Adachi, and K. Wasa, *Proceedings of the SPIE*, Santa Clara, 1989.

49. M. D. Ambersley and C. W. Pitt, *Thin Solid Films*, 80: 183 (1981).

50. J. E. Sundgren, B. O. Johansson, and S. E. Karlsson, *Thin Solid Films*, 80: 77 (1981).

51. N. Savides, D. R. McKenzie, and R. C. McPhedran, *Solid State Commun.*, 48: 189 (1983).
52. M. Scherer and P. Wirz, *Thin Solid Films*, 119: 203 (1984).
53. A. K. Dua, V. C. George, R. P. Agarwala, and R. Krishnan, *Thin Solid Films*, 121: 35 (1984).
54. A. J. P. Theuwissen and G. J. Declerck, *Thin Solid Films*, 121: 109 (1984).
55. M. K. Hibba, B. O. Johansson, J. E. Sundgren, and U. Helmersson, *Thin Solid Films*, 122: 115 (1984).
56. W. D. Wiggins, C. R. Aita, and F. S. Hickernell, *J. Vac. Sci. Technol.*, A2: 322 (1984).
57. I. Petrov, V. Orlinov, and A. Misiuk, *Thin Solid Films*, 120: 55 (1984).
58. P. K. Srivastava, T. V. Rao, V. D. Vankar, and K. L. Chopra, *J. Vac. Sci. Technol.*, A2: 1261 (1984).
59. R. B. Van Doner, D. D. Bacon, and W. R. Sinclair, *J. Vac. Sci. Technol.*, A2: 1257 (1984).
60. S. V. Krishnasway, W. A. Hester, J. R. Szedon, M. H. Francombe, and M. M. Driscoll, *Thin Solid Films*, 125: 291 (1985).
61. M. Milic, M. Milosavejevic, N. Bibic, and T. Nenadovic, *Thin Solid Films*, 126: 319 (1985).
62. G. Griffel, S. Ruschin, A. Hardy, M. Itzkovitz, and N. Croitoru, *Thin Solid Films*, 126: 185 (1985).
63. E. Leja, T. Stapinski, and K. Marszalek, *Thin Solid Films*, 125: 119 (1985).
64. T. D. Moustakas, H. Paul Maruska, and R. Friedman, *J. Appl. Phys.*, 58: 983 (1985).
65. N. Miyata and H. Kitahata, *Thin Solid Films*, 125: 33 (1985).
66. S. J. Jiang and C. G. Granqvist, *Proc. SPIE*, 562: 129 (1985).
67. M. Jachimowski, A. Brudnik, and H. Czternastek, *J. Phys. D. Appl. Phys.*, 18: L145 (1985).
68. D. Schalch, A. Scharmann, and A. Weib, *Thin Solid Films*, 124: 351 (1985).
69. K. Steenback, *Thin Solid Films*, 123: 239 (1985).
70. G. L. Harding, *Sol. Energy Mater.*, 12: 169 (1985).
71. J. Rostworaski and R. R. Parsons, *J. Vac. Sci. Technol.*, A3: 491 (1985).
72. R. Swanpoel, P. S. Swart, and H. Aharoni, *Thin Solid Films*, 128: 191 91985).
73. A. Nouhi and R. J. Stirn, *J. Vac. Sci. Technol.*, A4: 403 (1986).
74. W. D. Sproul and R. Rothstein, *Thin Solid Films*, 126: 257 (1985).
75. C. M. Gilmore, C. Quinn, E. F. Stelton, C. R. Gossett, and S. B. Qadri, *J. Vac. Sci. Technol.*, A4: 2598 (1986).
76. H. A. Jehn, S. Hofmann, V. E. Ruckborn, and W. D. Munz, *J. Vac. Sci. Technol.*, A4: 2701 (1986).
77. S. Nishikawa, *Thin Solid Films*, 135: 219 (1986).
78. D. S. Yee, J. J. Cuomo, M. A. Frisch, and D. P. E. Smith, *J. Vac. Sci. Technol.*, A4: 381 (1986).
79. J. B. Webb, C. Halpin, and J. P. Noad, *J. Vac. Sci. Technol.*, A4: 379 (1986).
80. N. Fortier and R. R. Parsons, *J. Vac. Sci. Technol.*, A4: 83 (1986).
81. (a) G. Hakansson, J. E. Sundgren, D. McIntyre, J. E. Greene, and W. D. Munz, *Thin Solid Films*, 153: 55 (1987). (b) K. H. Seidel, K. Reichelt, W. Schaal, and H. Dimigen, *Thin Solid Films*, 151: 243 (1987).
82. M. Kobayashi and Y. Doi, *Thin Solid Films*, 54: 67 (1978).

83. R. F. Bunshah, *Physical Vapor Deposition of Metals, Alloys, and Compounds: New Trends in Materials Processing*, American Society for Metals, Metals Park, Oh, 1976, p. 200.

84. Y. Murayama, *J. Vac. Sci. Technol.*, 12: 818 (1975).

85.. (a) M. Fukutomi, M. Fujitsuka, and M. Okada, *Thin Solid Films*, 20: 283 (1984). (b) J. Edlinger, J. Ramm, and H. K. Pulker, *Thin Solid Films*, 175: 207 (1989). (c) H. K. Pulker, W. Haag, and E. Moll, U.S. Patent 4,619,748 (1986). (d) D. Wang and T. Oki, *Thin Solid Films*, 185: 219 (1990).

86. J. N. Avaritsiotis and R. P. Howson, *Thin Solid Films*, 80: 61 (1981).

87. R. P. Howson and M. I. Ridge, *Thin Solid Films*, 77: 119 (1981).

88. M. Fukutomi, M. Fujitsuka, M. Kitajima, T. Shikama, and M. Okada, *Thin Solid Films*, 80: 271 (1981).

89. J. Machet, J. Guille, P. Saulnier, and S. Robert, *Thin Solid Films*, 80: 149 (1981).

90. K. H. Kloos, E. Broszeit, H. M. Gambriel, and H. J. Schroder, *Thin Solid Films*, 96: 67 (1982).

91. T. Shikama, M. Kitajima, M. Fukutomi, and M. Okada, *Thin Solid Films*, 117: 191 (1984).

92. A. J. Perry, L. Simmen, and L. Chollet, *Thin Solid Films*, 118: 271 (1984).

93. J. M. Molarius, A. S. Korhonen, and E. O. Ristolainen, *J. Vac. Sci. Technol.*, A3: 2419 (1985).

94. L. S. Wen, X. Jiang, and C. Y. Si, *J. Vac. Sci. Technol.*, A4: 2682 (1986).

95. B. H. Hahn, J. H. Jun and J. H. Joo, *Thin Solid Films*, 153: 115 (1987).

96. H. J. Erler, G. Reisse, and C. Weissmantel, *Thin Solid Films*, 65: 233 (1980).

97. D. L. Miller, H. Lutz, H. Weisman, E. Rock, A. K. Gosh, S. Ramamoorthy, and S. Strongin, *J. Appl. Phys.*, 49: 6192 (1978).

98. C. Weissmantel, K. Bewilogua, D. Dietrich, H. J. Erler, H. J. Hinneberg, S. Klose, W. Nowick, and G. Reisse, *Thin Solid Films*, 72: 19 (1980).

99. E. E. Chain, *J. Vac. Sci. Technol.*, A4: 432 (1986).

100. B. E. Cole, J. Moravec, R. G. Ahonen, and L. B. Ehlert, *J. Vac. Sci. Technol.*, A2: 372 (1984).

101. K. J. Bachmann, H. Schreiber, Jr., W. R. Sinclair, P. H. Schmidt, F. A. Thiel, E. G. Spencer, G. Pasteur, W. L. Feldmann, and S. Harsha, *J. Appl. Phys.*, 50: 3441 (1979).

102. R. N. Castellano and L. G. Feinstein, *J. Appl. Phys.*, 50: 4406 (1979).

103. G. P. Ceasar, S. F. Grimshaw, and K. Okumura, *Solid State Commun*, 38: 89 (1981).

104. J. Saraie, M. Kobayashi, Y. Fujii, and H. Matsunami, *Thin Solid Films*, 80: 169 (1981).

105. C. Sibran, R. Blanchet, M. Garrigues, and P. Viktorovitch, *Thin Solid Films*, 103: 211 (1983).

106. M. Garrigues, R. Blanchet, C. Sibran, and P. Viktorovitch, *J. Appl. Phys.*, 54: 2863 (1983).

107. V. A. Burdovitsin, *Thin Solid Films*, 105: 197 (1983).

108. P. J. Martin, R. P. Netterfield, W. G. Sainty, and D. R. McKenzie, *Thin Solid Films*, 100: 141 (1983).

109. J. M. E. Harper, H. T. G. Hensell, and J. J. Cuomo, *J. Vac. Sci. Technol.*, A2: 405 (1984).

110. T. Takami, T. Noguchi, K. Yokoyama, and K. Hamanaka, *International Superconductivity Electronics Conference*, Tokyo, 1989, p. 9.

111. (a) K. Narasimha Rao, M. A. Murthy, and S. Mohan, *Thin Solid Films*, 176: 181 (1989). (b) T. Fuji, M. Takano, R. Katano, and Y. Bando, *J. Appl. Phys.*, 66: 3168 (1989).

112. E. Fogarassy, C. Fuchs, A. Slaoui, and J. P. Stoquert, *Appl. Phys. Lett.*, 57: 664 (1990).

113. W. Posadowski, *Thin Solid Films*, 162: 111 (1988).

114. J. L. Wallace, *J. Appl. Phys.*, 64: 6053 (1988).

115.. M. Pinarbasi, N. Maley, A. Myers, and J. R. Abelson, *Thin Solid Films*, 171: 217 (1989).

116. F. C. Stedile, B. A. S. De Barros, Jr., C. V. Barros Leite, F. L. Freire, Jr., I. J. R. Baumvol, and W. H. Schreiner, *Thin Solid Films*, 170: 285 (1989).

117. T. Larsson, H. O. Blom, S. Berg, and M. Osling, *Thin Solid Films*, 172: 133 (1989).

118. H. Dintner, R. Mattheis, and G. Voghler, *Thin Solid Films*, 182: 237 (1989).

119. S. I. Shah, *Thin Solid Films*, 181: 157 (1989).

120. D. Wruck, S. Ramamoorthy, and M. Rubin, *Thin Solid Films*, 182: 70 (1989).

121. (a) M. Y. Al-Jaroudi, H. T. G. Hentzell, S. E. Hornstrom, and A. Bengton, *Thin Solid Films*, 182: 153 (1989). (b) K. Marzalek, *Thin Solid Films*, 175: 227 (1989).

122. (a) R. W. Simon, C. E. Platt, K. P. Daly, A. E. Lee, and M. K. Wagner, *IEEE Trans. Magn.*, MAG-25: 2433 (1989). (b) H. Wiesmann, D. H. Chen, R. L. Sabatini, J. Hurst, J. Ochab, and M. W. Ruckman, *J. Appl. Phys.*, 65: 1644 (1989).

123. (a) M. M. D. Romos, J. B. Almeida, M. I. C. Ferreira, and M. P. D. Santos, *Thin Solid Films*, 176: 219 (1989). (b) K. Kubota, Y. Kobayashi, and K. Fujimoto, *J. Appl. Phys.*, 66: 2984 (1989). (c) G. C. Xiong and S. Z. Wang, *Appl. Phys. Lett.*, 55: 902 (1989).

124. D. H. Mosca, P. D. Dionisio, W. H. Schreiner, I. J. R. Baumvol, and C. Achete, *J. Appl. Phys.*, 67: 7514 (1990).

125. Y. Nakagawa and T. Okada, *J. Appl. Phys.*, 68: 556 (1990).

126. H. Z. Wu, T. C. Chou, A. Mishra, D. R. Anderson, J. K. Lampert, and S. C. Gujrathi, *Thin Solid Films*, 191: 55 (1990).

127. R. C. Baumann, T. A. Rost, and T. A. Rabson, *J. Appl. Phys.*, 68: 2989 (1990).

128. K. Bewilogua, H. J. Heinitz, B. Rau, and S. Schulze, *Thin Solid Films*, 167: 233 (1988).

129. A. Ishida, K. Ogawa, T. Kimura, and A. Takei, *Thin Solid Films*, 191: 69 (1990).

130. A. F. Hebard, R. H. Eick, A. T. Fiory, A. E. White, and K. T. Short, *Proceedings of the Northeast Regional Meeting of the Metallurgical Society*, New Brunswick, NJ, 1988, p. 23.

131. S. Raud, J. Delage, J. P. Villain, and P. Moine, *Thin Solid Films*, 181: 333 (1989).

132. J. S. Lee, C. S. Park, J. W. Yang, J. Y. Kang, and D. S. Ma, *J. Appl. Phys.*, 67: 1134 (1990).

5

Ionized Cluster Beam Methods

If two men agree on everything, you may be sure that one of them is doing all the thinking.
Lyndon Johnson

In the thin film deposition technique discussed in this chapter microaggregates (clusters) from the deposit material are utilized instead of atomic or molecular state particles, as in thermal evaporation or sputtering. At the same time, ionized cluster beam (ICB) methods take advantage of the wide variety of effects caused by the chemical activity of the materials in an ionized state and by kinetic energy (imparted by accelerating the particles in an electric field). ICB, a relatively new deposition process for the preparation of thin films, has become a widely used thin film technique yielding a variety of metal, semiconductor, and insulator films. Epitaxial films of elemental and compound materials are obtained at low temperatures by ICB.

5.1 IONIZED CLUSTER BEAM DEPOSITION

Vaporized metal clusters are formed by adiabatic expansion of metal vapor into a high vacuum region without using inert gas as a carrier gas. This idea and the first reported data related to such vaporized metal clusters and the formation of films with such clusters were reported by Takagi et al. in 1972 [1]. Here adiabatic expansion of metal, semiconductor, or insulator vapor occurs when the vapor is ejected through the nozzle of a heated crucible into a high vacuum region of about 10^{-7}–10^{-5} torr and undergoes subsequent cooling to a supersaturated state. This leads to creation of clusters, which are atomic aggregates of 100–1000 atoms. The clusters are ionized by electron bombardment from an electron emitter situated just above the nozzle and are accelerated toward the substrate by a variable high potential (0–10 kV). Neutral clusters also move in the same direction. These are thus microparticles, with a small charge-to-mass ratio. The accelerated ionized clusters bombard the substrate together with the neutral clusters, which have some kinetic energy corresponding to the ejection velocity.

The basic configuration of an ionized cluster source is shown in Figure 5.1. The material is vaporized in a special crucible, and the vapor is ejected into the high vacuum region through a small aperture in the crucible. Clusters are formed [2] as the metal atoms in the crucible collide and transfer their energies to one another. Then a supersaturated state is formed by adiabatic expansion when the atoms are ejected through the nozzle into the high vacuum region. The atoms lose their energies by collision and start to aggregate to form nuclei. Nuclei smaller than a critical size are unstable and break up, whereas nuclei larger than the critical size and in the supersaturated state grow to form clusters. The rate of growth of the clusters is maximum near the nozzle and then becomes lower. In the high vacuum region the size of most of the clusters remains constant because there is very little collision between clusters and atoms. The clusters formed are ionized by electrons emitted from a filament coil in the acceleration–electrode assembly placed coaxially in front of the crucible and kept at a negative potential with respect to the crucible. The value of this potential is so large that the clusters are ionized by electron bombardment rather than by electron capture. The ratio of the ionized clusters to the

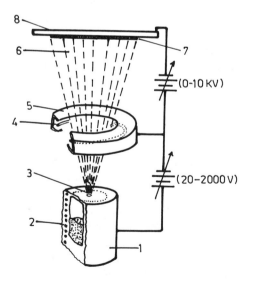

Figure 5.1   The basic configuration of an ionized cluster source: 1, crucible; 2, heating coil; 3, nozzle; 4, filament coil; 5, acceleration electrode; 6, ionized and neutral clusters; 7, substrate; 8, substrate holder.

total clusters can be adjusted by changing the electron beam current [3]. The cluster ions are accelerated opposite to the direction of flow of the injection electrons. A high negative potential is applied to the substrate to attract the ionized clusters. The energy of the ionized clusters impinging onto the substrate depends on the potential difference between the substrate and the crucible and can thus be adjusted.

Upon impact of the substrate, the kinetic energy of the accelerated ion clusters is converted to thermal energy, sputtering energy, and implantation energy. The main advantages of the ICB technique [4] are as follows.

1. In the metal phase, the substrate surface is sputtered and cleaned by ion bombardment and is deep etched.
2. The sputtered substrate material blends with the evaporant particles to form an interfacial layer.
3. Sputtering is continued during deposition.
4. Thermal energy converted from the kinetic energy of the bombarding ions is available at the surface of the substrate and the deposited layer, this gives a high surface temperature without bulk heating, resulting in increased chemical activity.
5. There is enhancement of interdiffusion and surface diffusion due to the high defect concentrations introduced (change in growth morphology.
6. Ion implantation is possible (enhancing interfacial layer formation).
7. Nucleation and deposition are influenced because of the migration of the depositing particles over the substrate surface, and this factor is an important characteristic of the ionized cluster beam deposition.

When ionized and neutral clusters are broken up into atomic state upon striking the substrate surface, their incident momentum is transformed into the surface diffusion energy for each atom. The atoms are scattered over the surface with high surface diffusion energy, and the enhanced adatom migration due to this surface diffusion energy contributes to the formation of good quality films. Enhanced adatom migration results in increased rates of island nucleation and growth [5,6].

Again in ICB the number of clusters of doubly or highly charged states is extremely small because the probability of electron collision with singly charged clusters is very small compared with the probability of collision with neutral clusters. Moreover it is assumed that doubly charged clusters are broken up into singly charged clusters by the Coloumb repulsion force [7]. This means that the charge-to-mass ratio is small for an ionized cluster (each cluster contains 500–1000 atoms but has a single charge), and this in turn reduces both space charge problems and the charging of an insulating substrate during deposition [4]. In the cluster beam, the cluster size can be taken to be about 1000 atoms; thus even if the clusters are accelerated to kilovolts, the energies of the individual atoms will be in the levels of a few electron volts. With such energies, deposition can be carried out at low temperatures, and films with good adhesion, high packing density, and good crystallinity can be prepared.

Since the first reported work on ICB, Takagi and his group have published several papers on the ICB deposition technique for preparating thin films, including epitaxial films and doped films. Several types of ICB apparatus have also been developed. Figure 5.2 is a schematic diagram of a laboratory ICB system consisting of a source with a single crucible and nozzle. The films can also be deposited at different substrate temperatures, by providing a heating arrangement [4]. Single-crucible systems, with multiple nozzles forming a curtain beam with high uniformity, have also been reported [8].

Details of the various kinds of film prepared by ICB and their salient features are summarized by Yamada and Takagi [9: Table 1]. Development of new materials by ICB technique is reported by Takagi [10]. A bibliography of recent work is given in Table 5.1.

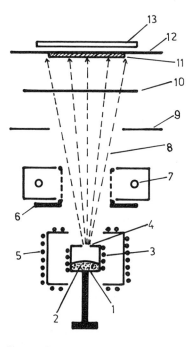

Figure 5.2 Schematic diagram of a laboratory ICB system: 1, crucible; 2, source material; 3, electron emitter for heating crucible; 4, nozzle; 5, cooling water; 6, electron accelerating electrode; 7, electron emitter for ionization; 8, ionized clusters and neutral clusters; 9, ionized cluster accelerating electrode; 10, shutter; 11, substrate; 12, substrate holder; 13, heater. (Data from Ref. 38.)

Table 5.1 Ionized Cluster Beam Deposition

Film	Substrate	Substrate Temperature (°C)	Acceleration voltage (kV)	Remarks, if any	Ref.
Al	Si (111) Si (100)	RT	0–5	Graphite crucible was used; substrate was not intentionally heated during deposition. Surface and interface of Al on Si (111) after 30 minute of annealing at 450°C were smooth. No annealing hillock and no alloy penetration were reported.	11
Cu	Glass Cr layer on glass	< 400	0–10	Acceleration voltage influenced the optical property of the films.	12
Al	Si (111) Si (100)	RT		Single-crystal Al films were obtained. Epitaxial Al films had excellent thermal and electrical stability. Epitaxial depositions of Al on GaAs, CaF$_2$, and sapphire substrates have also been done.	13
CdTe	Si (111)	250	5	Crystal properties of CdTe films showed that they are crystalline.	14
GaAs	GaAs (100)	< 400		Good crystallinity (single-crystal GaAs) was obtained with either both Ga and As beams ionized or with only the As beam ionized. Films with only Ga ionized were polycrystalline.	15
Al	Si (111) Si (100)	RT	0–5	Epitaxial Al films were obtained at low temperature. Interface between epitaxial Al film and the Si (111) was stable and well defined with no alloy penetration after annealing at 450°C for 30 minutes.	16

Table 5.1 (Continued)

Film	Substrate	Substrate Temperature (°C)	Acceleration voltage (kV)	Remarks, if any	Ref.
$Cd_{1-x}Mn_xTe$	Sapphire (0001)	300	0–7	Dual-source ICB system ionized CdTe cluster beam and neutral MnTe cluster beams. Crystalline quality was better for higher acceleration voltage of CdTe. When MnTe was ionized and accelerated, the crystalline quality improved, but composition remained the same.	17
Al	Si (111)	RT	5.0	No hillocks and valleys were found after annealing. Interface was abrupt. Crystalline quality remained stable after annealing.	18
CaF_2	Si (111)	700	1.0	Epitaxial CaF_2 films for MIS structure fabrication were obtained with CaF_2 as the insulator material.	18
Al	Si (100) n-type and p-type	RT		Graphite crucible heated to about 1550°C; substrate temperature rose as result of radiation heating of the source filament, but was below 90°C. Deposition rate was 3–5 Å. Resistivity for n-type wafer, 3–5 Ω cm, for p-type wafer, 0.07–0.13 Ω cm.	19
Au	(001) Si and (001) Si covered with thermally grown oxide		4 (Si) 1 (SiO$_2$)	Films were grown at low rates (203 Å/s) at a pressure of 3×10^{-6} torr. Deposition conditions such as ionization current, deposition rate, and substrate temperature were varied and the effect explored. Films were characterized by SEM, TEM and electron diffraction.	41

Al	SiO_2, 7000 Å thick	Unintentionally heated to a temperature <80°C	3	Pure Al (99.999%) was deposited from a carbon crucible with a nozzle of 2 mm diameter in a vacuum of 2×10^{-6} torr. An ionization voltage of 400 V and an ionization current of 100 mA were used.	42
GaAs	Si	550		Native oxide layer on the Si substrate removed at 550°C by the use of an accelerated arsenic beam. TEM and electron diffraction data suggest that single-crystal films of GaAs are formed.	43
$Hg_{1-x}Cd_xTe$	GaAs (100)	Not heated during deposition	0–5	CdTe, HgTe, and Te ingots were used as source materials, put in separate crucibles, and heated to 700, 250, and 470°C, respectively. Pressure during deposition was about 4×10^{-6} torr; deposition rate, 250–300 Å/min. Either CdTe or HgTe was ionized; band-gap could be controlled between 0.2 and 0.3 eV by adjusting the acceleration voltage for cluster ions.	44
Al	Si (100)			Structural study of the films deposied revealed a unique grain structure consisting of only two crystallographic orientations. Both were (110), but rotated through 90° with respect to each other about the (110) axis so that their (220) planes were perpendicular.	45

Wong et al. in 1985 [19] prepared SiO_2 films and later [20] prepared Al films on Si (100) wafers for Schottky barrier formation, using what they call nozzle jet beam deposition. The procedural method involved is the same as that of ICB deposition; however, the existence of clusters has not been proved in their experiments.

5.1.1 Ionized Cluster Assisted Film Deposition

Knauer and Poeschel have very recently reported [21] a deposition process which they believe to be the first truly cluster-based deposition process. They conducted cluster beam tests of the Takagi-type process, where silver and gold were evaporated in a graphite crucible. The vapor ejected through a nozzle was ionized and accelerated. Determination of the mass distribution in the resulting beam, under use of a retarding and also of a deflecting field analyzer, showed that the vapor emerging through the nozzle was constituted primarily of single atoms. Only in some cases a small quantity of clusters with up to 700 atoms was present, and these clusters never constituted more than 1% of the total vapor flux.

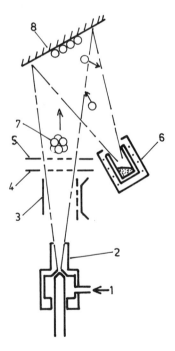

Figure 5.3 Schematic diagram of cluster-assisted film deposition: 1, argon gas; 2, nozzle; 3, ionizer; 4, mass separator; 5, accelerating electrode; 6, vapor source; 7, ionized cluster; 8, substrate. (Data from Ref. 21.)

A basic study of the cluster generation process and a theoretical evaluation was made by these authors. It was found that the clusters were formed on the crucible internal surfaces, not in the vapor stream ejecting through the nozzle (i.e., not by homogeneous nucleation and collisional growth in the vapor phase during passage through the nozzle). The results of the detailed analysis by Knauer [22] of the growth process, such as the formation of metal vapor nuclei on internal crucible surface, the growth of surface nuclei to larger clusters, and ejection of clusters from crucible surfaces confirm this heterogeneous cluster growth model. Also, only if the deposition material does not wet the crucible wall can cluster growth and ejection be expected to happen. The clusters occur consistently only when a temperature distribution is maintained within the crucible to favor cluster formation on surfaces near the crucible nozzle. Even with the most favorable temperature gradients, only about 1-2% of the vapor flux condenses into clusters. Clusters for metal deposition processes appear to be impracticable; instead an alternative process for the deposition of thin films under ionized cluster assistance has been proposed by Knauer and Poeschel [21].

A schematic diagram of the cluster-assisted film deposition is shown in Figure 5.3. Clusters are formed by homogeneous nucleation when a gas is expanded isotropically through a nozzle from a relatively high pressure into a vacuum [23]. Clusters formed are ionized by electron bombardment with currents of a few milliamperes and with energies in the range of 80-100 eV. The ion beam is then extracted and accelerated. All atomic ions (also small clusters) are removed from these beams, since these particles reach excessive levels of kinetic energy when accelerated. The deposition process involves noble gas clusters as an energizing assist to conventional vapor deposition.

Their deposition guidelines of Knauer and Poeschel were as follows.

Cluster energy	1-10 eV per atom
Size of clusters	≈ 2000 atoms
Acceleration voltage	2-20 kV
Deposition rate	10 Å/s
Impact rate	10 μA/cm^2
2000 atom clusters accelerated to 5 kV and pulsed with a duty cycle of 2%.	

The rate of arrival of energetic clusters is related to the film deposition rate, such that on an average, at least 1 eV is available for each arriving film atom.

Knauer and Poeschel have deposited fold films on GaAs substrates using this method with the assistance of ionized argon clusters, maintaining approximately the conditions stated above. They obtained films with excellent appearance and adherence, although sheet resistance did not show much difference from vacuum vapor deposition. To achieve significant improvements over vapor deposition, it is thought that the cluster kinetic energy must be raised from about 2–3 eV/atom to levels close to 10 eV/atom.

5.2 REACTIVE IONIZED CLUSTER BEAM (R-ICB) DEPOSITION

In R-ICB deposition, the standard ICB method is combined with a reactive gas at a relatively low pressure and the process described for ICB takes place. A reactive gas (e.g., oxygen, nitrogen, other reactive gases) is supplied from a nozzle near the metal vapor injection area through a controlled leak valve [24]. The reactive gas pressure is maintained at 10^{-5}–10^{-4} torr, so that the mean free path of the gas molecules is of the order of centimeters (longer than the distance between the nozzle and the substrate). Some reactive gas atoms or molecules are ionized together with the vaporized metal clusters by the electrons emitted from the tungsten filament in the ionization system above the crucible and are accelerated toward the substrate. Ionized and neutral particles of the metal vapors and the gas impinge on the substrate, and reaction takes place at the substrate surface aided, by enhanced migration effects, to form oxide, nitride, or other compound films. A schematic diagram of the R-ICB apparatus is shown in Figure 5.4.

Introducing reactive gas in the vicinity of the substrate increases the ratio of the arrival rate of reactive gas to metal atoms on the substrate, and the portion of the ionized reactive gas in the total flux will be very low. Stoichiometric Al_2O_3 films have been prepared by the R-ICB method with the oxygen introduced in the vicinity of the substrate, where the portion of the ionized oxygen in the total flux is estimated to be as low as 0.1% [25].

With a working reactive gas pressure of 10^{-5}–10^{-4} torr, R-ICB is a useful method for the preparation of good quality, highly reliable films. Low temperature deposition of crystalline insulators on semiconductor substrates of various kinds is of interest in the fabrication of metal-insulator-semiconductor (MIS) structures for integrated circuit application. Although plasma CVD deposition (discussed in Chapter 6) is a low temperature process, the surface of the compounds will be damaged by plasma bombardment [26], and a good semiconductor insulator interface could not be obtained in this method. Again, the R-ICB technique has been successfully developed for the low temperature deposition on silicon of single-crystalline insulating films [27].

Details of the early work on R-ICB are reported in references 9 and 10. A summary of the important materials from which films have been prepared recently by the R-ICB technique is given in Table 5.2.

5.2.1 Dual-beam Deposition Process

The simultaneous use of an ionized cluster beam system and a microwave ion source for the preparation of AlN and Al_2O_3 films has been reported [36,37]. This technique has the advantage of controlling the operation conditions separately for

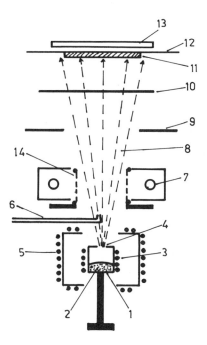

Figure 5.4 Schematic diagram of the R–ICB apparatus: 1, crucible; 2, source material; 3, electron emitter for heating crucible; 4, nozzle; 5, cooling water; 6, reactive gas; 7, electron emitter for ionization; 8, ionized and neutral clusters; 9, ionized cluster accelerating electrode; 10, shutter; 11, substrate; 12, substrate holder; 13, heater; 14, electron accelerating eleectrode. (Data from ref. 24.)

the ICB system and the microwave ion source. In addition, it has been possible to prepare oxide and nitride films at low substrate temperatures. The film properties can be controlled by adjusting the deposition conditions.

In principle, the setup is the same as that discussed earlier (Figure 4.22b), where the source material (produced by sputtering) and the reactive gas ions from an ion source strike the substrate simultaneously. Here the ionized clusters of the source material from the ICB system and the reactive gas ions from a microwave ion source are incident simultaneously on the substrate, to form the required film.

The deposition conditions given in Table 5.3 are those of Takaoka et al. [37]. Their results established that these films were chemically and thermally stable and also high electrical resistivity and breakdown voltage.

The recent status of work on film deposition by standard ICB and reactive ICB (R-ICB), as well as modifications of ICB using simultaneous ion deposition source, are given in Reference 38.

Table 5.2 Reactive Ionized Cluster Beam Deposition

Film	Material	Substrate	Substrate temperature (°C)	Reactive gas	Gas pressure	Remarks, if any	Ref.
Cu_2O	Cu	Glass	<400	O_2	1.4×10^{-4} torr	Acceleration voltage was 1–10 kV, copper of 99.999% purity. X-Ray diffraction indicated that film obtained is a mixture of Cu and Cu_2O.	12
SiO_2	SiO_2 grains	n-type Si(100) wafers	300	O_2	7.5×10^{-5} torr	Acceleration voltage can be varied from 0 to 10 kV, SiO_2 grains were 99.999% pure. Substrate chemically cleaned before deposition. Crucible was heated to 1500°C during deposition, which gave a deposition rate of 0.8 Å/s. Index of refraction and IR absorption spectrum of the films depended strongly on ionization current and acceleration voltage.	19
Al_2O_3	Al	p-type Si(111)	100–600	O_2	1×10^{-4} torr	Acceleration voltage was 0–5 kV, aluminum of 99.999% purity. Stoichiometry of the film was confirmed to be Al_2O_3 by Rutherford back-scattering spectroscopy. Oxygen was introduced near the substrate.	27
a-Si:H	Si	Glass	220	H_2	10^{-5}–10^{-4} torr	Acceleration voltage 1–10 kV. Hydrogenated amorphous silicon films were prepared. Doping was done in hydrogen gas containing PH_3 or B_2H_6. Deposited films had good adhesion, smooth surface finish, and thermally stable characteristics.	28

Film	Source	Substrate	Temp.	Gas	Pressure	Remarks	Ref.
SiO_2	SiO_2 powder	Si	200	O_2	3×10^{-5} torr	Films of SiO_2 had a refractive index of 1.46. The infrared spectrum peaked at 1000 cm^{-1}.	29
FeO_2	Fe	Glass	150	O_2	1×10^{-4} torr	Acceleration voltage was 0–3 kV, iron exceeded 99.9% purity. A heterostructure-type solar cell was fabricated using this method. Electrical and optical properties of iron oxide prepared under different deposition conditions were investigated.	30
SiC	Si	Fused quartz, Si(111) and Si(100) wafers	200–700	C_2H_2	10^{-5}–10^{-4} torr	Acceleration voltage was 0–8 kV, silicon exceeded 99.999% purity. Amorphous SiC and cubic β-SiC films were obtained.	31
FeO_x	Fe	n-type Si (100)	350	O_2	2×10^{-4} torr	Acceleration voltage was 0–1 kV, iron purity exceeded 99.9%, FeO_x/Si heterostructures fabricated.	32
TiO_2	Ti	Microscope glass slides and Si wafers	300–350	O_2	1×10^{-5} to 2×10^{-4} torr	Acceleration voltage, 0–6.6 kV, titanium purity 99.9%. Films prepared at an oxygen pressure of 1–2 $\times 10^{-4}$ torr gave TiO_2 stoichiometric films. By increasing the ionization current from 0 to 40 mA, the refractive index of the films was increased from 2.0 to 2.6.	33
Al_2O_3	Al	Si	100	O_2	5×10^{-5} torr	Acceleration voltage, 1 kV, aluminum 99.999% pure. Film composition examined by RBS. Oxygen introduced in the vicinity of the substrate. Al/Al_2O_3/Si MIS cells, p-type Si (111). Electrical characteristics were investigated.	34

Table 5.2 (Continued)

Film	Material	Substrate	Substrate temperature (°C)	Reactive gas	Gas pressure	Remarks, if any	Ref.
Al_2O_3	Al	Polished Si (100) wafers	100–600	O_2	5×10^{-5} torr	Acceleration voltage was 0.25-3 kV, aluminum was 99.99% pure. A protective oxide layer was first formed by an acid mixture and then removed in vacuum by direct current heatng of the sample at 850°C for 10 minutes to get a clean and ordered Si (100) substrate.	35
Y-Ba-Cu-O	Y, Ba, and Cu	MgO (100)	600–650	O_2	1.3×10^{-2} Pa	Activated ocygen generated by silent discharge and ejected to the substrate. Critical temperature ($R = 0$) of the as-grown film was above the boiling point of liquid N_2.	46
TiO_2	Ti (99.9% pure)	Quartz glass, n- and p-type, (100) Si	250–350	O_2	2×10^{-4} torr	Titanium dioxide films 100–450 nm thick were prepared. After deposition, films were annealed at 600°C in dry O_2 ambient. Electrical properties of films and TiO_2–Si interface were analyzed. Films exhibited a high breakdown voltage and low interface state density (1-2×10^{11} cm^{-2} eV^{-1}) at the midgap.	47

Table 5.3 Dual-Beam Deposition of AlN and Al₂O₃

Condition	AlN	Al₂O₃
Metal	Al (99.9999% pure)	Al (99.9999% pure)
Crucible temperature, °C	1520	1570
Reactive gas (introduced into the plasma discharge of the microwave ion source)	N_2 (99.999% pure)	O_2 (99.99% pure)
Gas pressure, torr	2×10^{-5}	2×10^{-5}
Acceleration voltage, kV	0.5–2	0.5–2
Incident energy of the gas ion, eV	100–500	100–500
Substrate	Sapphire (0001) and p-type Si (111)	Sapphire (0001) and p-type Si (111)
Substrate temperature, T_s – °C	100 (without substrate heating)	100

Two brief reviews of the ICB method of film deposition have been published very recently [39,40] and will be useful reading.

REFERENCES

1. T. Takagi, I. Yamada, and S. Kobiyama, *Proceedings of the 2nd International Conference on Ion Sources*, Vienna 1972, 1972, p. 790.
2. I. Yamada, H. Takaoka, H. Inokawa, H. Usui, S. C. Cheng, and T. Takagi, *Thin Solid Films*, 92: 137 (1982).
3. T. Takagi, Y. Imada, and A. Sasaki, *Inst. Phys. Conf. Ser.*, 38: 229 (1978).
4. T. Takagi, Y. Imada, and A. Sasaki, *Thin Solid Films*, 39: 207 (1976).
5. T. Takagi, *Thin Solid Films*, 92: 1 (1982).
6. T. Takagi, *J. Vac. Sci. Technol.*, A2: 382 (1984).
7. B. J. C. Burrows, P. G. Dawson, G. A. G. Mosson, E. S. Tay, and H. H. H. Watson, *5th Smposium on Fusion Technology* (UK), 1968, Cullem Laboratory Rep. 68/144, paper 51.
8. T. Takagi, I. Yamada, and H. Takaoka, *Surface Sci.*, 106: 544 (1981).
9. I. Yamada and T. Takagi, *Thin Solid Films*, 90: 105 (1981).
10. T. Takagi, *Mater. Res. Soc. Symp. Proc.*, 27: 501 (1984).
11. I. Yamada, H. Inokawa, and T. Takagi, *J. Appl. Phys.*, 56: 2746 (1984).
12. H. Takaoka, T. Yoshimura, and T. Takagi, *Proceeding of the 8th ISIAT Syumposium*, ISIAT '84, TOkyo, 1984, p. 263.
13. I. Yamada, H. Usui, H. Inokawa, and T. Takagi, *17th Conf. Solid State Devices and Materials*, Tokyo, 1985, p. 313.
14. T. Takagi, H. Takaoka, Y. Kuriyama, and K. Matsubara, *Thin Solid Films*, 126: 149 (1985).

15. P. R. Younger, *J. Vac. Sci. Technol.*, A3: 588 (1985).
16. I. Yamada, H. Ionokawa, and T. Takagi, *Thin Solid Films*, 124: 179 (1985).
17. T. Koyanagi, K. Matsubara, H. Takaoka, and T. Takagi, *Proceedings of the 9th ISIAT Symposium, ISIAT '85*, Tokyo, 1985, p. 397.
18. I. Yamada, H. Takaoka, H. Usui, and T. Takagi, *J. Vac. Sci. Technol.*, A4: 722 (1986).
19. J. Wong, T. M. Lu, and S. Mehta, *J. Vac. Sci. Technol.*, B3: 453 (1985).
20. J. Wong, S. N. Mei, and T. M. Lu, *Appl. Phys. Lett.*, 50: 679 (1987).
21. W. Knauer and R. L. Poeschel, *J. Vac. Sci. technol.*, B6: 456 (1988).
22. W. Knauer, *J. Appl. Phys.*, 62: 841 (1987).
23. O. F. Hagena, in *Molecular Beams and Low Density Gas Dynamics* (P. P. Wegener, ed.), Dekker, New York, 1974, p. 93.
24. T. Takagi, I. Yamada, K. Matsubara, and H. takaoka, *J. Cryst. Growth*, 45: 318 (1978).
25. H. Usui, T. Fujino, I. Takashita, I. Yamada, and T. Takagi, *Proceedings of the International Workshop on ICBT*, Tokyo-Kyoto, 1986, p. 207.
26. D. C. Cameron, L. D. Irving, C. R. Whitehouse, J. Woodward, G. T. Brown, and B. Cockayne, *Thin Solid Films*, 103: 61 (1983).
27. T. Fujino, M. Shibutani, H. Inokawa, H. Usui, I. Yamada, and T. Takagi, *Proceedings of the 9th ISIAT Symposium, ISIAT'85*, Tokyo, 1985, p. 307.
28. I. Yamada, I. Nagai, M. Horic, and T. Takagi, *J. Appl. Phys.*, 54: 1583 (1983).
29. Y. Minowa, K. Yamanishi, and K. Tsukamoto, *J. Vac. Sci. Technol.*, B1: 1148 (1983).
30. K. Hosono, K. Matsubara, H. Takaoka, and T. Takagi, *Proceedings of the International Ion Engineering Congress: ISIAT'83 and IPAT'83*, Kyoto, 1983, p. 1237.
31. K. Mameno, H. Takaoka, K. Matsubara, and T. Takagi, *Proceedings of the International Ion Engineering Congress: ISIAT'83 and IPAT'83*, Kyoto, 1983, p. 1233.
32. K. Fujino, K. Hosono, K. Matsubara, H. Takasha, and T. Takagi, *Proceedings of the 9th ISIAT Symposium, ISIAT'85*, Tokyo, 1985, p. 359.
33. K. Fukushima, I. Yamada, and T. Takagi, *J. Appl. Phys.*, 58: 4146 (1985).
34. H. Usui, T. Fujino, I. Takeshita, I. Yamada, and T. Takagi, *Proceedings of the International Workshop on ICBT*, Tokyo, Kyoto, 1986, p. 207.
35. H. Hashimoto, L. L. Levenson, H. Usui, I. Yamada, and T. Takagi, *J. Appl. Phys.*, 63: 241 (1988).
36. H. Takaoka, K. Fujino, J. Ishikawa, and T. Takagi, *Proceedings of the 11th ISIAT Symposium, ISIAT'87*, Tokyo, 1987, p. 351.
37. H. Takaoka, J. Ishikawa, and T. Takagi, *Thin Solid Films*, 157: 143 (1988).
38. T. Takagi, *Pure Appl. Chem.*, 60: 781 (1988).
39. S. E. Huq, R. A. McMahon, and H. Ahmed, *Semicond. Sci. Technol.*, 5: 771 (1990).
40. M. Sosnowski and I. Yamada, *Nucl. Instrum. Methods, Phys. Res. B*: B46: 397 (1990).
41. S. E. Huq, R. A. McMahon and H. A. Ahmed, *Thin Solid Films*, 163: 337 (1988).
42. R. E. Hummel and I. Yamada, *Appl. Phys. Lett.*, 53: 1765 (1988).
43. M. Y. Sung, K. C. Hseih, E. W. Cowell, M. S. Feng, and K. Y. Cheng, *J. Vac. Sci. Technol.*, A7: 792 (1989).
44. G. H. Takaoka, S. Murakami, and J. Ishikawa, *Appl. Phys. Lett.*, 54: 2550 (1989).
45. M. C. Madden, *Appl. Phys. Lett.*, 55: 1077 (1989).
46. K. Yamanishi, Y. Kawagoe, S. Yasunaga, K. Imada, and K. Sato, *International Symposium on New Developments in Applied Superconductivity*, Osaka, 1988, p. 135.
47. K. Fukushima and I. Yamada, *J. Appl. Phys.*, 65: 619 (1989).

6

Chemical Methods of Film Deposition

Light is good in whatever lamp it may burn, even a rose is beautiful in whatever garden it may bloom.

Dr. S. Radhakrishnan

Unlike the physical methods of preparation of thin films involving evaporation or ejection of material from a source (as in evaporation or sputtering, described in the earlier chapters), chemical methods of thin film deposition entail a definite chemical reaction. This may be due to thermal effects as in chemical vapor deposition and thermal growth or to the electrical separation of ions as in electrodeposition and anodic deposition.

Chemical methods of film deposition in general use simpler equipment and are more economical than physical approaches, although the former methods are complex and difficult to control.

This chapter discusses the different chemical methods of thin film deposition. The chemical vapor deposition method, which has became a very important technique as applied to modern technology such as solid state electronics, is treated here in considerable detail.

6.1 THERMAL GROWTH

A large number of films—oxides, carbides, nitrides—can be prepared by heating the metal substrates in the gases of the required type (e.g., in O_2 for oxygen). One common example of oxidation is the formation of thin films of oxide on aluminum (30–40 Å) at room temperature. The thickness of the film can be increased by increasing the temperature, but the total thickness is limited since the oxide growth rate generally diminishes with thickness.

Preparation of thin films by thermal growth is not a commonly used technique, but thermally grown oxides of metals and semiconductors have been widely investigated because the surface can be passivated and because insulating properties that are useful in the fabrication of electronic devices can be obtained. This section briefly discusses thermal oxidation.

All metals except gold will react with oxygen to form an oxide layer [1,2]. Several models have been proposed for thermal oxidation of metals [3,4]. Wilmsen's review of the growth models for the thermal oxidation of metals and alloys [5] may be referred for details regarding the nucleation and formation of thermal oxide films. In all these models, it is assumed that the metal cations or oxygen anions diffuse through the lattice of the oxide and not along grain boundaries or pores.

Thermal oxidation is usually carried out by conventional furnace oxidation, and thermal oxidation has been reported for several materials [6-10]. Quite a large number of reports have been published on the thermal oxidation of Si to form SiO_2, since SiO_2 is extremely important in silicon device technology, and it is not intended here to document all these works. Reviews of the work on the thermal oxidation of silicon [11-13] have been published, and Mott [14] has recently given a review of some current theories of oxidation process of silicon.

Two recent reports of the preparation of oxide films different from the conventional thermal oxidation are described below.

Rapid thermal heating of silicon in a dry oxygen ambient has been used by Ponpon et al. [15] to prepare very thin silicon oxide films. A growth rate higher than that of conventional furnace oxidation has been obtained, which is linearly dependent on the square root of the time. A rapid thermal cycle system capable of performing high temperature cycles in a few seconds [16] has been used, and an oxide layer up to 300 Å thick can be prepared in less than 1 minute. The temperature ranged from 1000 to 1200°C for steady state times between 1 and 64 seconds. This so-called high temperature, short time (HTST) method serves two purposes: (a) precise control of growth of thin silicon oxide ($\leq$ 300 Å) and (b) low charge and low interface states densities, which can be obtained only at high temperatures.

George et al. [17] have reported the preparation of Bi_2O_3 films by the oxidation of thin bismuth films in air and superheated steam, using the very simple setup shown in Figure 6.1. In this work, single-phase films of α-Bi_2O_3, β-Bi_2O_3, and γ-Bi_2O_3 films were prepared for the first time. It may be noted that the water molecules will not decompose into their components (oxygen and hydrogen) even at the maximum temperature (367°C) used in this experiment, and it is shown here that the high temperature steam has no part in the reaction other than to displace the air present in the reaction chamber, thereby changing the effective oxygen concentration there.

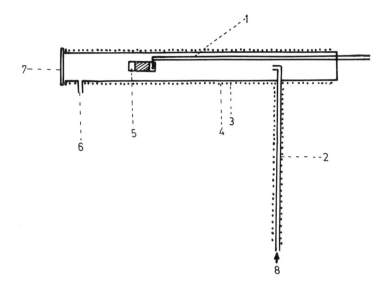

Figure 6.1 Experimental setup for the oxidation of thin bismuth films in air and super-heated steam: 1, thermocouple; 2, narrow glass tube; 3, heater winding; 4, glass tube (4cm diameter); 5, specimen; 6, outlet; 7, ground glass lid; 8, steam. (Data from Ref. 17.)

6.2 CHEMICAL VAPOR DEPOSITION

Chemical vapor deposition (CVD) is an important and popular technique for the preparation of thin films of a wide variety of materials—elements as well as compounds—on various substrates. CVD offers many advantages over other methods of thin film deposition. For example, films with a high degree of purity and good quality, with accurately controllable stoichiometric composition and doping levels, can be prepared. Because of the good throw power attainable with CVD, coating of complex shapes are possible. Since many reactions can be accomplished at ambient pressures, the need for expensive high vacuum equipment can be avoided. The higher deposition temperature (i.e., higher than in electrodeposition) leads to improved crystal perfection. Materials that decompose upon melting or evaporation can be prepared. The process can be adapted to large-scale multisubstrate operation.

An obvious disadvantage associated with CVD is that many coating reactions require high temperatures and the vapors react with substrates or apparatus. Also, the apparatus used in CVD may be complex, and there are many control variables.

There are several technological applications of CVD. Titanium nitride or silicon carbide coatings of cutting tools extend the articles useful life by enhancing wear resistance. The deposition of amorphous silicon by CVD on a large scale is used for low cost solar cells. CVD of titanium nitride is used to impart the gold color to low cost costume jewelry. The major application of CVD is in semiconductor integrated electronic circuits. Epitaxial deposition of silicon on silicon wafers and the deposition of dielectric films such as silicon dioxide and silicon nitride for passivation in integrated circuitry are made possible by CVD.

6.2.1 Common CVD Reactions

In CVD, the vapor/vapor mixture that enter a chamber containing the substrate, maintained at a suitable temperature, react to form a solid film on the substrate surface. The chemical reaction taking place is a very important characteristic in all CVD processes. Flow rate, gas composition, deposition temperature, pressure, and chamber geometry are the process variables by which thin film deposition is controlled. Thus a reactor-based CVD system for depositing thin films involves three fundamental processes: the transportation of the reactants, the supplying of heat to the reaction site, and the removal of the reaction by-products. Broadly speaking, CVD reactor designs can be classified into the atmospheric pressure and low pressure categories, the low pressure system having undergone significant development in the past few years. Disadvantages of atmospheric operation include the need for a large stream of carrier gas, large apparatus size, and high levels of contamination. But the low pressure CVD system allows removal of carrier gas and uses only small amounts of reactive gases at low partial pressures. Here the gas is injected from one end, with the vacuum pump at the other end. In what is called a "hot wall" reactor, the whole reactor reaches the process temperature, the substrate residing in an isothermal environment produced by uniform furnace heating. In a "cold wall" reactor, only the substrates have the necessary process temperature; in other words, the heated area is limited to the substrates or the substrate holder. It is not intended to catalog in detail all the designs of reactors in which thin films are produced and all the experimental approaches to the preparation of thin films. Some of the typical reactions involved in CVD are enumerated, however, with illustrative examples.

Thermal decomposition for metal films

The earliest approach to the preparation of silicon films was the thermal decomposition of silane when heated to a suitable temperature. The reaction is

$$SiH_4(g) \rightarrow Si(s) + 2H_2(g)$$

Many other gaseous compounds that are not highly stable can be dissociated to form the metal coatings:

$$Ni(CO)_4(g) \rightarrow Ni(s) + 4CO(g)$$

$$TiI_2(g) \rightarrow Ti(s) + 2I(g)$$

Reduction

In the hydrogen reduction of halides, silicon tetrachloride ($SiCl_4$) is a well-known example for the deposition of Si.

$$SiCl_4(g) + 2H_2(g) \rightarrow Si(s) + 4HCl(g)$$

Other examples involve tungsten and boron halides:

$$WCl_6(g) + 3H_2(g) \rightarrow W(s) + 6HCl\ (g)$$

$$WF_6(g) + 3H_2(g) \rightarrow W(s) + 6\ HF(g)$$

$$2BCl_3(g) + 3H_2(g) \rightarrow 2B(s) + 6HCl\ (g)$$

Chlorides are the more commonly used halides because of their greater volatility and because they can be readily purified by fractional distillation. Hydrogen reduction is not always suitable for the preparation of films of certain metals such as aluminum, and titanium because the halides of these elements are very stable.

Oxide formation

Silicon dioxide films are usually prepared by the oxidation of silane (SiH_4). Silane is mixed with oxygen together with an inert gas dilutant at atmospheric pressure. The reaction is as follows:

$$SiH_4(g) + O_2(g) \rightarrow SiO_2(s) + 2H_2(g)$$

The reaction can proceed at temperature as low as 450°C. Operation at atmospheric pressure is advantageous in the sense that it calls for a system simpler and less expensive than that required with a vacuum environment. When operating at atmospheric pressure, the gas phase nucleation is reduced by the degree of inert gas diluent used.

Other processes used for the deposition of SiO_2 are as follows:

$$SiH_4(g) + 2N_2O \rightarrow SiO_2(s) + 2H_2(g) + 2N_2(g)$$

$$SiH_2Cl_2(g) + 2N_2O \rightarrow SiO_2(s) + 2HCl(g) + 2N_2(g)$$

The temperature required for these reactions are approximately 850 and 900°C, respectively.

The direct oxidation of silicon tetrachloride and germanium tetrachloride also requires very high deposition temperatures:

$$SiCl_4(g) + O_2(g) \rightarrow SiO_2(s) + 2Cl_2(g)$$

$$GeCl_4(g) + O_2(g) \rightarrow GeO_2(S) + 2Cl_2(g)$$

An example of the oxide deposition of aluminum through the hydrolysis reaction also begins with a chloride:

$$Al_2Cl_6(g) + 3CO_2(g) + 3H_2(g) \rightarrow Al_2O_3(s) + 6HCl(g) + 3CO(g)$$

Nitride and carbide formation

Silicon nitride and boron nitride are two important examples of deposition of nitride films by CVD:

$$3SiH_4(g) + 4NH_3(g) \rightarrow Si_3N_4(s) + 12H_2(g)$$

High deposition rates are obtained with the following process:

$$3SiH_2Cl_2(g) + 4NH_3(g) \rightarrow Si_3N_4(s) + 6HCl(g) + 6H_2(g)$$

$$BCl_3(g) + NH_3(g) \rightarrow BN(s) + 3HCl(g)$$

The properties of the deposits are dependent on the nature of vapors and the deposition conditions (temperature, etc.). For example, for silicon nitride below certain temperatures, amorphous films are more likely to be formed.

Titanium carbide, boron carbide, and silicon carbide films can all be deposited by CVD using chloride reduction in the presence of hydrocarbon vapor:

$$TiCl_4(g) + CH_4(g) \rightarrow TiC(s) + 4HCl(g)$$

Thermal decomposition of trichloromethyl silane produces silicon carbide coatings:

$$CH_3SiCl_3(g) \rightarrow SiC(s) + 3HCl(g)$$

Compound formation

Deposition of compounds in Groups III-V from organometallic compounds proceeds as follows:

$$Ga(CH_3)_3(g) + AsH_3(g) \rightarrow GaAs(s) + 3CH_4(g)$$

If a temperature differential is set up in the system, a chemical transfer reaction takes place when the source material at temperature T_1 reacts with a transport agent to form a volatile species. When transported down the temperature gradient, this volatile material undergoes reverse reactions at temperature T_2 ($T_1 > T_2$) to form the original material in another part of the reactor:

$$6GaAs(g) + 6HCl(g) \underset{T_2}{\overset{T_1}{\rightleftharpoons}} As_4(g) + As_2(g) + 6GaCl(g) + 3H_2(g)$$

The material obtained after the reverse reaction is in a higher state of purity.

6.2.2 Methods of Film Preparation

A good, detailed account of the different aspects of CVD, the different reactor systems, and the earlier work on the preparation of films by CVD is presented by Kern and Ban [18], who give selected cross-references for further information. Table 6.1 outlines the salient features of the later work on films of several materials prepared by CVD.

A few interesting examples of later work on CVD are given below in somewhat more detail. Nakamura has reported a new CVD method of thin film preparation [31] that has been applied to the preparation of amorphous boron nitride films [56] using ammonia and decaborane reaction gases. The schematic diagram of his setup (Figure 6.2) indicates two adjustable leak valves, one for NH_3 and the other for $B_{10}H_{14}$. The film deposition conditions used by Nakamura are as follows.

Pressure of decaborane	2×10^{-5}
Pressure of ammonia	2×10^{-5} to 8×10^{-4} torr
$NH_3/B_{10}H_{14}$ ratio	1:40
Substrate temperature	300–1150°C
Deposition time	30–300 min
Substrates used	Ta, Si, sapphire

The crystal structures of the films prepared were studied by X-ray diffraction. It has been shown that the composition of the films can be closely controlled by regulating the pressure of the reaction gases. Stoichiometric boron nitride films were obtained at $NH_3/B_{10}H_{14}$ ratios of 20 or more at the substrate temperature of 850°C.

A vertical cold wall low pressure CVD system has been used by Fang and Hsu, for the preparation of WSi_2 films [57]. A quartz bell jar 24 cm in diameter was used as the reaction chamber and a graphite sheet (10 cm $\times$ 2 cm) coated with silica as the heater element. Tungsten hexafluoride (WF_6) and diluted SiH_4 (26% SiH_4/74% H_2) were used as the reactant gases. The substrates were (111) n-type silicon wafers, and the temperature was kept at 400°C. The chamber pressure was maintained

Table 6.1 Chemical Vapor Deposition

Film	Reactants	Deposition Temperature (°C)	Substrate	Remarks, if any	Ref.
ZnO	$(C_2H_5)_2$ Zn and O_2	200–500	Corning 7059 glass	Polycrystalline zinc oxide films highly oriented along the c-axis have been grown by the oxidation reaction of diethyl zinc. Highest degree of c-axis orientation was obtained around 400°C. Argon was used to transport the DEZ vapor to the reactor.	19
Ge	GeH_4	500–900	Si	Epitaxial Ge films were formed by the decomposition of GeH_4 in an H_2 ambient. Substrates of <100> Si were used. Germanium growth rate varied from 100 to 500 Å/s. Films with smooth surface morphologies and high crystalline quality are obtained.	20
SnO_2	$SnCl_2$ and O_2	350–500	Pyrex glass	Deposition was carried out in an open tube system. Stannous chloride vapor was driven through the furnace by maintaining an oxygen flow. Rates of deposition ranged from 300 Å/min at 350°C to 600 Å/min at 500°C. Electrical properties were studied.	21
Nb/Ge	$NbCl_5$ and $GeCl_4$	800 and 900	Alumina	Mixture of these chloride vapors was introduced into the coating chamber by the hydrogen carrier. Nb/Ge films were deposited by the hydrogen reduction of mixed chloride vapor. Superconducting transition temperature, crystal structures, and growth morphology were studied.	22

Material	Temperature	Reactants	Substrate	Remarks	Ref.
BN	600–1100	BCl_3 and NH_3	Fused silica and sapphire (0001) and ($1\bar{1}02$)	Films were prepared using reaction of BCl_3 and NH_3 in a flow system; BCl_3 diluted with argon; NH_3 was introduced directly into the center of the reaction tube to avoid premature reaction of BCl_3 with NH_3. After deposition, the substrates were cooled to RT in a stream of NH_3 and Ar. Growth rate was 1 μm/h. Transparent and smooth films were obtained between 1000 and 1100°C.	23
TiB$_2$	600–900	H_2, Ar, $TiCl_4$, and B_2H_6	Graphite disks	Argon was used as a carrier gas for $TiCl_4$ vapor. Deposition using the hydrogen reduction of the chlorides. Dense and adherent coatings were obtained: boron-rich at 600°C ($TiB_{3.12}$) and stoichiometric above 700°C.	24
BN	250–700	BCl_3 and NH_3	Copper	Vertical reaction tube (quartz). $BCl_3/H_2/Ar$ mixture and NH_3 were introduced (close to the substrate) separately through two lower gas inlets of the reaction tube and removed through an upper gas outlet. Films deposited below 450°C were unstable in moist atmosphere. Above 600°C, the films were very stable.	25
a-Si:H	380–475	Si_2H_6	Crystalline Si	Both hot wall and cold wall deposition methods were used. Final purified Si_2H_6 used contained no moisture or HCl, but possibly small amounts of SiH_4. Typical pressures ranged between 10 and 100 torr.	26
CdTe	550–650	CdTe and HCl	CdTe single-crystal (110) and nonoriented sample cut from	Deposition of CdTe layers was by closed tube CVD; vacuum was better than 10^{-6} torr. Source temperature always was 100°C higher than deposition temperature. Gaseous hydro-	27

Table 6.1 (continued)

Film	Reactants	Deposition Temperature (°C)	Substrate	Remarks, if any	Ref.
			CdTe ingots and polished	chloric acid obtained by thermal dissociation of ammonium chloride was used as the transporting agent. $CdTe(s) + 2 HCl\,(g) \rightleftharpoons CdCl_2\,(g) + H_2\,(g) + Te\,(g)$	
Polysilicon	SiH_4	570–640	<100> oriented Si wafers covered with SiO_2, 1000 Å thick	Polysilicon films were deposited as amorphous layers at temperature lower than 600°C. Crystallization of the deposited layer recurred during the continuous deposition of the films to the final thickness of 6000 Å The 640°C films had a very short nucleation period. Silane mixed with diluted phosphine was used to get doped films.	28
W	WF_6, Si, and H_2	300	Thermally oxidized Si wafers coated with 5000 Å polysilicon	Tungsten films were selectively deposited in a hot wall quartz reactor (reduced pressure). Total deposition time was 10 minutes. Films 400 Å thick were deposited by silicon reduction in the first minute; then hydrogen was introduced to achieve a total film thickness of about 1200 Å	29
Si_3N_4	SiH_2Cl_2/NH_3 mixture, 1:3	800	(111) oriented n-type Si	Deposition was carried out in a standard LPCVD reactor. Films were characterized by electrical measurements, AES, and RHEEDS.	30
B	$B_{10}H_{14}$	600–1200 350–700	Sapphire and Si Ta sheet	Sapphire and silicon heated by indirect heating and tantalum by direct resistance heating. Boron films were deposited on heated substrates by thermal decomposition of decaborane.	31

Material	Temperature (°C)	Reactants	Remarks	Ref.
			Pressure of decaborane kept between 1×10^{-5} and 2×10^{-4} torr. X-Ray and electron diffraction studies of the films indicated that the films are amorphous.	
Si	775	SiH_4	This low pressure CVD process deposits uniform specular epitaxial films. The wafer surface had to be sputter-cleaned in situ with an argon plasma before deposition. Predeposition in situ cleaning was essential for activating epitaxial growth at this temperature.	32
$TiSi_2$	650–700	SiH_4 and $TiCl_4$	An LPCVD reactor was used: a radiant heater kept outside the chamber heated the wafer from below the quartz substrate holder. Accuracy of temperature measurement was ± 25°C; pressure was 50–460 mtorr. Various $TiCl_4/SiH_4$ flow rate were used to deposit sequential films of polysilicon and titanium silicide. X-Ray diffraction studies showed that as-deposited films of titanium silicide were polycrystalline, the most stable phase (TiS_2) being predominant.	33
W	400	WF_6 and Si	Tungsten was deposited in an LPCVD reactor. When WF_6 came in contact with silicon, tungsten was deposited [WF_6 (g) is reduced by silicon]. Reaction was self-limiting. Total pressure of Ar and WF_6 was 0.3 torr; WF_6 flow rate, 12 cm³/min. Ultrathin tungsten filament cross section was 600 × 2000 Å.	34
BPSG (boro-	350–450	O_2, SiH_4 and PH_3/ SiH_4 and B_2H_6	Reactor with a vertical flow pattern was used. Mole ratio of the dopants was adjusted by the	35

Polycrystalline Si

Table 6.1 (continued)

Film	Reactants	Deposition Temperature (°C)	Substrate	Remarks, if any	Ref.
phospho-silicate glass)	SiH_4 mixtures			flow of various hydrides and mixing before injection into the reaction chamber. A separate injector is used for introducing oxygen. Total pressure was 0.10 torr. Concentration of boron and phosphorus in the films was determined by wet chemical calorimetric methods.	
SnO_2	SnI_4 and O_2	380–550	Glass and quartz	Films obtained were polycrystalline. Optical transmission in the visible range exceeded 80%. Sheet resistances were greater than 100 $\Omega/\square$.	36
SnO_2:F	$SnCl_4 \cdot 5H_2O$ and (99.9% pure) O_2	300–400	Quartz and glass	Hydrogen fluoride (48%) was used for doping, the required amount being mixed in the molten stannic chloride. Electrical, optical, SEM, and X-ray diffraction studies of the films showed good crystallinity.	37
Semi-insulating poly-crystalline Si	N_2O and SiH_4	565–623		Oxygen content in the film could be varied by altering the N_2O/SiH_4 input ratio. LPCVD process pressure varied from 0.13 to 0.44 torr.	38
$TaSi_2$	SiH_4 and $TaCl_5$	630–750	Silicon and silica	Hydrogen was used as the carrier gas. In this temperature range, the deposition rate was 700 Å/min. Composition of the film was determined by electron microprobe analysis; structure was identified by X-ray diffraction.	39

Material	Reactants	Temperature	Substrate	Description	Ref.
Si	100% Si$_2$H$_6$ and 1% PH$_3$ in N$_2$		<1Ī02>Sapphire and thermally oxidized Si wafers	Phosphorus-doped silicon films were prepared in a conventional LPCVD reactor. Three-zone heater produced a flat temperature profile in the reactor zone.	40
a-Si:H	SiH$_4$ and H$_2$			High quality hydrogenated amorphous silicon was obtained.	41
CdS	CdS and H$_2$	500–760	(111) CdTe single-crystal slices	Transport of CdS from the source to the deposition site was performed by reaction with hydrogen. Best quality films were obtained at a source temperature of 730°C and a deposition temperature of 630°C, with a hydrogen pressure of 250 torr. Epitaxial films were obtained.	42
Si	SiH$_4$ and H$_2$	550–725	Lightly doped Si wafers covered with thermally grown SiO$_2$ or LPCVD oxide	Growth of silicon films in a unique vertical flow reactor was compared with that in a conventional tube reactor. In the deposition of Si films from a hydrogen-carried silane process in the vertical reactor, the growth rate depends on both partial pressure and flow rate of silane. No dependence on hydrogen partial pressure was found.	43
B-H-N	B$_2$H$_6$ and NH$_3$	350, 400, and 440	Si wafers (100) polished on both sides	Conventional LPCVD system was used: B$_2$H$_6$ (6% mixture in N$_2$ or H$_2$ carrier gas) and NH$_3$ were introduced separately and mixed in the hot zone of the quartz reaction chamber. Ratio flow of gas (NH$_3$/B$_2$H$_6$) was between 0.1 and 1.5. Total pressure was 360 mtorr. Because chemical composition differs from stoichiometric BN, it is preferable to call the films obtained B-H-N (borohydronitride) films.	44

Table 6.1 (continued)

Film	Reactants	Deposition Temperature (°C)	Substrate	Remarks, if any	Ref.
TaSi$_2$	SiH$_4$ and TaCl$_5$	575	n^+-type poly-crystalline Si	Standard hot wall LPCVD reactor was used. Stoichiometry of the as-deposited film (Ta$_5$Si$_3$) changed to TaSi$_2$ by annealing at 900°C for 30 minutes. Poly-Si/tantalum silicide bilayer structures were fabricated.	45
W	WF$_6$ and H$_2$ or Ar	300	Self-aligned CoSi$_2$ structures and un-patterned CoSi$_2$ and Co layers, the substrates used being p-type <100> Si wafers	Tungsten was deposited by low pressure CVD in a vertical flow, hot wall reactor: WF$_6$ (99.98%), H$_2$ or Ar (99.998%). Pressure was 500 mtorr. WF$_6$/H$_2$ or WF$_6$/Ar flow ratio was = 44:1. Silicon wafers were sometimes covered with thermally grown SiO$_2$.	46
GaAs	GaAs and H$_2$O	750 ± 50	GaAs and Ge	Epitaxial films of GaAs were deposited on (100) oriented substrates, by the reversible vapor transport reaction (transport agent is water): $2GaAs (s) + H_2O (g) \rightleftharpoons Ga_2O (g) + As_2 (g) + H_2 (g)$. Source temperature was 800 ± 50°C.	47
Al	AlCl$_3$/H$_2$ mixture	700–1100	Pure Nickel (polished)	Kinetics of the formation of nickel aluminide was investigated.	48
a-Si:H	Si$_2$H$_6$	200–575		Doped and intrinsic a-Si:H films were deposited from 0.2 to 50 Å/s at 380–460°C. Amorphous silicon p-i-n solar cells were made.	49

3C-SiC	C_3H_8/SiH_4/H_2 gas system	1350	Si (100)	SiH_4 diluted with H_2, C_3H_8 diluted with H_2, and H_2 carrier gas purified by palladium–silver alloy cell introduced into the reaction tube (quartz), previously evacuated to 2×10^{-7} torr. Triethyl aluminum added to the reactant gases for Al doping.	50
TiO_2	Titanium alkoxides mixed with niobium ethoxide/ tantalum ethoxide/ t-butyl fluoride	400–600	Pyrex plates (Corning 7740 glass)	Carrier gas was 40% H_2 in N_2. Nibium/tantalum/ fluorine-doped films were prepared.	51
Fe	Ferrocene	490–600	Thin Ni strips	Films were deposited by thermal decomposition of the metallocene vapor in an all-glass vacuum system at a pressure of $\sim 4 \times 10^{-5}$ torr, with	52
Ni	Nickelocene	550	B16 stainless steel rods	3 hours of exposure to ferrocene and 3–6 hours nickelocene. Films were pure and had good adhesion to the substrates.	
Ti(C,N)	$TiCl_4$/H_2/N_2/CH_4	850–1150	WC with 6 wt % Co	Morphology and structure of titanium carbonitride films varied with deposition temperature, and composition. $TiCl_4$ vapor was carried with H_2 gas from the $TiCl_4$ saturater. Concentration of $TiCl_4$ was controlled by adjusting the temperature of the evaporator and the flow rate of H_2. Total pressure was 1 atm.	53
Poly-Si	Pure silane	630	Thermally oxidized (111) p-type Si	Undoped LPCVD of polysilicon films took place at a pressure of 80–180 mtorr. Growth rate was 48–91 Å/min; thickness, 7200–5500 Å	54

Table 6.1 (continued)

Film	Reactants	Deposition Temperature (°C)	Substrate	Remarks, if any	Ref.
SnO_2	$SnCl_4$ and O_2	300–500	Thermally grown SiO_2 layer on n-type <100> Si wafers	Liquid $SnCl_4$ (99.999% purity) kept at 25°C was vaporized with nitrogen gas flow and the gas flow was mixed with oxygen gas flow on the way to the reaction glass chamber, containing a quartz plate on a heater. Substrates were kept on the quartz plate. Conductivity, its dependence on temperature, and the Hall effect of the films were studied.	55
C	C_6H_6, Ar	1000	Quartz	Argon used as carrier gas. Carbon films with well-ordered graphite structure were obtained by the thermal decomposition of benzene. Film properties and their dependence on deposition conditions were investigated.	192
Amorphous Si	Undiluted silane	580–530	Thermally oxidized Si wafers (50 mm diameter)	Crystallization of undoped amorphous silicon films deposited (580–530°C) and annealed in the range of 550–950°C has been studied by TEM. A simple model is presented to explain the dependence of grain size on the annealing temperature.	193
TiC and TiN	$TiCl_4$, CH_4, H_2 gas mixtures; $TiCl_4/H_2$, /N_2 gas mixtures		Si_3N_4/TiC composite ceramics (hot pressed)	Auger electron spectroscopy survey was made to find the nonmetal-to-metal ratios of films under different deposition conditions. Microstructure and thermal shock resistance were also	194

Material	Temperature (°C)	Reactants	Substrate	Remarks	Ref.
Boron carbides	1027–1227	BCl_3 (99.99% pure), CH_4 (99.9995%), and H_2 (99.9995%) gas mixtures	α-Rhombohedral boron (prepared by depositing a 5–10 μm thick boron films on thin Mo)	investigated. TiC coatings had more resistance to thermal shock than TiN coatings. Different boron carbides were deposited using a cold wall reactor. The influences of various factors (vapor composition, temperature, total pressure) on the phase and chemical composition were studied.	195
W	475–750	Ar/WCl_6 and H_2/WCl_6	Patterned and unpatterned Si(100)	A conventional LPCVD reactor in a hot wall reactor was used for the selective deposition of tungsten. Selectivity of the chloride process was good when both Ar/WCl_6 and H_2/WCl_6 reactor gas mixtures were used. Thickness of the thermal oxide on the patterned wafers was 5000 Å.	196
ZrB_2	700–900	$ZrCl_4 + BCl_3 + H_2 + Ar$	Cu plate	Grain growth of the layer accelerated at temperatures above 750°C. Maximum deposition rate was obtained at a gas flow ratio ($BCl_3/ZrCl_4$) of 1.5; the rate decreased slowly above or below this value.	197
Si_{1-x} Ge	500–800	SiH_2Cl_2 and GeH_4 in H_2 carrier gas	Si in SiO_2	Low pressure CVD was carried out in a cold wall, lamp-heated rapid thermal processor. Selective growth occurred on <100> oriented silicon using windows in the oxide. An SiO_2 layer $\approx$ 2000 Å thick was thermally grown. Percentage of germanium in alloy depends on deposition temperature and flow ratio of gas.	198

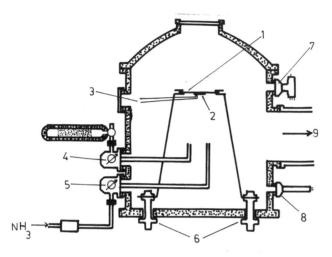

Figure 6.2 Schematic diagram of the apparatus for the preparation of amorphous boron nitride films: 1, heater; 2, substrate; 3, thermocouple; 4, variable leak valve (decaborane); 5, variable leak valve (ammonia); 6, electrode; 7, ionization gage; 8, Pirani gage; 9, to pump. (Data from Ref. 56.)

at 1 torr, and with the rates of flow rate 0.66 L/min (SiH_4) and 0.51 L/min (WF_6 gas), they obtained a deposition rate of 300 Å/min.

Metal oxide films of Al_2O_3, CuO, CuO/Al_2O_3, and In_2O_3 have been prepared by Ajayi et al. [58] by the pyrolytic decomposition onto glass substrates of the appropriate metal acetylacetonates. The schematic diagram of the experimental setup is shown in Figure 6.3. The starting material (the respective acetylacetonate) in the form of very fine powder is kept in an unheated receptacle and argon gas is bubbled through it. The argon flow rate was adjusted so that the fine argon-borne particles were transported into the working chamber (heated Pyrex tube). Some of the fine particles settled on the substrate on a holder (stainless steel) at the center of the furnace. At 420°C, the decomposition of the metal acetylacetonates carried on for about 2 hours results in films 10–20 nm thick, which were annealed overnight. The optical data the investigators obtained of these films compared well with data obtained by other methods, and this simple thermal decomposition method is useful for the preparation of good quality oxide films in the solar spectral range.

A new low pressure CVD (LPCVD) technique for preparing borophosphosilicate glass films by injection of miscible liquid mixtures of TEOS (tetraethylorthosilicate), TMB (trimethylborate), and TMP (trimethylphosphite) has been reported by Levy et al. [59]. Here the source materials (TEOS, TMB, and TMP) were measured volumetrically and mixed, and the mixed solution was delivered at a controlled rate into the reactor by a syringe pump through an adjustable leak valve. The exit side of the leak valve was heated to vaporize the liquid as soon as it passed through the valve orifice. There was also a second inlet for oxygen, and both gaseous inputs entered the reactor via porous glass diffusers to allow maximum mixing and to optimize the flow.

The substrates were single-crystal Si wafers (rectangular pieces ≈ 5 cm^2) positioned in a slotted fused quartz holder. The furnace provided a high temperature zone over about 50 cm. The following deposition parameters are typical.

Background pressure at temperature ($\approx 700°C$)	0.03 torr
Oxygen pressure	0.16 torr
Rate of input of organic mixture	5 mL/h
Pressure during the run	0.3 torr
Deposition time	1 hour

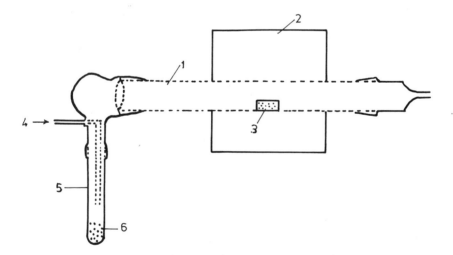

Figure 6.3 Schematic diagram of the experimental setup for the pyrolytic decomposition deposition of thin metal oxide films: 1, working chamber; 2, electrically controlled furnace; 3, substrate; 4, argon inlet; 5, receptacle; 6, fine powder. (Data from Ref. 58.)

For each run, 52 preweighted substrates were loaded back to back at 2 cm intervals.

Halide chemical vapor deposition has been successfully used by Engelhardt and Webb [60] for the preparation of Nb_3Ge films when metal chlorination and its hydrogen reduction proceed in independent gas flows of Cl_2 and H_2, respectively, and the metal film formed on the substrate is kept near the two jet streams. Sasaki et al. [61a] have developed a modified version; the significant feature of their apparatus is that the metal film is deposited on to an ultrasound vibrating substrate. Also there is a one-way type of reactant gas injection system in which metal chlorination and its hydrogen reduction occur successively along the quartz reactor, according to the reaction:

$$2Nb + 5Cl_2 \rightarrow 2Nb\ Cl_5$$

$$2NbCl_5 + 5H_2 \rightarrow 3Nb + 10HCl$$

A two-zone furnace has been used for metal chlorination (T^{Cl2} – 500°C) and for hydrogen reduction (T^d = 900°C), and the metal film is deposited onto a quartz (or silicon) substrate, attached to an ultrasound vibrating quartz rod driven by a barium titanate transducer. To avoid a backflow of reactant gases in the reactor, a small amount of excess H_2 gas is flowed through a bypass line. With the use of the ultrasound vibrating substrate, the thickness of the diffusion layer near the substrate is reduced.

The X-ray and electron diffraction analysis of the films obtained by Sasaki et al. [61a] showed that these Nb films have a face-centered cubic structure.

Vishwakarma et al. [61b] have reported the preparation of transparent conducting arsenic-doped tin oxide on glass substrates by a CVD technique. Clean soda glass substrates were heated to 673 K and $SnCl_2$ vapor was allowed to fall on the heated substrate, using oxygen as a carrier of the chloride vapor at a fixed flow rate of 1.35 L/min. For doping, $AsCl_3$ vapor was carried simultaneously over the substrate from a controlled amount of $AsCl_3$ solution using nitrogen as the carrier gas.

Films were deposited in the range of 523–723 K, and the investigators studied the electrical, thermal, and optical properties of these films. X-Ray and SEM studies were also made. It was found that the films deposited at 673 K were uniform and polycrystalline, with grain sizes in the range of 0.20–0.45 μm.

A new low temperature chemical vapor deposition method named catalytic CVD (cat- CVD or CTL-CVD), developed by Matsumura and his coworkers [61c], has been applied for the deposition of amorphous semiconductor films [61d]. Matsumura has recently reported the application of this technique for the low temperature deposition of silicon nitride [61e]. The deposition chamber, shown schematically in Figure 6.4, is made of steel, and the substrates are attached to the sample holder, which can be heated by a heater or cooled by an air jet behind the holder. A thermocouple is mounted on the substrate holder near the substrate to

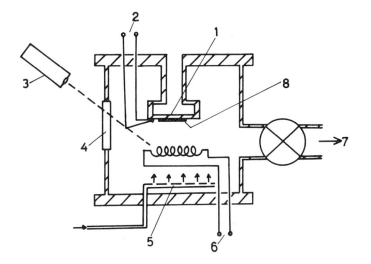

Figure 6.4 Schematic diagram of the setup for the low temperature deposition of SiN by cat-CVD: 1, substrate holder; 2, thermocouple; 3, IR thermometer; 4, window; 5, gas nozzles; 6, heater catalyzer; 7, to pump; 8, substrate. (Data from Ref. 61e.)

measure the temperature. The heated catalyzer, with a total surface area of 15–42 cm^2, is placed parallel to the substrate holder between the gas nozzles and the substrate holder, 3–4 cm from the holder. The catalyzer is 2% thoriated tungsten, coiled and pinned by molybdenum wire to keep it parallel to the substrate holder. The influence of the thermal radiation from the catalyzer was almost negligible for the temperatures used in this deposition. Hydrazine (N$_2$H$_4$), being a liquid at room temperature, it is bubbled by N$_2$ gas, and the gas mixture of hydrazine, N$_2$, and silane (SiH$_4$) was introduced into the chamber through the many nozzles.

The deposition conditions reported were as follows:

Catalyzer temperature	1200–1390°C
Substrate temperature	230–380°C
Gas pressure during deposition	7–1000 Pa
Partial gas pressure	(SiH$_4$), 7–24 Pa
Partial gas pressure ratio	0–10, (N$_2$H$_4$ + N$_2$)/(SiH$_4$)
SiH$_4$ flow rate	2–10 sccm

The properties of the films deposited under these conditions were investigated.

Here the mixture of gases (N_2H_4, N_2, and SiH_4) is decomposed by catalytic or pyrolytic reaction with the heated catalyzer, and the silicon nitride films can be deposited on the substrate at temperatures as low as 300°C with deposition rates near 1000 Å/min. Also it was shown that the resistivity (10^{14}–10^{16} Ω cm), the breakdown electric field (10^6 V/cm), and the hydrogen content of the films are almost equivalent to those of thermal CVD films prepared at about 700°C. Again the diffusivity of the depositing species in the cat–CVD method appears to be large, and the cat–CVD SiN_x films show high ability for step coverage.

Germane (GeH_4) was used for the first time as a reducing agent for WF_6 in selectively depositing tungsten on silicon by Vander Jengd et al. [61f]. Thermodynamic calculations showed that it should be possible to deposit tungsten under conditions comparable to the process in which SiH_4 is used as a reducer.

Vander Jengd et al. used a commercially available cold wall, single-wafer reactor for the deposition. The wafers were pressed against a graphite chuck heated by a lamp, and the temperature was controlled by a thermocouple in the chuck. The investigators used 100 mm p-type (100) Si wafers covered with thermally grown oxide, and after patterning, the total exposed fraction of bare Si was about 10%. The films deposited below 400°C consisted of the β-tungsten phase and at deposition above 400°C, the films consisted of a mixture of β- and α-tungsten phases.

6.2.3 Laser Chemical Vapor Deposition

Laser chemical vapor deposition is a recently developed deposition technique using a laser source. There are essentially two mechanisms by which a laser beam initiates a chemical reaction whereby thin films may be deposited. One is photolysis and the other is pyrolysis. In the photolytic process (laser photolytic photochemical deposition—LPD), photons of sufficient energy are used to dissociate the molecules to form a films or to react with other chemical species present in the reaction gases to form a compound film on an adjacent substrate. In the other process, the laser beam is used as a heat source at which pyrolytic decomposition due to the reaction of the ambient gases with the locally heated substrate takes place [62] and the temperature rise induced on the substrate controls the deposition reaction. The two predominant characteristics of the laser source—directionality and monochromaticity—are used to advantage in the deposition process. Directionality allows the beams to be aimed at very precise areas of very small dimensions, causing localized depositions, and the wavelength of the laser can be chosen to cause either photochemical or pyrolytic deposition. However it can be seen that pyrolytic and photolytic processes occur simultaneously in many instances. Although photolysis has been identified in many LCVD reactions, thermal effects are very often present, and the term LCVD is ordinarily used to describe reactions as mainly photolytic, or mainly pyrolytic.

This section discusses the pyrolytic (thermal) laser decomposition deposition of films, which is termed hereafter laser chemical vapor deposition (LCVD). Thus LCVD is a modification of conventional CVD and is analogous to it, despite reliance on a laser heat source. LPD techniques are presented in Section 6.2.4.

Although the reactant systems in LCVD may be similar to conventional CVD, the film growth characteristics may differ in several ways, for a variety of reasons [63]. Much higher reaction temperatures can be used in LCVD because the heated area is very localized. Large volume preheating of the reactants is possible in LCVD, as well as the use of much higher reactant concentrations with no significant gas phase reaction. Again contamination from heated surfaces other than the substrate is minimal. For film nucleation, surface defects serve not only their usual role as nucleation sites but may be more strongly absorptive, thus generating a higher surface temperature upon laser heating. Increased diffusion of the reactants to the reaction zone due to the "point" geometry of LCVD as compared to the planar diffusion geometry in conventional CVD allows access to higher temperatures and can produce deposition rates many orders of magnitude greater than that can be obtained in conventional CVD [64]. It may be noted that the high local surface temperature attainable in LCVD is in a small region for a short time, and as such the LCVD deposition rate will in effect be limited by the diffusion of reactants as well as by convection. These factors limiting the deposition rate are functions of the initial concentration of the reactants and inert species, the surface temperature, the gas temperature, and also the geometry of the reaction zone. Allen et al. [63] have measured the deposition rate in real time for LCVD of several metal films on fused silica substrates using visible optical monitoring.

The many published reports on laser chemical vapor deposition feature, for example, aluminum [65], nickel [66], gold [67], silicon [68], silicon carbide [69a], polysilicon [69b], and aluminum and gold [69c].

6.2.4 Photochemical Vapor Deposition

Photochemical vapor deposition (photo CVD) is an attractive vapor deposition technique for the preparation of high quality, damage-free films for many technological application. Other advantages of this technique include lower temperature, faster deposition rate, growth of metastable phase, and abrupt junction formation. Again in contrast to plasma-enhanced CVD (described in Section 6.2.5, below), photo CVD has no highly energetic species bombarding the growing film surface. Also the photons causing molecular dissociation of the reactant molecules are not energetic enough to result in significant ionization. And this technique can produce quality films as well as good interfaces.

As mentioned earlier, in the photo CVD process, the photochemical deposition takes place when high energy photons selectively excite states in the surface-absorbed or gas phase molecules leading to bond rupture and the production of free chemical species to form the films or to react to form compound films on the adjacent substrate. This process is highly dependent on the wavelength of the incident radiation. Photochemical deposition can be achieved by laser (laser photochemical deposition) or by ultraviolet lamp. Apart from direct photolytic decomposition processes, high quality films are also prepared by mercury-sensitized photo CVD. It may be noted that dissociation and nucleation in photo CVD are controlled by the photon source, and then the substrate temperature becomes an independent variable to be selected with regard to the substrate and the film requirements.

Films of many different materials have been prepared by the photo CVD process: various metals [70–72], dielectrics and insulators [73,74], and compound semiconductors [75,76]. Amorphous Si (a-Si) and related alloys such as a-SiGe, a-SiGe:H are promising classes of materials for photovoltaic applications. For example, high quality a-Si films have been prepared by mercury-sensitized [77] and direct photo CVD processes [78]. The mercury photosensitization of monosilane and disilane was investigated earlier by Pollock et al. [79]. The main step for this process is

$$Hg^* + SiH_4 \rightarrow Hg + 2H_2 + Si$$

where Hg^* indicates the excited state of mercury atom due to UV radiation. This reaction takes place through several intermediate silyl radical reactions [80].

Konagai [81] recently developed a mercury-sensitized photo CVD system consisting of four reaction chambers and one loading chamber, partitioned by gate valves for solar cell structure fabrication. The major limitation, namely the unintended deposition of silicon on the UV transparent window was overcome by coating the quartz window with perfluoropolyether. Undoped a-Si films are prepared in one chamber using monosilane or disilane. Konagai has obtained typical deposition rates of 1–3 Å/s; for more details of the preparation of a-Si film and other related alloys, refer to the paper [81].

Very recently Kim et al. [82a] prepared for the first time undoped a-Si:H films by atmospheric pressure mercury-sensitized photo CVD. The schematic diagram of their setup is shown in Figure 6.5. SiH_4 gas was led into the chamber using Ar as a carrier gas. A low pressure mercury lamp with 2537 and 1849 Å resonance lines was used. Low vapor pressure fluorinated oil was painted on the internal surface of the quartz window, to retard the deposition of film on the window. Mercury vapor was introduced into the reaction chamber above the substrate (T_s = 200–350°C) as shown. A deposition rate of 45 Å/min was obtained by optimizing the mercury reservoir temperature (20–200°C) and the flow rate (SiH_4, 1–30 sccm; Ar, 100–700 sccm). The investigators have observed that the concentration of oxygen in the films prepared was 2×10^{18} cm^{-3}, which is much lower than that of films grown by low pressure photo CVD.

The first reported direct photo CVD of SiO_2 from a mixture of N_2O_3 and Si_2H_6 used an external deuterium lamp with magnesium fluoride as the vacuum UV

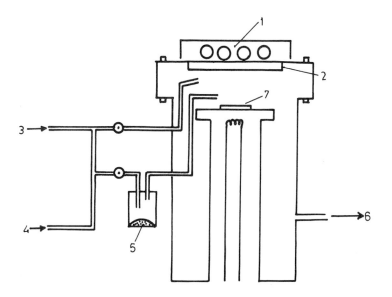

Figure 6.5 Schematic diagram of the setup for atmospheric pressure mercury sensitized photo CVD preparation of undoped a-Si: H films: 1, mercury lamp; 2, quartz glass; 3, Ar inlet; 4, SiH₄ inlet; 5, mercury; 6, exhaust; 7, substrate (Data from Ref. 82a.)

source [82b]; n-type <100> 1–5 Ω cm crystalline silicon was used as substrates, and the deposition was shown to be photon-limited to a maximum deposition rate of about 1 Å/s. Infrared adsorption studies indicated that the films obtained were nearly stoichiometric, with no detectable Si—N bonds and only trace amounts of Si—H bonds. In the light of their experiments with NO, NO$_2$, and N$_2$O$_3$, these investigators are of the view that with Si$_2$H$_6$ and N$_2$O$_3$, NO and its photodissociated fragments play the dominant role in the deposition process.

Yoshida et al. [82c] have reported the deposition of high quality hydrogenated amorphous silicon films by the direct photolysis of disilane with vacuum ultraviolet light from a windowless discharge. The schematic diagram of their setup is shown in Figure 6.6. A hydrogen discharge tube excited by microwave (2.45 GHz) power was used for the vacuum UV light source in this experiment. Disilane gas diluted with helium ($\leq 5\%$) served as the reactant source and was introduced into the reaction chamber close to the substrate. The flow rates of Si$_2$H$_6$/He and He were kept at 50 and 150 sccm, respectively. The total pressure in the reaction chamber during deposition was 2 torr. Glass or Si wafers were used as the substrates. The deposition was carried out at a substrate temperature of 50–350°C for 5 hours. The deposition rate was found to be independent of T_s implying that there was no thermal decomposition.

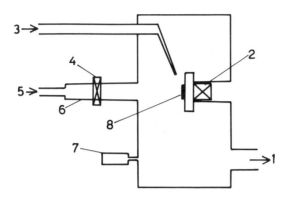

Figure 6.6 Schematic diagram of the photo CVD setup for the deposition of high quality a-Si: H films by the direct photolysis of disilane: 1, to pump; 2, heater; 3, Si_2H_6 inlet; 4, microwave cavity; 5, H_2 inlet; 6, quartz tube (straight); 7, vacuum gage; 8, substrate. (Data from Ref. 82c.)

Yoshida et al. had investigated the electrical and optical properties of the films obtained and found that the photosensitivity (σ_{ph}/σ_d) was about 10^7 in the films prepared at 250°C.

The advantages of this system are as follows.

1.   The vacuum UV light can be introduced into the chamber without any absorption, at a window.
2   The problem of deposition of the film material at the window is avoided.
3.   No light reaches the substrate directly, but the excited hydrogen enters the chamber in the same way as the straight tube.

Items 1 and 2 are very serious problems encountered in conventional photo CVD.

The application of laser photo CVD for the preparation of thin films is a potentially important processing approach. A few interesting examples are discussed below.

Shirafugi et al. [83] have reported the preparation of a-SiO$_x$ film using an excimer laser (repetition rate, 50 Hz; average output power, 1.5 W) operated at the ArF line (1930 Å). The source gases were 100% Si_2H_6 and N_2O, both of which will decompose by direct reaction with incident laser photons (6.24 eV). The schematic diagram of the setup (Figure 6.7) indicates that the laser beam passing through a synthetic quartz cylindrical lens was aligned parallel and close to the substrate surface. The inner surface of the quartz entry window was purged with argon. The mixture of source gases was showered down to the substrate surface through a multiholed plate. The substrate (silicon/thin glass plates) temperature was kept at

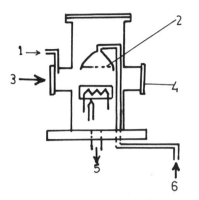

Figure 6.7   Schematic diagram of the system for the laser photo CVD of amorphous SiO$_x$ film using an excimer laser: 1, Ar inlet; 2, multiholed plate; 3, laser beam; 4, quartz window; 5, to pump; 6, source gases. (Data from Ref. 83.)

300°C. The N$_2$O-to-Si$_2$H$_6$ flow rate ratio and the pressure of the gas mixture were varied, and it was found that an N$_2$O/Si$_2$H$_6$ ratio larger than 200 gave a-SiO$_x$ films close to stoichiometric in composition. The investigators also found that annealing at 450°C was useful for improving the breakdown field strength and reducing the oxide charge density.

Amorphous silicon nitride thin films are of very great interest now as useful materials in microelectronic circuit applications. Jasinski et al. [84] have reported the preparation of silicon nitride films showing useful electrical properties with higher breakdown fields and lower interface state trap densities. Here ArF excimer laser photolysis has been performed in a low pressure, hot wall reactor, using ammonia as the nitrogen source and either silane or disilane as the silicon source. In this setup (Figure 6.8), the laser beam was directed horizontally along the length of the reactor and passed 5-10 mm above the horizontally placed substrate. The quartz furnace was evacuated and purged continuously with He to prevent the backstreaming of pump oil. A 5:1 mixture of ammonia and silane or disilane was allowed to flow at a total pressure of 0.25 torr, and deposition was initiated when the laser (repetition rate, 20 Hz) beam was admitted into the furnace tube. The films were deposited at temperatures in the range 225-625°C. The film stoichiometry could be varied from nitrogen rich to stoichiometry to silicon rich by changing the ammonia/silane ratio, the deposition temperature, or the gas (i.e., from silane to disilane).

Epitaxial InP films deposited from organometallic precursors at a steady state temperature below the incongruent decomposition temperature of InP ($\approx$ 350°C) were first reported by Donnelly et al. [85], using an excimer laser (1930 Å, 20 Hz repetition rate) InP films have been deposited on substrates of several types (Si,

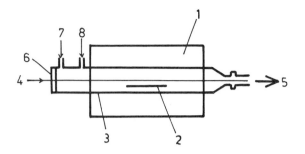

Figure 6.8 Schematic setup of the laser photo CVD deposition apparatus for the deposition of silicon nitride films with high breakdown fields: 1, furnace; 2, substrate holder; 3, quartz; furnace tube; 4, excimer laser; 5, to pump; 6, Suprasil window; 7, He inlet; 8, SiH₄, NH₃ inlet. (Data from Ref. 84.)

InP, InGaAs, GaAs) by photochemical decomposition of trimethyl indium and trimethyl phosphorus gas phase precursors. The deposition apparatus [86,87] consisted of a glass reaction cell with windows for laser transmission and fluorescence detection. The carrier gas, purified He, was bubbled through liquid $(CH_3)_3InP$ $(CH_3)_3$ at 50°C. Hydrogen purified with a Pd diffuser was used as an additional diluent and radical scavenger. To prevent condensation, the gases were mixed in a heated tubing and a second He purge was used to prevent deposition on the laser input window. The substrates could be heated to 400°C, and the system was operated at the pressure of 0.5–10 torr.

Deposition resulted from gas phase photochemistry; the surface irradiation stimulated removal of carbon and crystallization. Films ranged from amorphous to epitaxial, depending on conditions. The investigators found no film deposition in the absence of light, establishing that this was a 100% photochemical deposition.

West et al. [88] have reported a new method for the deposition of titanium silicide films using a CO_2 laser to heat gas mixtures of titanium chloride and silane above a silicon substrate whose temperature can be controlled separately from the gas mixture. Here gas mixtures of SiH_4 and $TiCl_4$ enter a cold wall CVD reactor through the mixing nozzle and argon enters through a separate port located near the laser windows, to prevent window deposits. A CO_2 (160 W) laser beam (944.195 cm^{-1}) enters the reactor parallel to the substrate and is absorbed slightly off resonance in the 944.213 cm^{-1} band of SiH_4, causing the gas temperature to increase. Argon is required to pressure broaden the absorption line width sufficiently to induce chemical vapor deposition. Argon and $TiCl_4$ do not absorb the CO_2 laser energy directly; rather, molecular collisions thermally equilibrate the reaction mixture during chemical vapor deposition. The deposition is controlled at a rate of 200 Å/min for a total pressure of 6.5–7.5 torr. The deposited film morphology has been found to depend to a large extent to the substrate temperature T_s; indeed, for T_s = 400°C, the films obtained are seen to be amorphous. It is reported that these films compare well with the films obtained by other silicide deposition methods.

Illustrative examples of photo CVD are compiled in Table 6.2, along with their salient features.

6.2.5 Plasma-enhanced Chemical Vapor Deposition

In recent years there has been a growing interest in plasma-assisted deposition technology to meet the demand for novel and better materials in microelectronics, modern optics, photovoltaic applications, and so on. Plasma-enhanced chemical vapor deposition (PECVD) is a versatile technique for depositing a wide variety of film materials. Dielectric and semiconductor films such as SiO_2, Si_3N_4, amorphous hydrogenated silicon, polycrystalline silicon, and silicon carbide have become increasingly important, since many such elemental and compound material films can be prepared at much lower temperatures than are possible with conventional CVD. In the majority of the work reported, the plasma is produced by an rf field, although dc and microwave fields have also been used. The primary role of the plasma is to promote chemical reactions, the average electron energies (1–20 eV) in the plasma being sufficient to ionize and dissociate most types of gas molecule. An important aspect in this substitution of electron kinetic energy for thermal energy is that the substrate degradation consequent to excessive heating is avoided and a variety of thin film materials can be formed on temperature-sensitive substrates such as polymers. Although electrons are the ionizing source, collisions involving excited species can be lead to the formation of free radicals and also can assist the ionization process. From a study on the formation of Si_3N_4 upon the dissociation of $SiCl_4$ and NH_3, Ron et al. [105] have concluded that the deposition is by the free radicals produced in the glow discharge. Yet it may be noted that for each systems, one must examine the role in the deposition of films of glow discharge electrons, ions, photons, and other excited species. For details on the basic phenomena and plasma chemistry, the interested reader is referred to various papers in the literature [106–108]. Hess [109] has reviewed plasma surface interaction in PECVD, and this paper makes interesting reading.

PECVD of thin film materials involves many reactor designs and different experimental conditions, and it is neither intended nor possible to discuss or catalog here all the published work.

Since the early work on silicon nitride films by PECVD in the 1960s [110,111], many different dielectric, metal, and semiconductor films have been prepared by PECVD and characterized for microelectronic, photovoltaic, and other applications. Plasma deposition of inorganic films has been discussed in detail by Hollahan and Rosler [112], and PECVD of thin films was extensively reviewed in a later work by Ojha [113]. Tungsten [114], SiO_2 [115], Si [116], GaAs and GaSb

Table 6.2 Photochemical Vapor Deposition

Film	Reactants	Source	Substrate and substrate temperature (°C)	Remarks, if any	Ref.
Zn, Se, and and ZnSe	$Zn(CH_3)_2$ and $Se(CH_3)_2$	1000 W Hg/Xe arc lamp emitting ≈ 1 W of total UV radiation < 2400 Å	Quartz RT	Vacuum of 10^{-6} torr was achieved prior to filling the test cells with $Zn(CH_3)_2$ and $Se(CH_3)_2$. Zinc selenide films obtained were nonstoichiometric. No pyrolysis occurred.	89
Mo, W, and Cr	Respective hexacarbonyls	Excimer laser output can be operated at 1570, 1930, 2480, or 3080 Å	Pyrex and quartz plates; also Si wafers RT	Substrates could be held either parallel or normal to incident laser beam. Films characterized were obtained at normal incidence. The most carbon-free films with the highest deposition rates were obtained with a laser of 2480 Å. Deposition rates for Mo, W, and Cr were 0.25, 0.17, and 0.2 μm/min, respectively, over 2.5 × 2.5 cm².	90
SiO_2	5% SiH_4 in N_2 and N_2O	Pulsed ArF laser (1930 Å), 100 Hz repetition rate 40 W/cm² average power	3 in. <100> *n*-type Si wafers 20–600	Reactant gases were mixed before flowing through the cell, maintained at 8 torr by a vacuum pumb. Laser beam held parallel to the substrate. Film quality was better at temperatures above 200°C. Films obtained were stoichiometric and contained only <5% nitrogen.	91
ZnO	Dimethyl zinc and NO_2 or N_2O	Excimer laser, either 1930 Å (ArF) or	Si wafers or quartz	Highly stoichiometric films were obtained, 97% pure DMZ and commercial	92

Material	Reactants	Energy source	Substrate	Temperature (°C)	Remarks	Ref.
		or 2480 Å (KrF)		RT to 220	grade NO_2 or N_2O used; NO_2 or N_2O together with DMZ introduced into the cell. At 2480 Å, with N_2O as the oxygen donor, clear films of ZnO were obtained at a deposition rate of 200 Å/min. There was a slight increase in deposition rate with 1930 Å. Results of 2480 Å with NO_2 not any better. Highest deposition rate (3000 Å/min.) was obtained with NO_2 irradiated with 1930 Å. Both perpendicular and parallel substrate irradiation was done.	93
Ti	$TiCl_4$	5 mW UV laser, 5145 Å	Li NbO_3		Typical average UV laser beam intensities at the substrate were less than 50 kW/cm²; temperature rise at beam focus was less than 30°C. Photodeposited films were characterized by solvent tests, X-ray fluorescence, and SEM.	
Ge	GeH_4 in He	KrF laser; 2480 Å	Amorphous SiO_2 (quartz) and (100) NaCl		On NaCl at a substrate temperature of 120°C, the films were epitaxial; growth rates up to 3.6 μm/h were reported.	94
a-Si:H	10% disilane (S_2H_6) diluted with 90% He	Low pressure Hg lamp, 2537 Å	Si	< 300	Films were prepared by mercury photosensitized decomposition. Reactant gases S_2H_6/He were premixed with a small amount of mercury vapor in a temperature-controlled mercury vaporizer. Substrate temperature was not	95

Table 6.2 (continued)

Film	Reactants	Source	Substrate and substrate temperature (°C)	Remarks, if any	Ref.
				very important in determining the deposition rate, but rate could be enhanced by raising the mercury vaporizer temperature. Conductivity strongly depended on the substrate temperature during deposition. *n*-type doping has been achieved by adding phosphine to disilane during deposition.	
W	WF_6 H_2	ArF excimer laser, 1930 Å; average power, 4–7 W	Si and Si covered with thermal SiO_2 240–440	Deposition rates have exceeded 1000 Å/min. Film resistivities as low as twice the bulk value have been obtained at 440°C. Properties were compared with those obtained by conventional thermal deposition.	96
Phosphorus nitride	NH_3 (100%) and PH_3 (2% pure H_2)	ArF excimer laser, 1930 Å	InP 100–300	Fused quartz cylindrical lens provided a rectangualr beam parallel to the substrate. Films exhibited a much lower leakage current than that of thermal CVD. Best results were obtained at 300°C. Deposition rate was 400 Å/min.	97
a-SiC:H	Methylsilanes or acetylene and disilane	Low pressure Hg lamp at 1849Å	Corning glass 200	Disilane gas was diluted in helium (He/Si_2H_6 = 9:1). Total gas pressure was 5 torr; deposition rate, 1 Å/s. Undoped and boron-doped films were prepared.	98

Material	Precursor	Lamp/UV source	Substrate / Temperature	Description	Ref.
$Cd_x Hg_{1-x}$ Te (CMT)	Diethyl telluride, dimethyl cadmium and mercury	3 kW Hg arc lamp	CdTe, InSb 250	CMT was prepared by the nonthermal decomposition of metal organics using UV photolysis. UV source was a 3 kW mercury arc lamp. Horizontal flow system was operated at 1 atm total pressure. Metal organics flowed over Hg reservior and mixed with Hg vapor before passing over the substrate; carrier gas was H_2/He. Substrate was heated by infrared lamp.	99
Si	$Si_2H_6 + SiH_2F_2 + H_2$ or $SiH_4 + SiH_2F_2 + H_2$	Low pressure Hg lamp at 1849 Å, 2537 Å	(100) Si 100–300	Epitaxial films were prepared.	100
Zn_3P_2	Dimethyl zinc and PH_3	Low pressure Hg lamp, 1849 and 2537 Å	Stainless steel, Si (111), ITO/glass, and ordinary glass plates RT to 250	DMZ and PH_3 were introduced together into the reactor. Deposition of the films on the different substrates was carried out in the same chamber at the same time. Deposition rates were very much affected by UV light intensity, the density of gases, PH_3/DMZ molar ratio, and substrate temperature.	101
GaAs	10% mixture of arsine in 99.995% pure H_2 and 99.999% pure triethyl gallium	1000 W Hg-Xe arc lamp	Synthetic fused silica ≈ 240	Film deposition was carried out in an LPCVD system. This is the first report of a GaAs deposition process using this source with no deposition occurring in the absence of light. X-Ray diffraction, energy-dispersive spectrometry, and optical transmittance were used to analyze the films.	102

Table 6.2 (continued)

Film	Reactants	Source	Substrate and substrate temperature (°C)	Remarks, if any	Ref.
Si_3N_4	SiH_4 and NH_3	Quartz Hg arc lamp	50–250	Optical, electrical, and morphological properties were studied. Film deposition on the quartz window was avoided by having thin layer of special oil to prevent adhesion and deposition. Chamber pressure was controlled by bleeding nitrogen at the intake of the rotary pump. During the first 10 minutes the deposition rate was about 56 Å/min and it decreased with time.	103
$a\text{-}SiO_2$	O_2, Si_2H_6 and Si_2F_6 or SiF_4	Deuterium (D_2) lamp	n-type Si wafers 200	Deuterium lamp was set to irradiate vacuum UV light through an MgF_2 window perpendicular to the substrate. Mixing Si_2F_6, defects, such as—H and -OH in the films, were effectively eliminated; growth rate was enhanced; and a slight doping of fluorine was found in the films. SiF_4 had no effect except a slight doping of fluorine.	104
SiO_2	Si_2H_6 and O_2	Kr resonance lamp (123.6 nm)	Si 145	Si_2H_6 was selected excited but not O_2. Amount of Si—OH bonding in the films was much lower than found with other vacuum UV lamps.	199

Silicon oxynitride	Si_2H_6, NH_3, and NO_2	Vacuum UV light of deuterium lamp	≈ 330	Film deposition rate varied with the NO_2 flow rate and also with the excitation source spectrum, which was varied by a low pass filter of a synthetic or fused silica plate. Film composition was sensitive to NO_2 flow rate.	200
Ge	GeH_4	193 nm laser, ArF	Cr or Si doped (100) GaAs RT to 415	Growth process consisted of photodissociating GeH_4 in a parallel geometry; films grown at different substrate temperatures were examined by TEM. This work indicates that the laser directly produces a species (radical) and is collisionally transformed en route to the substrate into a more stable species. Upon reaching the substrate surface, this species (Ge_2H_6) is pyrolyzed to form the epitaxial layer.	201
TiN	$TiCl_4/NH_3$ (or N_2)/H_2	D_2 lamp	SUS 302, 0.5 × 10 × 10 mm³	Temperature could be lowered by 50–100°C by irradiation (vs. without irradiation). Deposition rate increased by 35-300% with irradiation.	202
SnO_2	$SnCl_4$ and N_2O	ArF (193 nm)	Fused silica RT	Optical bandgap of 3.20 eV and transmission cutoff wavelength of 330 nm compared favorably with films deposited by other methods. Selective area photodeposition of SnO_2 with resistivities as low as 0.04 Ω cm has been demonstrated.	203

Table 6.2 (continued)

Film	Reactants	Source	Substrate and substrate temperature (°C)	Remarks, if any	Ref.
TiC	$TiCl_4/CH_4$ (or C $Cl_4)/H_2/Ar$	UV light from D_2 lamp (160 nm)	Graphite or Cu plate, $0.5 \times 10 \times 10$ mm^3 800–900	Deposition temperature and deposition rate compared with that without irradiation by the D_2 lamp. Deposition rate increased 1.5–2.3 times by irradiation. No lowering of deposition temperature with irradiation could be observed.	204
a-C:H	C_2H_2 (5%) diluted in Ar (95%)	ArF laser (193 nm)	Single-crystal wafers of Si or GaAs 150–350 Also quartz glass, 350	Laser beam was directed parallel to and above the substrate at a distance of 3 mm. Reaction gas mixture was supplied through a flat nozzle, which directed the gas through the light path to the substrate at an angle of 20°. Properties including growth rate and microhardness were studied as a function of substrate temperature.	205

W, C, and W/C multi-layers	WF_6/C_6H_6	ArF excimer laser 193 nm	B-doped (100) silicon crystals RT to 300	Total pressure of WF_6 (or C_6H_6) and He gas mixture was kept at 20 torr. Dependence of tungsten and carbon film characteristics on substrate temperature was studied. Carbon films were deposited directly on the substrates and then W films on them. Amorphous W and C films with very smooth surfaces obtained at substrate temperatures of 100–200 and 100–300°C for W and C, respectively.	206
TiB_2	$TiCl_4/BCl_3/H_2/Ar$	D_2 lamp 160 nm	Cu plate (0.2 × 10 × 10 mm³) 600–800	With irradiation, deposition could be obtained at a temperature as low as 600°C. Deposition temperature was lowered by 50°C by irradiation; deposition rate increased 1.5–2.5 times with irradiation.	207

[117], titanium silicide-TiS_x [118], and many other materials have been prepared by PECVD. Nguyen [119] has reviewed plasma-assisted chemical vapor deposition of thin films for microelectronic applications. Table 6.3 gives the salient features of PECVD of thin films of certain important materials reported recently.

Also discussed below are certain specific designs of plasma CVD systems of recent interest.

Sakai et al. [142] have reported the preparation of a-Si:H films using what they call an "interdigital vertical electrode deposition apparatus" design for high throughput deposition of a-Si:H on rigid substrates such as glass or thick sheet metal. The apparatus consists of a heating chamber, three deposition chambers, and a cooling chamber, the substrates being placed vertically in the deposition chamber. An interdigitally arranged vertical electrode configuration is shown schematically in Figure 6.9. Since the configuration provide four plasma sections, a-Si:H films can be deposited simultaneously on four substrates. Sakai et al. used the following deposition conditions in their experiment.

Gas mixture ratio, $SiH_4/(SiH_4 + H_2)$	10–100%
RF power density	10–20 mW/cm^2
Total gas pressure	0.1–2.0 torr
SiH_4 flow ratio	60 sccm
Substrate temperature	200–300°C

Sakai et al. have evaluated films thus prepared by terms of thickness, uniformity, and optical and electrical properties.

The PECVD technique has been used to prepare SiO_2 and Si_3N_4 films at low substrate temperatures from gas mixtures of SiH_4 and either O_2 or N_2, respectively. The usual approach is to have a parallel plate reactor and use a capacitive coupling of rf power into the glow discharge containing the gas mixtures. The films produced by this PECVD process contains H incorporated in the form of SiH and OH groups in oxides and SiH and NH in the nitrides [143]. But Richard et al. [144], who have reported a new low temperature process, remote plasma-enhanced CVD (RPECVD), have been able to deposit silicon dioxide and silicon nitride films at substrate temperatures between 350 and 500°C [145], with no measurable IR absorption associated with the bonding groups. Their process involves the rf excitation of only one of the constituent gases—the oxygen- or nitrogen-containing molecules or a mixture of this gas with a rare gas, which is then mixed with SiH_4. The deposition chamber of their design (Figure 6.10) provides two separate gas feed lines, one located at the top of the chamber and the other coupled into the dispersal ring at the center. The gas, $NH_3/N_2/O_2$ (or mixture of the gases with He or

Table 6.3 Plasma-Enhanced Chemical Vapor Deposition

Film	Input Materials	Discharge Data	Substrate and substrate temperature (°C)	Remarks, if any	Ref.
Si	$SiCl_4$, H_2 and Ar	Rf, 27.12 MHz	Stainless steel	Plasma was initiated in a quartz reactor inductively coupled to rf generator. Silicon films were formed by decomposition of $SiCl_4$ in the plasma and their reduction by hydrogen (B_2H_6 or PH_3 added for doping). Chemical and electrical characteristics of microcrystalline Si films deposited were studied using XPS and conductivity measurements.	120
B-C-N-H	B_2H_6 (4.8 vol % in N_2), C_2H_6, and Ar	Rf discharge power level of 20 W, 13.56 MHz	Corning 7059 glass and NaCl crystals	Films were deposited at RT without heating the substrate. Films were transparent at wavelengths of 2000–10,000 Å. Deposition rate was 200 Å/min. Transparency of the films increased with decreasing deposition rate.	121
a-Si	SiH_4	Rf hollow cathode discharge	Glass and Cu; no external substrate heating	Preliminary results gave a deposition rate of 2nm/s. Deposition rates appeared to be constant, with roughly equal rates at 20 and 6 Pa SiH_4 pressures. Films were hard and adherent, and showed moderate photoconductivity. From dark conductivity measurements, the films were confirmed to be amorphous.	122

Table 6.3 (continued)

Film	Input Materials	Discharge Data	Substrate and substrate temperature (°C)	Remarks, if any	Ref.
Si	SiH$_4$	Rf	2 in. Si wafer 650	System used CVD of silane at very low pressures ($< 10^{-2}$ torr), with in situ cleaning of the surface. Adatom surface mobility at pressures of 10^{-2} torr was high enough to allow epitaxial growth at temperatures as low as 650°C.	123
Diamondlike carbon	CH$_4$, CH$_4$/H$_2$ CH$_4$/Ar and CH$_4$/He	Microwave discharge; microwave power at a frequency of 2450 MHz from a 1 kW magnetron	(111) oriented boron doped p-Si wafers 700 (in CH$_4$ and CH$_4$/H$_2$); 800 (inCH$_4$/Ar and CH$_4$/He plasma)	Substrate was kept in a quartz tube placed in the discharge tube. Total pressure in the discharge tube after introduction of the gas was 133 Pa. Effect of the dilution gases on the deposition was examined.	124
BN	B$_2$H$_6$ in H$_2$ + NH$_3$	13.56 MHz rf		Effect of low intensity magnetic field on the deposition rate in a parallel plate plasma reactor was examined. Deposition rate was greater with the applied magnetic field at 400 mtorr pressure, whereas above this value the deposition rate was higher without magnetic field enhancement. Total pressure was 100–700 mtorr; diborane, 5.32% in H$_2$.	125
Si:H	SiH$_4$/H$_2$ gas mixture	Rf power, 6 W	200	Total gas pressure was 1 torr. Silane fraction, defined by SiH$_4$/SiH$_4$ + H$_2$	126

Material	Source gases	Method	Substrate / Temperature	Remarks	Ref.
$a\text{-}Si_{1-x}Ge_x\text{:}H$	SiH_4/GeH_4 mixture	Rf power, 3 W, 13.56 MHz	200–400	in the silane/hydrogen gas mixture, was varied from 100% to 2%. Flow rate also was changed depending on silane fraction: Amorphous to crystalline state was observed at a silane fraction of about 10%. Optoelectronic properties of the deposited films were studied.	127a
TiB_2	$TiCl_3$, BCl_3, and H_2	Rf power, 20 W, 15-MHz	Sapphire, Pyrex, quartz, p-type Si, and thermally oxidized Si 480–650	Chamber pressure kept at 0.25 torr; x values ranging from 0.14 to 0.74 were grown at 300°C. Unalloyed a-Si:H and a-Ge:H films were also prepared. SiH_4 and GeH_4 were 99.4% pure. Structural, electrical, and optical properties were investigated.	127b
TiN	$TiCl_4$, H_2 N_2, and Ar	Dc glow discharge	Tool steel 500	High quality titanium boride films that are smooth, shiny, crackfree, and highly conductive may be used as diffusion barriers for integrated circuit.	128
Ti (O, C, N)	$Ti(OC_3H_7)_4$ (titanium isopropoxide), H_2, N_2, and Ar			A rotating gas supply was used to feed the gas mixture over almost the entire length of the reactor, directly toward the cathode fall around the substrates. The advanatge of using this organic titanium compound is the absence of any chloride products in the deposited layers.	

Table 6.3 (continued)

Film	Input Materials	Discharge Data	Substrate and substrate temperature (°C)	Remarks, if any	Ref.
SiO_x (Si sub-oxide) $x<2$	$SiH_4 + N_2O$ mixture diluted with He	Rf power, 20–60 W	350	A conventional PECVD system was used. Total pressure was 75–125 mtorr; N_2O/SiH_4 ratio, 5–10; He dilution, 0–60 vol %; 50% gave film composition close to $SiO_{1.9}$. Local atomic structure was studied using IR spectroscopy.	129
a-$Si_{1-x}Ge_x$:H, F	SiF_4, GeF_4, and H_2	Rf glow powered by 300 W rf power generator	Glass 400	Deposition pressure typically was 0.11 torr, base pressure about 10^{-7} torr. An ultra high vacuum system has been designed and constructed for the preparation of the films. Composition, structural, electrical, and optical properties of the alloy films were reported.	130
Silicon Nitride	SiH_4, N_2, H_2 gas mixture	Rf glow discharge in a capacitive reactor, 13.56 MHz	Highly resistive Si wafers and Mo-coated glass 300	SiH_4 (10% diluted in H_2) and N_2 were introduced to the reaction chamber through a mass flow controller system. Film composition and the concentration of bonded hydrogen depended strongly on the conditions of deposition. Higher rf power was needed because of the higher dissociation energy compared with that of NH_3.	131

BN	B_2H_6, NH_3, and H_2	Rf power supply at 13.56 MHz	300	Effect of a low intensity magnetic field on the growth of films in a low pressure parallel plate plasma reactor was studied, along with structural and optical properties. Optical properties were identical with and without the application of a low magnetic field.	132
SiO_2	SiH_4 and NO_2	Rf discharge power, 10–50 W	n-type crystalline Si <100> wafers	Films of near stoichiometric character were formed in a parallel plate system; base pressure, 10^{-6} torr. Gases were premixed before being admitted to the chamber. Ratio of NO_2 to SiH_4 varied from 5 to 30. Flow rate of SiH_4 was 6 sccm (fixed); deposition rate, 5–9 Å/s. Physical and electrical properties were comparable with films produced by N_2O/SiH_4 mixture.	133
a-Si:H	$SiH_4 + B_2H_6$	Rf glow discharge	Glass 250	p-type films produced by mixing 1×10^{-3}, 5×10^{-3}, and 1×10^{-2} parts by volume of diborane in silane. Influence of boron doping on the transport properties was studied.	134
a-SiC:H	$SiH_4 + CH_4 + B_2H_6$	Rf glow discharge	Glass 250	p-type SiC:H films have been prepared with 1×10^{-3} and 1×10^{-2} parts of diborane in silane + methane gas. Effect of boron doping was investigated.	134

Table 6.3 (continued)

Film	Input Materials	Discharge Data	Substrate and substrate temperature (°C)	Remarks, if any	Ref.
Silicon Nitride	SiH_4 (2% in N_2) and NH_3; 100% SiH_4 and NH_3	Rf glow discharge, 50 kHz	Si 325	Used parallel plate plasma reactor: pressure, 0.25 torr with NH_3/SiH_4 (2% in N_2) and 0.20 torr with SiH_4/NH_3 mixture. Physical, chemical, and structural propreties of the films prepared in the two cases were compared.	135
Mo	$Ar + Mo(CO)_6$ or $H_2 + Mo(CO)_6$ mixture	Rf discharge power at 100 W	Thermally grown SiO_2 over Si substrate 100 and 300	Films were obtained from the first mixture contained large quantities of carbon and oxygen before and after annealing. Films deposited from the other mixture contained carbon, but oxygen was present in the form of hydroxide only. Annealing removed oxygen.	136
AlN	$AlBr_3 + N_2 + H_2$ + Ar gas mixture	Rf discharge, 50–500 W power, 13–56 MHz	Graphite 200–800	Morphology of AlN films was examined by SEM. Fine-grained polycrystalline AlN films were obtained at 700°C under a total pressure below 10 torr. Translucent polycrystalline films with <001> preferred orientations were grown at a total pressure of 10–40 torr.	137
TiN	$TiCl_4$, N_2, and H_2	Rf discharge power, 0–200 W	M_2 steel 350–500	Total pressure was 1 torr; $TiCl_4$ flow rate maintained constant. Ratio of	138

H_2 to N_2 varied from 0.7 to 4.5. Applied bias from −250 to 0 V (both grounded and floating conditions used). At 500°C, highly crystalline stoichiometric TiN films were deposited. Plasma was essential for TiN formation for 500°C and below.

Material	Reactants	Method	Substrate (temp.)	Description	Ref.
$SiO_x N_y H_z$	SiH_4 in He, N_2O, and NH_3	Rf discharge, 13.56 MHz	Si, polished C and glass 200	Total pressure was 500 mtorr. Silicon oxtnitride film composition was changed by varying the ratio of N_2O to NH_3 and by alternating the total pressure. Deposition rate was found to be proportional to the concentration of N_2O. Reactor configuration was such that bombardment of ions and other particles was reduced to a minimum.	139
Silicon Nitride	Mixture of NH_3 and SiH_4 (10% SiH_4 in Ar)	Rf discharge, 0.44 W and 13.56 MHz	Single-crystal Si plates (100) 300–450	Parallel plate plasma reactor was used. Ratio NH_3 to SiH_4 concentration varied from 1.2 to 10. Dependence of the deposition rate, electrical resistivity, and RI on temperature, pressure, and gas mixture was investigated.	140
Fluorinated $SiH_x H_y$	SiH_4/He/NF_3/NH_3 also SiH_4/He/ NH_3/F_2 mixtures	Rf power density, 0.4 W/cm^{-2} 13.56 MHz	p-type Si <100> wafers 300	Films were deposited in a stainless steel, axial flow, capacitively coupled reactor. Electrical properties were investigated and compared with unfluorinated silicon nitride films deposited from SiH_4/He/	141a

Table 6.3 (continued)

Film	Input Materials	Discharge Data	Substrate and substrate temperature (°C)	Remarks, if any	Ref.
Si	SiH_4	Rf, 2.5–20 W; 13.56 MHz	(100) oriented n-type wafers 700–800	NH_3 plasmas. With fluorine added, the dielectric strength increased. Epitaxial Si films were deposited in pure silane at a pressure of 6 mtorr. Films were n type, with a carrier concentrations of 1–10×10^{15} cm^{-3}. In contrast to ultra low pressure CVD, a low power PECVD (2.5–5 W) improved the electrical quality and lowered the minimum epitaxy temperature. High power (20 W) degraded the electrical quality in comparison to a strictly thermal deposition.	141b
$Si_{1-x} C_x{:}F{,}H$	SiF_4, CF_4, and H_2	13.56 MHz power density 150,500 or 800 mW/cm^2	Corning glass/ polished Si wafers	Films with a wide range of structural and optoelectronic properties were formed. Dark conductivity and activation energy were seen to be very sensitive to the proportion of hydrogen.	208
a-C:H	CH_4 and H_2	Low frequency (50 Hz) plasma	Si and glass RT	Films obtained were transparent, very uniform, and also highly resistive ($> 10^{14}$ Ω cm), with high breakdown field strength (10^6 V/cm).	209
a-Si:H,F	SiF_4 and H_2	Rf glow discharge (27 MHz)	Corning 7059 glass	Feeding mixture composition was kept constant and other deposition para-	210

				meters changed. Discharge pressure of 20–80 Pa and discharge power of 4–40 W were used for different samples. Films were deposited under ionic bombardment conditions by placing the substrate on the rf-powered electrode.	
SiN$_4$	Silane (diluted in N$_2$), NH$_3$, and N$_2$	Rf power 20 W	p-type Si (111)	A conventional PECVD system with parallel plates (35 cm dia.) was used; temperature of the lower plate was 350°C. Total flow rate of the reactant gases in the chamber was 162 sccm; ammonia-to-silane flow ratio, 18; pressure kept at 107 Pa. Under these conductions, the films had low hydrogen content with near stoichiometric composition.	211
Boron-doped a-Si:H	Silane mixture (5% SiH$_4$ in 95% H$_2$) mixed with different amounts of B$_2$H$_6$	Rf (13.56 MHz) power 5–30 W	Corning 7059 glass 200–280	Films prepared with different amounts of diborane and their electrical properties were studied.	212
a-SiC$_x$:H	SiH$_4$ + a hydro- carbon precursor	Rf glow discharge 13.56 MHz	Variety of substrates	Mixing ratio of silane and the hydro- carbon gas (molar gas concentration M) was 0–0.85. Deposition rate was found to be higher with C$_2$H$_2$ than with C$_2$H$_6$, C$_2$H$_4$, or CH$_4$. By choos- ing the appropriate molar gas con- centration, a-SiC$_x$:H films of ident- ical composition could be prepared.	213a

Table 6.3 (continued)

Film	Input Materials	Discharge Data	Substrate and substrate temperature (°C)	Remarks, if any	Ref.
TiN and TiC	N_2 (for TiN), CH_4 (for TiC), $TiCl_4$, and Ar	Dc voltage, 350–500 V	M_2 high speed steel and pure Fe 425–600	Micro-Vickers hardnesses of the coatings were determined. Hardness decreased with increasing chlorine content. Hardness of TiC was reduced with excess carbon. Wear and galling resistances were also determined.	213b
a-SiN:H	NH_3 and SiH_4	440 kHz, 20 W rf power	Si wafers 360	Semiinsulating a-SiN:H films were deposited in a hot wall horizontal tube plasma reactor; NH_3/SiH_4 ratio varied from 5 to 0.5. Electrical, physical, and chemical characteristics of the film for use as a resistance field shield to passivate high voltage integrated circuits were studied. Film properties were found to be controlled by NH_3/SiH_4 flow ratio.	214
Diamond	CH_4 (5%) H_2 (99.5%)	Dc plasma discharge voltage, 700 V Current density, 1.8 Å/cm²	C-BN {111} 900	During the initial stages of film growth, island structure was observed. Continuous film was obtained at a thickness of about 2000Å, giving a smooth surface.	215a
a-C:H	C_2H_2, 99.95%	13.56 MHz rf	Si	Using conventional vacuum equipment with a stainless steel chamber, the substrate was applied directly onto the	215b

				water-cooled cathode. Bias voltage of substrate was -300 V. During preparation, the C_2H_2 gas pressure was changed from 2×10^{-1} to 5.0×10^{-3} mbar to prepare films with different properties. Mechanical properties of the films were investigated.	
Diamond	CH_4 and H_2	Arc discharge voltage 200–300 V. Corresponding current 5–4 A	Single-crystal Si M wafers 800–1000	Diamond films were prepared by arc discharge plasma CVD. In situ optical emission spectroscopy (OES) measurements were carried out during the growth process. Results indicate the presence of a number of atomic hydrogens in the dc arc discharge plasma, which is the key factor in preparing diamond films of high quality at high deposition rates. A typical condition was a CH_4/H_2 ratio not exceeding 1%.	216
μc-Si	Silane and H_2	Rf power varied	Corning 7059 glass 50–350	Studied the effects on the growth of microcrystalline silicon films of substrate temperature and rf power. Added 1% diborane in silane to make p-type films; H_2/SiH_4 ratio was fixed at 50. Optimum conditions: rf power, 15 W; substrate temperature, 230°C. For the deposition process, H_2/SiH_4 passed into the downstream reactor and helium through the plasma generating region.	217

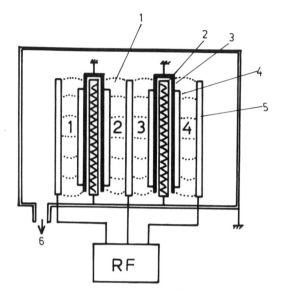

Figure 6.9 Schematic diagram of an interdigitally arranged vertical electrode configuration: 1, plasma; 2, grounded electrode; 3, substrate holder; 4, substrate; 5, electrode; 6, to pump. (Data form Ref. 142.)

Ar), is supplied to the center of the feed line and is inductively excited at the top of the chamber. This RPECVD is a four-step process in which the films are deposited outside the plasma region.

1. Rf excitation of the gas or gas mixture
2. Transport of the excited nitrogen or oxygen species (other atoms, molecules, radicals, etc.) that contained in the gas or gas mixture out of the plasma region.
3. Mixing and gas phase reaction of the excited nitrogen or oxygen with neutral silane or disilane in the chamber outside the plasma region.
4. Final process of CVD reaction at the heated substrate.

The first three steps in the RPECVD process are meant to generate gas phase precursors, either molecules or radicals that include the bonding group in the deposited films. For example, one such precursor molecule is disiloxane [146,147]. This molecule contains the planar Si—O—Si group, the structural building block for stoichiometric SiO_2. The deposition parameters used for the deposition of SiO_2, $Si(NH_2)$, and Si_3N_4, reported in the work of Richard et al. are given in Table 6.4.

The gas flow rates and operating pressures are such that the flow of gas molecules back into the plasma region from other parts of the chamber is unlikely.

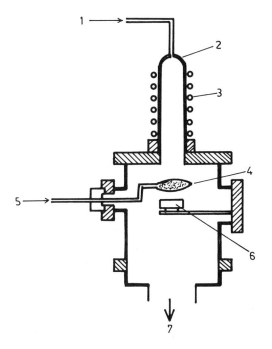

Figure 6.10 Schematic diagram of the deposition chamber for remote plasma-enhanced CVD: 1, gas mixture feed line; 2, Pyrex tube; 3, rf coil; 4, gas ring; 5, silane inlet; 6, substrate heater block; 7, to pump. (Data from Ref. 145.)

Table 6.4 Deposition Parameters for the RPECVD Process

Base pressure	$\approx 5 \times 10^{-8}$ torr
Operating pressure	300 mtorr
Gas mixtures	10% silane in Ar, 20% O_2 in He, 20% N_2 in He

Deposited Film	Rf power (W)	Gas mixture A	Flow rate (sccm) A	Flow rate (sccm) Silane, Ar	Substrate temperature (°C)	Average deposition (Å/s)
SiO_2	10	O_2,He	75	11.6	350	2.5
$Si(NH)_2$	25	NH_3	75	11.6	100	0.5
Si_3N_4	25	NH_3	75	11.6	550	0.4
		N_2, He	65	11.6	400	0.1

Source: Ref. 145.

An experimental study of the deposition conditions that lead to the different types of nitride films in RPECVD also was reported [148a].

RPECVD has been successfully used to deposit low interface state density SiO_2 at 300°C [148b,c].

The nitride deposition process (RPECVD) discussed earlier uses either N_2 or NH_3 (mixed with inert gas) discharge to create nitrogen species, which react with SiH_4 downstream from the plasma tube to form the film. Usually N_2 is preferred as a gas source for nitride deposition, because NH_3 presents problems of gas purity and also results in increased incorporation of bonded hydrogen in the films [148a].

Hattangady et al. [148d] have recently reported the deposition of low hydrogen content silicon nitride at low temperature by a variant of the RPECVD technique [148e] in which an inert gas (Ar) discharge excites nitrogen and silane, both introduced downstream from the plasma region. Here the nitrogen is excited in the immediate vicinity of the silane species and close to the substrate surface. Thus the nitrogen species do not have to be transported from the remote discharge, and the loss due to collisional deexcitation, both with ground state nitrogen molecules and the tube wall, is avoided.

Mito and Sekiguchi [149] have reported the deposition of silicon nitride (SiN) films on silicon substrates by induction-heated (IH) plasma-assisted CVD. Their experimental setup is shown schematically in Figure 6.11. The IH plasma was generated in an inductively coupled quartz tube. Nitrogen was introduced into this discharge tube, and a high rf power (3–4 kW) source applied at 1 torr pressure. The gas in the IH plasma is thermally excited so that it contains radicals with long lifetimes, and intense photoemissions are radiated from the IH plasma. The SiH_4 in the depo-

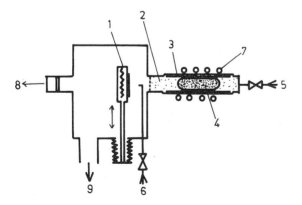

Figure 6.11   Schematic diagram of the experimental setup for induction-heated plasma-assisted CVD of silicon nitride films on silicon substrates: 1, substrate holder; 2, glowlike plasma; 3, quartz tube; 4, IH plasma; 5, N_2 inlet; 6, SiH_4 inlet; 7, rf coil; 8, Vacuum UV spectroscope; 9, to pump. (Data from Ref. 149.)

sition chamber is dissociated by the vacuum ultraviolet emission without ionic collisions. There are two deposition modes for the preparation of films.

Mode 1, photo (vacuum UV emission) and radical assisted CVD, gave a deposition rate of 60 Å/min. Here the plasma does not come into contact with the substrate.

Mode 2 is plasma-assisted CVD in which the glowlike plasma generated around the IH plasma comes into contact with the substrate. A deposition rate of 500 Å/min was reported [149].

This method is almost similar to that of RPECVD discussed earlier. But in the method above the rf power supply is very high; the plasma in the discharge tube radiates intense photoemission (mode 1), and the deposition is a photo + radical-assisted CVD. Deposition mode 2 is a versatile high rate PECVD.

In conventional PECVD, which usually features an rf discharge plasma, the substrates are normally introduced into the discharge region and therefore become exposed to the plasma containing highly energetic particles (electrons, ions, etc.), resulting in radiation damage. Moreover, we cannot avoid the incorporation of impurities sputtered from electrodes into the growing films.

The nonexposure of the substrate to energetic species and charged particles is turned to advantage in thin film growth methods by using microwave excited plasma [150,151]. Zaima et al. [152a] have reported the low temperature thin film deposition of dielectric films such as SiN_x, using microwave-excited plasma, where the discharge region can be separated from the reaction region. In setup (Figure 6.12), the microwave exciting plasma chamber and the reaction chamber are separated. A 2.45 GHz microwave generated by the magnetron is guided through

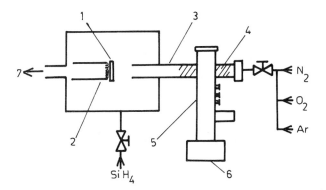

Figure 6.12 Schematic diagram of the microwave-excited apparatus for the deposition of thin PECVD dielectric films: 1, substrate; 2, heater; 3, quartz tube; 4, plasma; 5, waveguide; 6, magnetron; 7, to pump. (Data from Ref. 152a.)

the rectangular waveguide to the 32 mm diameter quartz tube serving as the plasma exciting chamber. The substrate is kept in the deposition (reaction) chamber (300 mm from the center of the discharge region). The substrate can be heated to 600°C with a substrate heater. The chambers can be evacuated to 1×10^{-7} torr. Nitrogen gas excited in the plasma chamber diffuses toward the reaction chamber and reacts with the unexcited SiH_4 (20% diluted in N_2) and the microwave-excited N_2 (or N_2/Ar), depositing SiN_x films. Zaima et al. have found the silicon nitride films formed to have a nearly stoichiometric composition in a wide range of experimental conditions, with excellent dielectric properties .

CVD by microwave electron cyclotron resonance (ECR) is a new technology for the deposition of thin films. The electron cyclotron discharge, with its ability to create a high density of charged and excited species at low pressures (10^{-5}-10^{-3} torr), is made use of in the CVD of thin films, and the unique plasma environment available through ECR condition [152b] offers distinct advantages over the conventional plasma-enhanced CVD process.

Numerous reports on ECR CVD of thin films [152c-e] have been published since the first report [152f] of the development of the application of an ECR discharge for the CVD of thin films in 1983. This newly developed ECR plasma deposition system (Figure 6.13) consists of two chambers, the plasma chamber and the deposition chamber. The plasma chamber receives 2.45 GHz microwave power being connected to the microwave power supply through a rectangular waveguide and window made of fused quartz plate. The plasma chamber operates as microwave cavity resonator, and two coaxial magnet coils are mounted around the periphery of the plasma chamber for ECR plasma excitation. Electron cyclotron resonance occurs at a magnetic field of 875 G, and a highly activated plasma is obtained.

In this deposition system the ions are extracted from the plasma chamber to the deposition chamber and flow out of the diverging magnetic field to deposit a film on the substrate.

For silicon nitride film deposition, nitrogen gas is introduced into the plasma chamber, and silane (SiH_4) is introduced into the deposition chamber. For SiO_2, oxygen gas is introduced into the plasma chamber in place of nitrogen. The investigators were able to obtain high quality films without substrate heating.

High quality a-Si:H films have been prepared by microwave ECR plasma CVD using SiH_4 gas with excited argon or hydrogen plasma at deposition temperatures below 150°C [152g,h]. In a continuation of their earlier report [152i] Kitagawa et al. [152j] have reported the influence of deposition conditions on the properties of a-Si:H films prepared by the ECR plasma CVD method using SiH_4 gas without intentional substrate heating.

The deposition conditions reported were as follows.

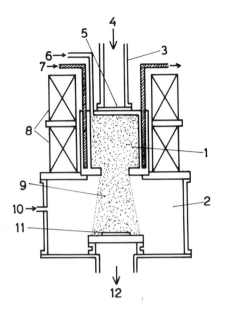

Figure 6.13 Schematic diagram of the ECR plasma deposition system: 1, plasma chamber, 2, process chamber; 3, rectangular waveguide; 4, 2.45 GHz microwave power; 5, quartz window; 6, gas inlet; 7, cooling water; 8, magnetic coils; 9, plasma stream; 10, gas inlet; 11, substrate; 12, to pump. (Data from Ref. 152d.)

Substrates	Corning 7059 glass, crystalline silicon wafers
Substrate temperature	< 60°C, without intentional heating
Backpressure	2×10^{-6} torr
SiH$_4$ flow rate	4–30 sccm
Pressure	1×10^{-4} to 2×10^{-3} torr
Microwave frequency	2.45 GHz
Microwave power	50–350 W
Magnetic field	875 G

The films prepared at pressures below 4×10^{-4} torr exhibited higher photoconductivity and a wider optical gap than the films deposited at pressures above 4×10^{-4} torr.

Pool and Shing [152k] have reported recently the deposition of hard a-C:H films using ECR microwave plasma decomposition of CH4 diluted with H2 gas. The two coaxial coils mounted approximately 33 cm apart supply the confining magnetic field in a magnetic mirror configuration. The plasma generation gas (H2) is introduced into the chamber through an annular distribution ring located just below the quartz window. The substrate stage (stainless steel) is located 15 cm below the aperture connecting the plasma chamber and the deposition chamber. The substrate stage can be biased by the application of an external dc or rf voltage. The reaction gas (CH4) is introduced through a circular gas ring in the deposition chamber.

The general deposition data are as follows.

Microwave frequency	2.45 GHz
Microwave power	360 W
Magnetic field (ECR)	875 G
Substrates	Corning 7059 optical glass, quartz and p-type silicon (0.1 Ω cm)

All depositions were made at ambient temperature. The films studied were deposited without an external bias, with an applied negative bias, and with a 13.56 MHz rf external bias applied to the substrate stage. It has been found that the properties of the films were strongly dependent on an rf-induced negative self-bias of the substrate stage, as well as on the magnetic field profile used during deposition.

The preparation of amorphous boron films by ECR plasma CVD using B2H6/H2 gas mixtures has been reported by Shirai and Gonda [152l]. The films were deposited at very low pressures (0.1 Pa) on quartz and p-type silicon at substrate temperatures below 600°C. The films were prepared over a wide range of parameters, and the deposition characteristics were studied by observing the effects of the deposition rate on various parameters (e.g., substrate temperature during deposition, diborane concentration in the source gases, gas flow rate, gas pressure).

Asmussen [152m] has reviewed the basic physics of ECR discharges and the associated microwave system and applicator technologies, and this paper will be useful for interested readers. As has been pointed out here, the fact that microwave discharges can be produced without high plasma potential is a definite advantage over rf parallel plate discharge because of the absence of surface damage. There are, however, other sources of damage, such as high energy electrons and ions produced and accelerated in mirror magnetic fields, direct microwave radiation, and ultraviolet and other plasma radiations. These areas must be evaluated for each specific application.

Diamond thin films have been grown by dc plasma chemical vapor deposition from methane and hydrogen gases on Si and a-Al2O3 substrates [153a], as shown

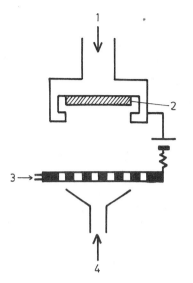

Figure 6.14 Schematic diagram of the dc plasma CVD experimental setup for the preparation of thin diamond films: 1, water inlet; 2, substrate; 3, water inlet; 4, gas inlet. (Data from Ref. 153a.)

schematically in Figure 6.14. The reaction gas ($CH_4 + H_2$) at a pressure of 200 torr is fed through the cathode (2 cm in. diameter). The water-cooled anode (1 cm diameter) is set above the cathode, and the substrate is mounted on it. The spacing between the anode and the cathode is about 2 cm. The ratio of CH_4 to H_2 is varied from 0.3 to 4%, and the flow rate is fixed at about 20 sccm. The typical discharge conditions used are 1 kV and 4 A/cm². The substrate temperature rises to 800°C and can be varied by varying the flow rate of the water through the water-cooled anode. Suzuki et al. [153a] have characterized the deposited films by RHEEDS and X-ray diffraction and have reported that the interplanar spacing, the lattice constant, and the Vickers hardness all agreed with that of the natural diamond. They have also briefly discussed the influence on diamond synthesis of the dc discharge in a low vacuum ($\approx$ 200 torr).

The application of hollow cathodes to the deposition of thin films has been reported by Jansen et al. [153b], and deposition rates up to 30μ/min were observed for a-Si:H. Here the condensable radicals in the PECVD process are carried in the forced gas flow through hollow cathodes toward the substrate to condense. The electrode arrangements used are shown in Figure 6.15. Figure 6.15(1) is actually a plasma diode configuration with a cylindrical metal cathode; in Figure 6.15(2), the substrate does not function as the anode and the discharge current is drawn by a

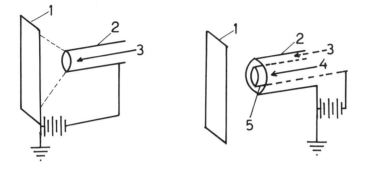

Figure 6.15 Schematic diagram of the electrode arrangements for the hollow cathode deposition of thin films. (1) Single tube structure (for, e.g., a-Si:H): 1, substrate; 2, cathode; 3, process gas. (2) Coaxial tube structure (for, e.g., a-SiN): 1, substrate; 2, anode; 3, gas 2; 4, gas 1; 5, cathode. (data from Ref. 153b).

grounded coaxial tube of larger diameter than the hollow cathode. For stable operation, the ground tube extends a little farther toward the substrate than the cathode.

The cylindrical electrodes are made of stainless steel and are sealed with compression O-ring fitting against insulators constituting the wall of the vacuum system. To prevent a dc discharge outside the metal tube in the configuration shown in Figure 6.15(1), a quartz tube surrounds the metal tube. This arrangement is useful for the deposition of electrically conductive films such as amorphous silicon. For the deposition of a-Si:H from silane, undiluted silane gas was introduced at a prescribed flow rate and the chamber pressure maintained at 500 mtorr. Films were deposited on grounded substrate (silicon wafers or glass coated with ITO) kept at 230°C, held 40 mm from the end of the cathode tube and perpendicular to the tubes. The discharge power was varied by changing the dc currents (25–150 mA). The high rates of deposition observed were attributed to the higher power density in the plasma, combined with the efficient transport of radicals toward the substrate.

The arrangement in Figure 6.15(2), with the coaxial electrode, was used for the deposition of amorphous silicon nitride. Here the anode function is assumed by the grounded outer tube. To ensure that the insulating material did not build up on the anode tube, a shroud gas was flowed through the annular space between the electrodes.

Two different flow patterns were used to deposit silicon nitride films. Nitrogen gas were flowed through the central tube, while an Ar and silane flow was maintained through the annular space. The other pattern uses an auxiliary gas inlet for silane gas well away from the substrate holder, while N_2 and Ar, which flow through the electrode assembly, were kept the same. Also in this configuration, the flow of argon gas could be replaced with nitrogen without affecting the results very

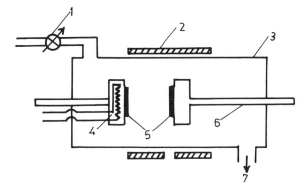

Figure 6.16 Schematic diagram of the setup of the pulsed electromagnetic inductive system for the deposition of amorphous carbon films: 1, variable leak valve for admitting silane/argon mixture; 2, one-turn coil, 3, discharge tube; 4, heater; 5, substrate; 6, pedestal; 7, to pump. (Data from Ref. 155a.)

much. Amorphous silicon and silicon nitride films produced by these electrode structures were studied, and the unusual properties of these films are discussed in the report [153b].

Ebihara and Maeda have reported a method for the growth of thin films of a-Si with pulsed inductive discharge is silane gas [154]. The pulsed plasma was generated by supplying pulsed current of 70 kA to a solenoid coil (i.d., 80 mm; axial length, 200 mm). The discharge tube (Pyrex, o.d., 800 mm) was filled with a silane/argon mixture (20% SiH_4/80% Ar). The films were deposited on the substrate plates set normal to the discharge tube at the middle and end of the coil. Using this pulsed electromagnetic inductive (PEI) system, Ebihara et al. [155a] recently prepared amorphous carbon films. Their experimental setup is shown in Figure 6.16. Pure methane was introduced at the rate of 10 cm³/min. The films were deposited on Corning 7059 glass and crystalline Si wafer (<100> orientation) substrates set normal to the axis at the middle and end. The optical energy gap and electrical conductivity of the amorphous carbon films were investigated and compared with rf glow discharge PECVD. Optical energy gaps were found to decrease with increase of substrate temperature and charging voltage. The deposited films were transparent in the IR and adhered well to room temperature substrates.

Rahman et al. [155b] have prepared hydrogenated amorphous silicon carbide (a-SiC:H) using a novel rf plasma-enhanced CVD system in which a longitudinal dc electric field is applied independently of the inductively coupled rf field (Figure 6.17). An rf power supply at 13.56 MHz is inductively coupled to the reactor. The quartz reactor contains two parallel metal plates that serve as electrodes, with one being the substrate holder. Quartz and p-Si (111) were used as substrates. Source

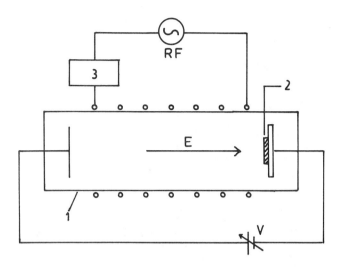

Figure 6.17 Schematic Diagram of the setup of the field-enhanced PECVD system for the preparation of amorphous hydrogenated SiC: 1, quartz tube; 2, substrate; 3, tuning circuit. (Data from Ref. 155b.)

gases SiH4 and CH4 each diluted to 5% in H2, were supplied through mass flow controllers, which maintained a flow rate ratio CH4/(SiH4 + CH4) between 0 and 75%, and a reactor pressure of about 0.5 torr. Substrate temperature ranged from 200 to 500°C, the rf power being fixed at 50 W. The dc voltage ranged from –300 to +250 V. The average growth rate, optical bandgap, and photoconductivity for a-SiC:H were seen to be changed substantially by the dc field; generally, the films formed were improved by applying an electric field directed out of the substrate surface.

Bausch et al. [155c] have recently reported a novel technique for the preparation of pyrite films using rf plasma and the pure elements iron and sulfur as source materials. The films were prepared by a two-step process of plasma-assisted sulfurization of thin iron films. The iron films were first evaporated on quartz substrates and then exposed to sulfur plasma for the formation of pyrite films by the plasma-assisted thermal reaction of iron films with sulfur gas at temperatures below 450°C.

6.3 ELECTRODEPOSITION

Electrodeposition is the process of depositing a substance by the passage of electric current through the conducting medium (called the electrolyte), producing a chemical change (electrolysis). The system used for electrodeposition consists of an anode and a cathode immersed in a suitable electrolyte. When the electric current is passed, the material is deposited on the cathode. The properties of electrodeposited films depend on the electrolyte, the electrodes, and the current density. Early work on electrodeposition has been discussed by Campbell [156] and Lowenheim [157], and in the references cited by these authors.

In recent years there has been considerable interest in the electrodeposition of semiconductor films for photovoltaic applications. Table 6.5 gives a summary of the principal materials studied recently.

6.4 ELECTROLESS DEPOSITION

Deposition of films directly by chemical reaction, without the application of any electrode potentials, is treated here as electrodeless deposition. The chemical reaction can occur with or without a catalyst. Catalytic reactions using activators also are taken as electroless deposition. Silver coating is a typical example of a non-catalytic reaction: here silver mirrors are deposited on glass by the use of a reducing agent such as formaldehyde in a solution of silver nitrate. On the other hand, there is a process of reduction taking place only on certain surfaces (catalytic surfaces), as in the case of reduction of $NiCl_2$ by sodium hypophosphite. Here the metal will be deposited on the surface of nickel itself (or cobalt/iron/aluminum), the metal acting as a catalyst. Not all metal surfaces will catalyze deposition, and the number of potential catalysts is limited. However it is seen that the surfaces of noncatalystic metals can be activated so that deposition will take place on these surfaces. For example, copper dipped in a very dilute solution of $PdCl_2$ will catalyze deposition. The role of the activator here is to lower the activation energy for the reduction reaction so that deposition will take place on the surface of the metal.

Electroless deposition is a simple technique that does not require high temperatures and is very economical. Large-area depositions are also possible with this technique.

Several metals (e.g., Ni, Co, Pd, Au) have been deposited by electroless deposition, and the interested reader can have more details from Reference 157. The electroless deposition technique has also been used to prepare oxide films. The basic principle is to have a controlled homogenous precipitation of the metal hydroxide, and the corresponding oxide film is obtained by annealing these films in vacuum or air. Freezing agents such as NH_3/NH_4F and a small amount of silver ions (Ag^+) acting as catalyst are mixed with the solution of metal ions. For example, PbO_2 [168], Tl_2O_3 [169], In_2O_3 [170], and SnO_2 and Sb-doped SnO_2 [171] films have been prepared using this technique.

Raviendra and Sharma [172] have prepared transparent conducting films of cadmium stearate, zinc oxide, and Al-doped ZnO using the electroless deposition technique.

Chemical ferrite plating is an electroless deposition technique [173] by which spinal films of various compositions [$(Fe, M_3)O_4$, where M = Fe, Ni, Co, Mn, Zn,

Table 6.5 Electrodeposition: Bath Composition and Operating Conditions

Film	Bath composition	Substrate	Remarks, if any	Ref.
$MoSe_2$	H_2MoO_4 + NH_4OH + SeO_2 + H_2O	Ti	Electrolyte was prepared by mixing ammoniacal solution solution of molybdic acid and an aqueous solution of SeO_2. Final pH of the mixture was 9.7. Deposition was carried out at 40°C. Current density, 2.4 mA/cm². Structural characteristics of the films investigated using EDAX, EM, and SEM.	158
$AgInSe_2$	Aqueous solution of $AgNO_3$ $In(NO_3)_3$:H_2O and SeO_2		Polycrystalline $AgInSe_2$ films were prepared. As-deposited films consisted mainly of $AgInSe_2$ and excess selenium. Effect of annealing the as-deposited films in a nitrogen atmosphere in the approximate temperature range of 500–600°C, on the electrical, optical, and structural properties was investigated.	159
$CuInS_2$	Aqueous solution containing $InCl_3$ (10 and 6 mM, respectively), 0.2 vol % triethanolamine and 0.25 vol % ammonia, acidified to pH ≈ 2 by dilute HCl, followed by addition of thiourea to 0.125 M	Ti	Plating current density was 6–8 mA/cm². Deposited film subsequently annealed in H_2S (or H_2S/Ar) at about 500°C for 15–60 minutes to get $CuInS_2$ films. Films were characterized by SEM and microphobe analysis.	160
CdS	Solution containing 2×10^{-3} M $CdSO_4$, 0.1 M Na_2SO_3	Al	pH was adjusted to 2.3 with H_2SO_4. Structural, optical, and morphological characteristics of the films were studied.	161

	Substrate	Solution	Description	Ref
CdTe	Ti sheets	Aqueous H_2SO_4 solution containing 1 M $CdSO_4$ and 1 mM TeO_2	Deposition was carried out at room temperature. Reagent grade $CdSO_4$ (99.5%) and TeO_2 (99%) were used. Three electrode cells were used for the deposition, with a platinum sheet and a Ag/AgCl electrode as counter and reference electrodes, respectively.	162
CdSe	Ti (foil, 1 cm²)	30 mL of aqueous selenosulfite (0.75 M Na_2SO_3 + 0.05 MSe) and 10 mL of aqueous cadmium nitrilotriacetate [0.1 M $N(CH_2\ CO_2H)_3$ + 0.088 M CdCl]	Substrate was polished and etched before deposition. Electrodeposition took place under constant current at pH 9 and at room temperature. Deposition was monitored by measuring electrode potentials simultaneously, also the intensity of light specularly reflected by the electrode.	163
Cu_2O	Stainless steel 0.05 mm thick	Solution of cupric sulfate , (0.4 M), lactic acid (2.7 M), and NaOH ($\approx$ 4 M)	Solution pH was about 9. Purities were higher than 99% (cupric sulfate), 90% (lactic acid), and 97% (sodium hydroxide). Cell had copper as anode and substrate as cathode; reference electrode was of the 0.1 M $CuSO_4$-Cu type. Distance between anode and cathode (working electrode) was about 4 cm. Deposition area was 10–15 cm².	164
CdS	Vacuum-deposited Cr + Ag glass, tin oxide coated conducting glass, polished stainless steel	AR grade 1.0g $CdCl_2$ and 0.6 g sulfur powder in 100 mL of dimethyl sulfoxide	Deposition was carried out at constant current for various times (medium kept at 140°C), giving films of different thicknesses. Films with oriented crystalline structure have been obtained. Microstructural and X-ray diffraction studies of the films were made.	165
$CuInSe_2$	Polished Ti	Aqueous solution containing 10 mM CuO_4, 25 mM In_2 $(SO_4)_3$, and 30 mM SeO_2 adjusted to pH 1 with dilute H_2SO_4	Plating solution was kept constant at 50–55°C. Film close to stoichiometric composition was exclusively deposited at –0.8 V (vs. a saturated calomel electrode).	166

Table 6.5 (continued)

Film	Bath composition	Substrate	Remarks, if any	Ref.
CdTe	Aqueous electrolyte containing Cd and Te complexes	Si doped n^+ GaAs with (100) orientation	A 0.5 M aqueous solution was prepared using reagent grade CdSO$_4$ in deionized water. Impurities were plated out. The solution pH was adjusted to 1.7 by adding H$_2$SO$_4$, and 30 ppm of Te was added to the electrolyte by passing a current between a pure tellurium anode and platinum cathode. Deposition was made at constant cathode potential. Electrolyte was kept at 95°C. CdTe lasers with strong (111) orientation were grown.	167
Cu$_2$O	Lactic acid (3.25 M), anhydrous cupric sulfate (0.39 M) and sufficient NaOH to adjust the pH in the 4–12 range	0.5 mm stainless steel	T_s was kept at about 60°C. Films of different grain size and with different preferential orientations could be prepared by controlling the deposition parameters. Effects of pH, bath temperature, and deposition rate on grain size and orientation were discussed.	218
AgInSe$_2$	5 mM Ag$_2$SO$_4$, 30 mM In$_2$(SO$_4$)$_3$, 30 mM H$_2$SeO3 adjusted to pH 1 with dilute H$_2$SO$_4$	Ti polished and etched in 10% HF for 3 seconds, and rinsed in distilled water	Films obtained under optimum conditions revealed n-type conduction, optical bandgap of 1.2 eV.	219
CuInSe$_2$	3 · 7 mM CuCl$_2$, 22 mM InCl$_3$, and 3.6 mM SeO$_2$ adjusted to pH ≈ 1.5	Ti plate polished and chemically etched	Films were heat treated at 500°C in atmosphere containing both O$_2$ and H$_2$. As-grown and treated samples were examined by X-ray diffraction, SEM, and electron microphobe analysis.	220

etc.] are prepared on various substrates. Abe et al. [174] have also reported a method of preparing polycrystalline Co-ferrite film on organic compound substrates without an intermediate layer to enhance the adhesion of the ferrite film to the surface. Prior to ferrite plating, the investigators exposed the surface to an rf (13.56 MHz, 120 W) plasma glow of air; this served to improve the wettability of the surface, which enhanced the adhesion of the ferrite film.

Electroless deposition of certain compound, alloy, and metal films has been reported recently. Examples include CdS [175], NiP [176], Co/Ni/P [177], Co/P [177a], ZnO [177b], Ni/W/P [178a], C/Ni/Mn/P [178b], Cu/Sn [178c], Cu/In [178d], Ni [178e], Cu [178f], and Sn [178g].

6.5 ANODIC DEPOSITION

The electrodeposition process discussed above is concerned with cathodic reactions, whereas the anodization process relies on anodic reactions. In anodization the metal is made the anode in a suitable electrolyte and either a metal or graphite serves as the cathode. When the electric current is passed, the surface of the metallic anode is used up and forms its oxide coating; in other words, an oxide grows on the metal–anode surface. In earlier reports, this type of oxide was limited to a few metals to give useful films (e.g., Al, Nb, Ta, Si, Ti, Zr), but aluminum was by far the most important. Anodic oxide films on semiconductors have been reported recently. An anodic sulfidization process for forming native sulfide films on Hg_{1-x} Cd_x Te also has been reported [187a]. Here native sulfide films, instead of native oxides, terminate the lattice and passivate the surface.

This simple method produces amorphous continuous films, but film thickness is limited. Anodic films can be grown under constant current or constant voltage conditions.

Table 6.6 presents a few typical examples of the work done on anodic deposition.

Table 6.6 Anodic Deposition

Substrate with anode	Electrolyte	Remarks	Ref.
Al on glass	3% tartaric acid made up to pH 5.5 with NH_4OH	Aluminum films were evaporated onto glass slides and these films were anodized, Anodized Al_2O_3 films were cleaned in distilled water and dried for 15 minutes in a desiccator; then Al was again evaporated on to make $Al/Al_2O_3/Al$ sandwich structure.	179
Single Si crystal polished plates	0.04 N solution of KNO_3 in CH_3OH	SiO_2 films ranging from 100 to 1000 Å thick for Al/p-Si /SiO_2/Al structures were obtained. Two successive conditions, constant current and constant potential, were used. Constant current (45 mA/cm^2) was maintained in the initial stages; upon reaching the required potential, the anodization process automatically transferred to the constant potential condition, which was maintained until the current reached 2 or 3 mA/cm^2.	180
InP	Tartaric acid/propylene gylcol electrolytes, pH 2–12	Anodic oxide of InP, with composition varying from P_2O_5 rich to In_2O_5 rich, depending on the pH, is obtained. Subsequent annealing was necessary to dehydrate the oxides formed.	181a
InP	Mixture of 40% boric acid and 2% NH_3 in ethylene glycol	InP (n type) was anodized under constant voltage conditions (5–125 V) using a platinum cathode (3.75 cm^2). Anodizing bath is commercially available. InP used was Czochrolaski-grown (111) slice, and both slides	181b

288

Material	Electrolyte	Description	Ref.
		were polished mechanochemically. A mechanism for the anodization of InP has been proposed.	
Ni	0.1 M KOH	Anodic oxide film on nickel is a poorly conducting oxide at lower potentials. At higher potentials, results indicated conversion into a good conducting oxide, as evidenced by changes in ellipsometric parameters.	182
Mo	Acedic acid	Anodic oxidation of Mo has been studied using a self-nulling ellipsometer to follow field-induced changes in the optical characteristics of the oxide film formed. At constant field, the oxide film showed large optical anisotropy.	183
$Pb_{1-x}Sn_xSe$ ($x \approx 0.068$)	N-Methyl acetamide, water, and propylene glycol	Constant current source was used. Reproducible films were obtained, with pH of the bath adjusted to 6.6 ± 0.2 by adding oxalic acid and ammonium hydroxide. Anodic oxide was a compound of PbO, SnO_2, and an oxide of Se.	184
$Al_xGa_{1-x}As$	N-Methyl acetamide, adjusted to pH 8.0–8.5 by NH_4OH	Platinum wire (dia. 1 mm) was used as the cathode. Anodic oxidation was carried out at a constant current density of 0.5 mA/cm². Very homogeneous oxide film was obtained.	185
W (99.999%) pure foil	Aqueous solution containing 0.4 M KNO_3 and 0.04 M HNO_3	Thin films of tungsten oxide were obtained. Photo-conductivity of the oxide was investigated. Optical bandgap measured and gave a value corresponding to the anodic WO_3 films.	186

Table 6.6 (continued)

Substrate with anode	Electrolyte	Remarks	Ref.
$Hg_{1-x}Cd_xTe$	Nonaqueous basic solution of Na_2S	In this novel sulfidization process for obtaining native sulfide films on $Hg_{1-x}Cd_xTe$, the resulting films terminate the lattice and passivate the surface instead of the native oxides. Native CdS films were grown on $Hg_{1-x}Cd_xTe$, the layer identified by Auger electron spectroscopy. Constant current densities of 50–100 $\mu A/cm^2$, with a carbon counter electrode, were used. Films from 100 to several hundred angstroms thick were obtained at the rate of about 10 Å/min.	187a
Al (99.999% pure) rectangular samples	Ammonium tetraborate, concentration, 41.3 g/L, pH 9.0; 0.2 M H_3PO_4, adjusted to pH 7.6 with NaOH	Aluminum oxide films were prepared anodically at 20°C using a constant current density of 1 mA/cm². At a predetermined voltage, the power supply crossover made an automatic change to constant voltage operation with the voltage constant; the current was allowed to decay for 10 minutes before anodization was stopped. Specimens were then washed in distilled water and dried in air. A spectrophotometric technique to determine refractive index of the oxide formed has been decribed.	187b
$Cd_xHg_{1-x}Te$ (x = 0.23)	1:9 solution of 1 N KOH in	Bulk crystals of n-type $Cd_xHg_{1-x}Te$ were mechanically polished and used. Electrolyte temperature varied from –25 to 95°C. Anodic films had optimum	188

Material	Electrolyte	Remarks	Ref.
		dielectric properties at 50°C. Oxides at lower temperatures were very unstable. Using MOS structures, the effect of electrical properties of the coatings was investigated.	
Si	Water	Thin silicon dioxide films were produced by anodization of Si in pure water with no salt added. Oxide resistivity was in the 10^{17}–10^{18} Ω cm range.	189
Ti	Neutral phosphate and acidic sulfate solutions	Anodic films were formed on titanium at various potentials. Films changed from amorphous to crystalline beyond a certain critical potential.	190
InP, n type (100)	3% phosphoric acid propylene glycol (1:2 by volume; pH 2)	Anodic oxide films were prepared under constant current conditions, 0.1–0.2 mA/cm². Oxide obtained could be considered to be a double-layered structure: outer indium-rich layer and inner phosphorus-rich layer.	10a
Si (p-doped single-crystal wafer <100> oreientation)	Ethylene glycol + 0.04 N NH$_4$NO$_3$	Oxide films were anodically prepared at constant voltage. A model has been proposed for the growth kinetics of thin oxides of silicon.	191
Nb (99.9% pure)	0.1 N oxalic acid	Anodization was carried out at different current densities and temperatures.	221
Ti rod (99.98%) fixed in Teflon holder; exposed surface, disk of 0.50 cm² geometrical area	0.5 M H$_2$SO$_4$	Oxide films were prepared by applying anodic potentials in steps to get films of different types. Each potential the was kept constant for 15 minutes; during this period, current value decreased to almost zero. During oxide formation, the electrode rotated. Behavior of anodic oxide films grown at voltage higher than 10 V was studied.	222

Table 6.6 (continued)

Substrate with anode	Electrolyte	Remarks	Ref.
Superpure Al	40 g/L oxalic acid	Anodic films were formed at constant current density of 600 A/m² at 313 K; time of deposition, 50 minutes. Films were characterized morphologically and with respect to composition.	223
Spec-pure Cd disks	0.01 M NaOH + yM Na₂S ($0 \leq y \leq 0.03$) and xM NaOH + 0.01 M Na₂S ($0.01 \leq x \leq 1$) solutions at 25°C under purified N₂ gas saturation conditions	Anodization at low potentials produced a thin layer of CdS and at high potentials a thin layer of cadmium sulfide/oxide/hydroxide.	224
Polished n-type <100> oriented Si	0.04 N solution of potassium nitrate in ethylene gylcol containing about 0.35 vol %	Current density was 7 mA/cm². As the films formed, potential difference, across it increased; when it reached a certain predetermined value, (depending on the film thickness required), the voltage was maintained at that value for 30 minutes. Electrophysical properties were studied for MOS structures.	225
SiNi (thickness 120 nm) deposited on Si (111) substrates	0.2 N KNO₃ solution in ethylene glycol; bath kept at 20°C	Samples were anodized at different constant currents (0.1–10 mA). Only a circular area (0.29 cm²) was allowed to be anodized. Nitride films showed good anodization. Oxidation rate about 3.8 times greater than p-type silicon. Oxynitride exhibited a double layer structure with a silicon oxide film on top of a less oxygen-rich oxynitride film.	226

Ta	Aqueous solution {0.1 N H_2SO_4 (4% by vol. in deionized distilled water)}	Sample anodically oxidized using a constant current density of 1 mA/cm² followed by a constant voltage until the current dropped to a very low level (10 µA). Oxide thickness was proportional to final anodizing voltage.	227
Ta (99.9% purity)	Different aqueous acids: citric, oxalic, acetic, and tartaric 10 mol m⁻³/ 100 mol m⁻³. Temperature, 298 K	Tantalum substrates were cleaned and chemically polish-ed, etched, washed, and dried. Anodization was carried out under constant current conditions. Current density was 50.0 A/m². Formation and breakdown characteristics of the anodic oxide films, their dielectric properties, and effects of heating were studied.	228
InP	Aqueous solution (1:3 wt %) of the complexing agents, citric acid or oxalic acid or tartaric acid diluted with 20 parts of ethylene glycol, made to pH 6.6 with 25% NH_4OH solution	Both n- and p-InP (001) were used. Anodization was per-formed galvanostatically at 0.85 mA/cm² in daylight until a preset voltage of 80 V was reached; the InP electrode was kept for 1–2 minutes at this potential be-fore rinsing with water and spin-drying. Anode was rotated continuously during anodization. It is shown that choice of complexing agent is important in the final composition and properties of the anodic oxide formed.	229

REFERENCES

1. O. Kubaschewski, *Oxidation of Metals and Alloys*, Butterworths, London, 1962.
2. K. Hauffe, *Oxidation of Metals*, Plenum Press, New York, 1965.
3. N. Caberra and N. F. Mott, *Rep. Prog. Phys.*, 12:163 (1949).
4. H. H. Uhlig, *Acta Metall.*, 4:541 (1956).
5. C. W. Wilmsen, *Thin Solid Films*, 39:105 (1976).
6. S. K. Sharma and S. L. Pandey, *Thin Solid Films*, 62:209 (1979).
7. N. P. Sinha and M. Misra, *Thin Solid Films*, 62:209 (1979).
8. C. D. Fung and J. J. Kopanski, *Appl. Phys. Lett.*, 45:757 (1984).
9. D. Raviendra, Sudeep, and J. K. Sharma, *Phys. Stat. Sol.a*, 88:PK 83 (1985).
10. (a) Y. Robach, A. Gagnaire, J. Joseph, E. Bergignat, and G. Hollinger, *Thin Solid Films*, 162:81 (1988). (b) S. Mitra, S. R. Tatti, and J. P. Stark, *Thin Solid Films*, 177:171 (1989). (c) M. A. Mohammed and D. V. Norgan, *Thin Solid Films*, 176:45 (1989).
11. R. B. Fair, *Applied Solid State Science*, Suppl. 2B (R. Wolfe, Ed.),Academic Press, NY 1981.
12. N. F. Mott, *Proc. R. Soc. London A*, 376:207 (1981).
13. A. Atkinson, *Rev. Mod. Phys.*, 57:437 (1985).
14. N. F. Mott, *Phil. Mag.*, B55:117 (1987).
15. J. P. Ponpon, J. J. Grob, A. Grob, and R. Stuck, *J. Appl. Phys.*, 59:3921 (1986).
16. J. Nulman, J. P. Krusius, and A. Gat, *IEEE Electron Device Lett.*, EDL hy 6; 205 (1985).
17. J. George, B. Pradeep, and K. S. Joseph, *Thin Solid Films*, 144:255 (1986).
18. W. Kern and V. S. Ban, in *Thin Film Process* (J. L. Vossen and W. Kern, Eds.), Academic Press, New York, 1978, p. 257.
19. S. K. Gandhi, R. J. Field, and J. R. Shealy, *Appl. Phys. Lett.*, 37:449 (1980).
20. T. F. Kuech, M. Maenpaa, and S. S. Lau, *Appl. Phys. Lett.*, 39:245 (1981).
21. K. B. Sundaran and G. K. Bhagarat, *J. Phys. D: Appl. Phys.*, 14:333 (1981).
22. M. Suzuki, H. Onodera, T. Anayama, G. Oya, and Y. Onodera, *Appl. Phys. Lett.*, 39:354 (1981).
23. M. Sano and M. Aoki, *Thin Solid Films*, 83:247 (1981).
24. H. O. Pierson and A. M. Mullendore, *Thin Solid Films*, 72:511 (1982).
25. Y. Tamura and K. Sugiyama, *Thin Solid Films*, 88:269 (1982).
26. M. Akhtar, V. L. Dalal, K. R. Ramaprasad, S. Gau, and J. A. Cambridge, *Appl. Phys. Lett.*, 41:1146 (1982).
27. A. M. Mancini, P. Pierini, A. Quirini, A. Rizzo, and L. Vasanelli, *J. Cryst. Growth*, 62:34 (1983).
28. E. Kinsbron, M. Sternheim, and R. Knoell, *Appl. Phys. Lett.*, 42:835 (1983).
29. W. A. Metz, J. E. Mahan, V. Malhotra, and T. L. Martin, *Appl. Phys. Lett.*, 44:1139 (1984).
30. L. I. Popova, B. Z. Antov, A. V. Shopov, M. S. Sotirova, G. D. Beshkov, and E. N. stefanov, *Thin Solid Films*, 122:153 (1984).
31. K. Nakamura, *J. Electrochem. Soc.*, 131:269 (1984).
32. T. J. Donahue, W. R. Burger, and R. Reif, *Appl. Phys. Lett.*, 44:346 (1984).
33. P. K. Tedrow, V. Ilderem, and R. Reif, *Appl. Phys. Lett.*, 46:189 (1985).

34. H. H. Busta, A. D. Feinerman, J. B. Ketterson, and G. K. Wong, *J. Appl. Phys.*, 58:987 (1985).
35. A. J. Learn and B. Baerg, *Thin Solid Films*, 130:103 (1985).
36. R. D. Tarey and T. A. Raju, *Thin Solid Films*, 128:181 (1985).
37. A. K. Sexana, R. Thankaraj, S. P. Singh, and O. P. Agnihotri, *Thin Solid Films*, 131:121 (1985).
38. B. Verstegen, F. H. P. M. Habraken, W. F. Vander Weg, J. Holsbrink, and J. Snijder, *J. Appl. Phys.*, 57:2766 (1985).
39. C. Weiczorak, *Thin Solid Films*, 126:227 (1985).
40. W. Ahmed and B. Meakin, *J. Cryst. Growth*, 79:394 (1986).
41. H. Matsumura, *Jpn. J. Appl. Phys.*, 25:L949 (1986).
42. A. M. Mancini, L. Vasanelli, and C. DeBlasi, *J. Cryst. Growth*, 79:734 (1986).
43. D. W. Foster and A. J. Learn, *J. Vac. Sci. Technol.*, B4:1182 (1986).
44. S. S. Dana and J. R. Maldonado, *J. Vac. Sci. Technol.*, B4:235 (1986).
45. A. E. Widmer and R. F. Fehlmann, *Thin Solid Films*, 138:131 (1986).
46. P. vander Putte, D. K. Sadana, E. K. Broadbent, and A. E. Morgan, *Appl. Phys. Lett.*, 49:1723 (1986).
47. B. A. Lombos, D. Cote, J. P. Dodelet, M. F. Lawrence, and J. I. Dickson, *J. Cryst. Growth*, 79:455 (1987).
48. W. P. Sun, H. J. Lin, and M. H. Hon. *Thin Solid Films*, 146:55 (1987).
49. S. Steven, R. E. Hegedus, W. Rocheleau, W. Buchman, and B. N. Baron, *J. Appl. Phys.*, 61:381 (1987).
50. M. Yamanaka, H. Daimon, E. Sakuma, S. Miswa, and S. Yoshida, *J. Appl. Phys.*, 61:599 (1987).
51. S. R. Kurtz and R. G. Gordon, *Thin Solid Films*, 147:167 (1987).
52. G. T. Stauf, D. C. Driscoll, P. A. Dowben, S. Barfuss, and M. Grade, *Thin Solid Films*, 153:421 (1987).
53. D. J. Cheng, W. P. Sun, and M. H. Hon, *Thin Solid Films*, 146:45 (1987).
54. C. A. Dimitriadis and P. A. Coxon, *J. Appl. Phys.*, 64:1601 (1988).
55. M. Kojima, H. Kato, and A. Imai, *J. Appl. Phys.*, 64:1902 (1988).
56. K. Nakamura, *J. Electrochem. Soc.*, 132:1757 (1985).
57. Y. K. Fang and S. L. Hsu, *J. Appl. Phys.*, 57:2980 (1985).
58. O. B. Ajayi, M. S. Akanni, J. N. Lambi, H. D. Burrows, O. Osasona, and B. Podor, *Thin Solid Films*, 138:91 (1986).
59. R. A. Levy, P. K. Gallegher, and F. Schrey, *J. Electrochem. Soc.*, 143:430 (1987).
60. J. J. Engelhardt and G. W. Webb, *Solid State Commun.*, 18:837 (1976).
61. (a) M. Sasaki, M. Koyano, H. Negishi, and M. Inoue, *Thin Solid Films*, 158:123 (1988). (b) S. R. Vishwakarma, J. P. Upadhyay, and H. C. Prasad, *Thin Solid Films*, 176:99 (1989). (c) H. Matsumura, H. Ihara, and H. Tachibana, in *Proceedings of the IEEE Photovoltaic Specialist Conference* (A. M. Barnet, Ed.), Las Vegas, 1985, p. 1277. (d) H. Matsumura, *Appl. Phys. Lett.*, 51:804 (1987). (e) H. Matsumura, *Jpn. J. Appl. Phys.*, 28:2157 (1989). (f) C. A. Vander Jengd, G. I. Lensink, G. C. A. M. Janssen, and S. Radelaar, *Appl. Phys. Lett.*, 57:354 (1990).
62. S. D. Allen, *J. Appl. Phys.*, 52:6501 (1981).
63. S. D. Allen, R. Y. Jan, S. M. Mazuk, and S. D. Vernon, *J. Appl. Phys.*, 58:327 (1985).
64. S. D. Allen, A. B. Trigubo, and R. Y. Jan, *Mater. Res. Soc. Symp. Proc.*, 17:207 (1983).

65. R. W. Bigelow, J. G. Black, C. B. Duke, W. R. Salaneck, and H. R. Thomas, *Thin Solid Films*, 94:233 (1982).
66. T. R. Jervis, *J. Appl. Phys.*, 58:1400 (1985).
67. T. H. Baum and C. R. Jones, *Appl. Phys. Lett.*, 47:538 91985).
68. M. Hanubasa, S. Moriama, and H. Kikuchi, *Thin Solid Films*, 107:227 (1983).
69. (a) F. Shaapur and S. D. Allen, *J. Appl. Phys.*, 60:470 (1986). (b) D. Tonneau, G. Auvert, and Y. Pauleau, *Thin Solid Films*, 155:75 (1988). (c) T. H. Baum, *Proc. SPIE Int. Soc. Opt. Eng.*, 1190:188 (1990).
70. C. R. Jones, F. A. Houle, C. A. Kovac, and T. H. Baum, *Appl. Phys. Lett.*, 46:97 (1985).
71. A. E. Adams, *Photochemical Processing of Semiconductor Materials and Devices Synposium*, Wembly, UK, July 1985.
72. M. S. Chiu, Y. G. Tseng, Y. K. Ku, *Opt. Lett.*, 10:113 (1985).
73. R. Solanki, W. H. Richie, and G. J. Collins, *Appl. Phys. Lett.*, 43:454 (1983).
74. (a) K. Hamano, Y. Numazawa, and K. Yamazaki, *Jpn. J. Appl. Phys.*, 23:1209 (1984). (b) A. Tate, K. Jinguji, T. Yamada, and N. Takato, *J. Appl. Phys.*, 59:932 (1986). (c) S. Motojima and H. Mizutane, *Appl. Phys. Lett.*, 54:1104 (1989).
75. R. W. Schwartz, *Mater. Lett.*, 4:370 (1986).
76. (a) V. M. Donnelly, V. R. McGrary, A. Applebaum, D. Brasen, and W. P. Lowe, *J. Appl. Phys.*, 61:1410 (1987). (b) U. Sudarsan, N. W. Cody, T. Dosluoglu, and R. Solanki, *Appl. Phys. Lett.*, 55:738 (1989). (c) P. K. York, J. G. Eden, J. J. Coleman, G. E. Fernandez, and K. J. Beernik, *Appl. Phys. Lett.*, 54:1866 (1989). (d) D. P. Norton and P. K. Ajmera, *J. Electron. Mater.*, 19:367 (1990).
77. Y. Tarui, K. Sorimachi, K. Fujii, and K. Aota, *J. Noncryst. Solids*, 59, 60:711 (1983).
78. K. Kumata, U. Itoh, Y. Toyoshima, N. Tanaka, H. Anzai, and A. Matsuda, *Appl. Phys., Lett.*, 48:1380 (1986).
79. T. L. Pollock, H. S. Sandhu, A. Jodhan, and O. P. Strausz, *J. Am. Chem. Soc.*, 95:1017 (1973).
80. H. Niki and G. J. Mains, *J. Phys. Chem.*, 68:304 (1964).
81. M. Konagai, *Tech. Digest Int.*, PVSEC-3:15 (1987).
82. (a) W. Y. Kim, M. Konagai, and K. Takahashi, *Jpn. J. Appl. Phys.*, 27:L948 (1988). (b) Y. K. Bhatnagar and W. I. Milne, *Thin Solid Films*, 168:345 (1989). (c) A. Yoshida, K. Inone, H. Ohashi, and Y. Saito, *Appl. Phys. Lett.*, 57:484 (1990).
83. J. Shirafugi, S. Miyoshi, and H. Aoki, *Thin Solid Films*, 157:105 (1988).
84. J. M. Jasinski, B. S. Meyerson, and T. N. Nguyen, *J. Appl. Phys.*, 61:431 (1987).
85. V. M. Donnelly, D. Brasen, A. Appelbaum, and M. Geva, *J. Vac. Sci. Technol.*, A4:716 (1986).
86. V. M. Donnelly, M. Geva, J. Long, and R. F. Karlicek, *Appl. Phys. Lett.*, 44:951 (1984).
87. V. M. Donnelly, D. Brasen, A. Applebaum, and M. Geva, *J. Appl. Phys.*, 58:2022 (1985).
88. G. A. West, A. Gupta, and K. W. Beeson, *Appl. Phys. Lett.*, 47:476 (1985).
89. W. E. Johnson and L. A. Schile, *Appl Phys. Lett.*, 40:798 (1982).
90. R. Solanki, P. K. Boyer, and G. J. Collins, *Appl. Phys. Lett.*, 41:1048 (1982).
91. P. K. Boyer, G. A. Roche, W. H. Ritchie, and G. J. Collins, *Appl. Phys. Lett.*, 40:716 (1982).
92. R. Solanki and G. J. Collins, *Appl. Phys. Lett.*, 42:662 (1983).

93. J. Y. Tsao, R. A. Becker, D. J. Ehrlch, and F. J. Leonberger, *Appl. Phys. Lett.*, 42:559 (1983).

94. J. G. Eden, J. E. Greene, J. F. Osmundsen, D. Lubben, C. C. Abele, S. Gorbatkin, and H. D. Desai, *Laser Diagnostics and Photochemical Processing of Semiconductor Devices, Symposium Proceeding*, Boston, 1982, p. 185.

95. T. Inoue, M. Kanagai,and K. Takahashi, *Appl. Phys. Lett.*, 43:774 (1983).

96. T. F. Deutsch and D. D. Rathman, *Appl. Phys. Lett.*, 45:623 (1984).

97. Y. Hirota and O. Mikami, *Electron Lett.*, 21:77 (1985).

98. A. Yamada, J. Kenne, K. Konagai,and K. Takahashi, *Appl. Phys. Lett.*, 46:272 (1985).

99. S. J. C. Irvine, J. Giess, J. B. Mullin, G. W. Blackmore, and O. D. Dosser, *J. Vac. Sci. Technol.*, B3:1450 (1985).

100. A. Yamada, S. Nishida, M. Konagai, and K. Takahashi, *18th International Conference on Solid State Devices and Materials*, 1986, p. 217.

101. Y. Kato, S. Kurita, and T. Suda, *J. Appl. Phys.*, 62:3733 (1987).

102. D. P. Norton and P. K. Ajmera, *Appl. Phys. Lett.*, 53:595 (1988).

103. M. Berti, M. Meliga, G. Rovai, S. Stano, and S. Tamagno, *Thin Solid Films*, 165:279 (1988).

104. H. Nonaka, K. Arai, Y. Fujino, and S. Ichimura, *J. Appl. Phys.*, 64:4168 (1988).

105. Y. Ron, A. Revh, U. Carmi, A. Inspektor, and R. Avni, *Thin Solid Films*, 107:181 (1983).

106. S. C. Brown, *Introduction to Electrical Discharges in Gases*, Wiley, New York, 1966.

107. F. K. McTaggard, *Plasma Chemistry in Electrical Discharges*, Elsevier, Amsterdam, 1967.

108. J. R. Hoolhan and A. T. Bell, Eds., *Techniques and Applications of Plasma Chemistry*, Wiley, New York, 1974.

109. D. W. Hess, *Annu. Rev. Mater. Sci.*, 16:163 (1986).

110. H. F. Sterling and R. W. Warren, *Solid state Electron.*, 8:653 91965).

111. R. G. G. Swan, R. R. Mehta, and T. P. Cauge, *J. Electrochem. Soc.*, 114:713 (1967).

112. J. R. Hollahan and R. S. Rosler, in *Thin Film Processes* (J. L. Vossen and W. Kern), Eds. Academic Press, New York, 1978 p. 335.

113. S. M. Ojha, in *Physics of Thin Films* Vol. 12 (G. Hass, M. H. Francombe, and R. W. Hoffman), Academic Press, New York, 1982, p. 237.

114. C. C. Tang and D. W. Hess, *Appl. Phys. Lett.*, 45:633 (1984).

115. A. C. Adams, F. B. Alexander, C. D. Capio, and T. C. Smith, *Solid State Technol.*, 24:135 (1983).

116. P. E. Vanier, F. J. Kampas, R. R. Coderman, and G. Rajeswaran, *J. Appl. Phys.*, 56:1812 (1984).

117. K. Matsushita, T. Sato, Y. Sato, V. Sugiyama, T. Hariu, and Y. Shibata, *IEEE Trans. Electron Devices*, ED hy 31:1092 (1984).

118. R. S. Rosler and G. E. Engle, *J. Vac. Sci. Technol.*, B2:733 (1984).

119. S. V. Nguyen, *J. Vac. Sci. Technol.*, B4:1159 (1986).

120. E. Grossman, A. Grill, and M. Polak, *Thin Solid Films,*, 119:349 (1984).

121. K. Montasser, S. Hattori, and S. Morita, *Thin Solid Films*, 117:311 (1984).

122. C. M. Horwitz and D. R. Mckenzie, *Appl. Surface Sci.*, 22/23:925 (1985).

123. T. J. Donahue and R. Reif, *J. Appl. Phys.*, 57:2757 (1985).

124. O. Matsumoto, H. Toshima, and Y. Kanzaki, *Thin Solid Films*, 128:341 (1985).

125. T. H. Yuzuriha, W. E. Mlynko, and D. W. Hess, *J. Vac. Sci. Techol.*, A3:2135 (1985).
126. J. Shirafuji, S. Nagata, and M. Kuwagaki, *J. Appl. Phys.*, 58:3661 (1985).
127. (a) K. D. Mackenzie, J. R. Eggert, D. J. Leopold, Y. M. Li, S. Lin, and W. Paul, *Phys. Rev.*, B31:2198 (1985). (b) L. M. Williams, *Appl. Phys. Lett.*, 46:43 (1985).
128. P. Mayr and H. R. Stock, *J. Vac. Sci. Techol.*, A4:2726 (1986).
129. P. G. Pai, S. S. Chao, Y. Takagi, and G. Lucovsky, *J. Vac. Sci. Technol.*, A4:689 (1986).
130. J. Kolodzey, S. Aljishi, r. Schuwarz, D. Slobodin, and S. Wagner, *J. Vac. Sci. Technol.*, A4:2499 (1986).
131. H. Watanabe, K. Katoh, and S. Imagi, *Thin Solid Films*, 136:77 (1986).
132. T. H. Yuzuriha and D. W. Hess, *Thin Solid Films*, 140:199 (1986).
133. D. E. Eagle and W. I. Milne, *Thin Solid Films*, 147:259 (1987).
134. R. Banerjee and S. Ray, *J. Non cryst. Solids*, 89:1 (1987).
135. V. S. Dharmadhikari, *Thin Solid Films*, 153:459 (1987).
136. N. J. Ianno and J. A. Plaster, *Thin Solid Films*, 147:193 (1987).
137. H. Itoh, M. Kato, and K. Sugiyama, *Thin Solid Films*, 146:255 (1987).
138. M. R. Hilton, G. J. Vandentop, M. Salmeron, and G. A. Somorjai, *Thin Solid Films*, 154:377 (1987).
139. J. E. Schoenholtz and D. W. Hess, *Thin Solid Films*, 148:285 (1987).
140. L. N. Aleksandrov, I. I. Belousov, and V. M. Efimov, *Thin Solid Films*, 157:337 (1988).
141. (a) R. V. Livengood and D. W. Hess, *Thin Solid Films*, 162:59 (1988). (b) S. Ohi, W. R. Burger, and R. Beif, *Appl. Phys. Lett.*, 53:891 (1988).
142. H. Sakai, K. Maruyama, T. Toshida, Y. I. Chikawa, M. Kamiyama, T. Ichimura, and Y. Uchida, *Proceedings of the 17th IEEE Photovoltaic Specialists Conference*, Florida, 1984 p. 76.
143. A. C. Adams, *Solid State Technol.*, 26:135 (1983).
144. P. D. Richard, R. J. Markunas, G. Lucovsky, G. G. Fountain, A. N. Mansour, and D. V. Tsu, *J. Vac. Sci. Technol.*, A3:867 (1985).
145. G. Lucovsky, P. D. Richard, D. V. Tsu, S. Y. Lin, and R. J. Markunas, *J. Vac. Sci. Technol.*, A4:681 (1986).
146. P. A. Longeway, R. D. Estes, and H. A. Weakliem, *J. Phys. Chem.*, 88:73 (1984).
147. H. Yasuda, *Plasma Polymerization*, Academic Press, Orlando, FL, 1985.
148. (a) D. V. Tsu and G. Lucovsky, *J. Vac. Sci. Technol.*, A4:480 (1986). (b) R. A. Rudder, G. G. Fountain, and r. J. Markunas, *J. Appl. Phys.*, 60:3519 (1986). (c) G. g. Fountain, R. A. Rudder, S. V. Hattangady, and P. S. Lindorme, *J. Appl. Phys.*, 63:4744 (1988). (d) S. V. Hattangady, G. G. Fountain, R. A. Rudder, and R. J. Markunas, *J. Vac. Sci. Technol.*, A7:570 (1989). (e) R. A. Rudder, G. G. Fountain, and R. J. Markunas, *J. Appl. Phys.*, 60:3519 (1986).
149. H. Mito and A. Sekiguchi, *J. Vac. Sci. Technol.*, A4:475 (1986).
150. I. Kato, S. Wakana, S. Hara, and H. Kezuka, *Jpn. J. Appl. Phys.*, 21:L470 (1982).
151. S. Kimura, E. Murakami, K. Miyake, T. Warabisako, H. Sunami, and T. Tokuyama, *J. Electrochem. Soc.*, 132:1460 (1985).
152. (a) S. Zaima, Y. Yasuda, S. Takashima, T. Nakamura, and A. Yoshida, *18th (1986 International) Conference on solid State Devices and Materials*, Tokyo, A-7-2, 1986, p. 249. (b) S. R. Mejia, R. D. McCleod, K. C. Kao, and H. C. Card, *Rev. Sci. Instrum.*, 57:493 (1986). (c) S. Matsuo, *Abstracts, 16th International Conference on Solid State Devices and materials*, Kobe, 1984, p. 459. (d) T. Ono, C. Takanashi, and S. Matsuo,

Jpn. J. Appl. Phys., 23:L534 (1984). (e) K. Wakita and S. Matsuo, *Jpn. J. Appl. Phys.*, 23:L556 (1984). (f) S. Matsuo and M. Kiuchi, *Jpn. J. Appl. Phys.*, 22:L210 (1983). (g) T. Watanabe, K. Azuma, M. Nakatani, K. Suzuki, T. Sonobe, and T. Shimada, *Jpn. J. Appl. Phys.*, 25:1805 (1986). (h)K. Kobayashi, M. Hayama, S. Kawamoto, and H. Miki, *Jpn. J. Appl. Phys.*, 26:202 (1987). (i) K. Kitagawa, S. Ishihara, K. Setsune, Y. Manabe, and T. Hirao, *Jpn. J. Appl. Phys.*, 26:L231 (1987). (j) M. Kitagawa, K. Setsune, Y. Manabe, and T. Hirao, *Jpn. J. Appl. Phys.*, 27:2026 (1988). (k) F. S. Pool and Y. H. Shing, *J. Appl. Phys.*, 68:62 (1990). (l) K. Shirai and S. Gonda, *J. Appl. Phys.*, 67:6281 (1990), (m) J. Asmussen, *J. Vac. Sci. technol.*, A7:853 (1989).

153. (a) K. Suzuki, A. Sawabe, H. Yasida, and T. Inusuzka, *Appl. Phys. Lett.*, 50:728 (1987). (b) F. Jansen, D. Kuhman and C. Taber, *J. Vac. Sci. Technol.* A7: 3176 (1989).

154. K. Ebihara and S. Maeda, *J. Appl. Phys.*, 57:2482 (1985).

155. (a) K. Ebihara, S. Kanazawa, Y. Yamagata, and S. Maeda, *J. Appl. Phys.*, 64:1440 (1988). (b) M. M. Rahman, C. Y. Yang, D. Sugiarto, A. S. Byrne, M. Ju, K. Tran, K. H. Lui, T. Asano, and W. F. Sticle, *J. Appl. Phys.*, 67:7065 (1990). (c) S. Bausch, B. Sailer, H. Keppner, G. Willeke, E. Bucher, and G. Frommeyer, *Appl. Phys. Lett.*, 57:25 (1990).

156. D. S. Campbell in *Handbook of Thin Film Technology* (L. I. Maissel and R. Glang), (Eds.), McGraw-Hill, New York, 1970, Ch. 5.

157. F. A. Lowenheim, in *Thin Film Processes* (J. L. Vossen and W. Kern, Eds.), Academic Press, New York, 1978, p. 209.

158. S. Chandra and S. N. Sahu, *J. Phys. D: Appl. Phys.*, 17:2115 (1984).

159. D. Ravienda and J. K. Sharma, *Phys. Stat. Sol. a*, 88:365 (1985).

160. G. Hodes, T. Engelhard, D. Cahen, L. L. Kazmerski, and C. Harrington, *Thin Solid Films*, 128:93 (1985).

161. E. Fates, P. Herrasti, F. Arjona, E. G. Camerero, and M. Leon, *J. Mater. Sci. Lett.*, 5:583 (1986).

162. M. Takahashi, K. Vosaki, and H. Kite, *J. Appl. Phys.*, 60:2046 (1986).

163. M. Fracastoro-Decker, J. L. S. Ferreira, N. V. Gomes, and P. Decker, *Thin Solid Films*, 147:291 91987).

164. A. E. Rakshani and J. Varghese, *Thin Solid Films*, 157:87 (1988).

165. K. S. Balakrishnan and a. C. Rastogi, *Thin Solid Films*, 163:279 (1988).

166. Y. Ueono, H. Kawai, T. Sugiura, and H. Minoura, *Thin Solid Films*, 157:159 (1988).

167. P. Sircar, *Appl. Phys. Lett.*, 53:1184 (1988).

168. W. Mindt, *J. Electrochem. Soc.*, 118:93 (1971).

169. R. N. Bhattacharya and P. Pramanik, *Bull. Mater. Sci.*, 2:287 (1980).

170. R. P. Goyal, D. Raviendra, and B. R. K. Gupta, *Phys. Stat. Sol. a*, 87:79 (1985).

171. D. Raviendra and J. K. Sharma, *J. Phys. Chem. Solids*, 46:945 (1985).

172. D. Raviendra and J. K. Sharma, *J. Appl. Phys.*, 58:838 (1985).

173. M. Abe and Y. Tamaura, *J. Appl. Phys.*, 55:2614 (1984).

174. M. Abe, Y. Tanno, and Y. Tamaura, *J. Appl. Phys.*, 57:3795 (1985).

175. P. K. Nair and M. T. S. Nair, *Solar Cells*, 22:103 (1987).

176. S. V. S. Tyagi, V. K. Tandon, and S. Ray, *Bull. Mater. Sci.*, 8:433 (1986).

177. (a) E. L. Nicholson and M. R. Khan, *J. Electrochem. Soc.*, 133:2342 (1986). (b) M. Ristov, G. J. Sinadinovski, I. Grozdanov, and M. Mitreski, *Thin Solid Films*, 149:65 (1987).

178. (a) K. Koiwa, M. Usuda, and T. Osaka, *J. Electorchem. Soc.*, 135:1222 (1988). (b)L.

Zhihui, Y. Chen, C. Haoming, H. Yusheng, and L. Ke, *Thin Solid Films*, 182:255 (1989). (c) G. T. Duncan and J. C. Banter, *Plat. Surface Finish*, 76:54 (1989). (d) A. Gupta and A. S. N. Murthy, *J. Mater. Sci. Lett.*, 8:559 (1989). (e) M. Lambert and D. J. Duquette, *Thin Solid Films*, 177:207 (1989). (f) J. E. A. M. van den Meerakker, *Thin Solid Films*, 173:139 (1989). (g) A. Molenaar and J. W. G. de Bakker, *J. Electrochem. Soc.*, 136:378 (1989).

179. A. Akschi, *Thin solid Films*, 80:395 (1981).
180. A. G. Abdullayev, A. M. Karnaukhov, and S. K. Khanjanov, *Thin Solid Films*, 79:113 (1981).
181. (a) A. Yamanoto, M. Yamaguchi, and C. Vemura, *J. Electrochem. Soc.*, 129:2795 (1982). (b) D. De Cogan, G. Eftekhart, and B. Tuck, *Thin Solid Films*, 91:277 (1982).
182. W. Visscher and E. Barendrecht, *Surface Sci.*, 135:436 (1983).
183. D. J. Desmet, *J. Electrochem. Soc.*, 130:280 (1983).
184. S. C. Gupta and H. J. Richter, *J. Electrochem. Soc.*, 130:1469 (1983).
185. J. Yu, Y. Aoyagi, K. Toyoda, and S. Namba, *J. Appl. Phys.*, 56:1895 (1984).
186. F. M. Nazar and F. Mahmood, *Int. J. Electron.*, 56:57 (1984).
187. (a) Y. Nemirovsky and L. Burstein, *Appl. Phys. Lett.*, 44:443 (1984). (b) G. F. Pastore, *Thin Solid Films*, 123:9 (1985).
188. E. Bertagnolli, *Thin Solid Films*, 135:267 (1986).
189. F. Gaspard and A. Halimaoni, *Proceedings of the INFOS 85 International Conference*, North Holland, 1986, p. 251.
190. T. Ohtsuka, J. Guo, and N. Sato, *J. Electrochem. Soc.*, 133:2473 (1986).
191. S. K. Sharma, B. C. Chakravarty, S. N. Singh, and B. K. Das, *Thin Solid Films*, 163:373 (1988).
192. Y. Yoshimoto, T. Suzuki, Y. Higashigabi, and S. Nakajima, *Thin Solid Films*, 162:273 (1988).
193. M. K. Hatalis and D. W. Greve, *J. Appl. Phys.*, 63:2260 (1988).
194. D. W. Kim, Y. J. Park, J. G. Lee, and J. S. Cun, *Thin Solid Films*, 165:149 (1988).
195. U. Janson, J. O. Carlsson, B. Stridh, S. Soderberg, and M. Olson, *Thin Solid Films*, 172:81 (1989).
196. A. Harsta and J. Carlsson, *Thin Solid Films*, 176:263 (1989).
197. S. Motojima, K. Funahashi, and K. Kurosawa, *Thin Solid Films*, 189:73 (1989).
198. Y. Zhong, M. C. Ozturk, D. T. Grider, J. J. Wortman, and M. A. LittleJohn, *Appl. Phys. Lett.*, 57:2092 (1990).
199. K. Inoue, M. Okuyama, and Y. Hamakawa, *Jpn. J. Appl. Phys.*, 2, Lett:2152 (1988).
200. J. Watanabe and M. Hanabusa, *J. Mater. Res.*, 4:882 (1989).
201. C. J. Kiely, T. Tavitian, and J. G. Eden, *J. Appl. Phys.*, 65:3883 (1989).
202. S. Motojima and H. Mizutani, *Appl. Phys. Lett.*, 54:1104 (1989).
203. R. R. Munz, M. Rothschild, and D. J. Ehrlich, *Appl. Phys. Lett.*, 54:1631 (1989).
204. S. Motojima and H. Mizatani, *Thin Solid Films*, 186:L17 (1990).
205. B. Discheler and E. Bayer, *J. Appl. Phys.*, 68:1237 (1990).
206. K. Mutoh, Y. Yamada, T. Iwabuchi, and T. Miyata, *J. Appl. Phys.*, 68:1361 (1990).
207. S. Motojima and H. Mizatani, *Appl. Phys. Lett.*, 56:916 (1990).
208. G. Ganguly, J. Dutta, S. Ray, and A. K. Barua, *Phys. Rev. B, 40:3830 (1989)*.
209. M. Shimozuma, G. Tochitani, H. Ohno, H. Tagashira, and J. Nakahara, *J. Appl. Phys.*, 66:447 (1989).

210. R. Murri, L. Schiavulli, G. Bruno, P. Capezzuto, and G. Grillo, *Thin Solid Films,* 182:105 (1989).
211. I. Montero, O. Sanchez, J. M. Albella, and J. C. Pivin, *Thin Solid Films,* 175:49 (1989).
212. O. S. Panwar, P. N. Dixit, A. Tyagi, T. Seth, B. S. Satyanarayanan, R. Bhattacharyya, and V. V. Shah, *Thin Solid Films,* 176:79 (1989).
213. (a) D. Kuhman, S. Grammatica, and F. Jansen, *Thin Solid Films,* 177:253 (1989). (b) D. W. Kim, Y. J. Park, J. G. Lee, and J. S. Chun, *Thin Solid Films,* 165:149 (1989).
214. J. W. Osenbach, J. L. Zell, W. R. Knolle, and L. J. Howard, *J. Appl. Phys.,* 67:6830 (1990).
215. (a) S. Koizumi, T. Murakame, T. Inuzuka, and K. Suzuki, *Appl. Phys. Lett.,* 57:563 (1990). (b) X. Jiang, K. Reichelt, and B. Stritzker, *J. Appl. Phys.,* 68:1018 (1990).
216. F. Zhang, Y. Zhang, Y. Yang, and G. Chen, *Appl. Phys. Lett.,* 57:1467 (1990).
217. S. C. Kim, M. H. Jung, and J. Jang, *Appl. Phys. Lett.,* 58:281 (1991).
218. A. E. Rakshani and J. Varghese, *J. Mater. Sci.,* 23:3847 (1988).
219. Y. Ueno, Y. Kojima, T. Sugiura, and H. Minoura, *Thin Solid Films,* 19:91 (1990).
220. N. Khare, G. Razzine, and L. P. Bicelli, *Thin Solid Films,* 186:113 (1990).
221. R. K. Nigam, K. C. Singh, and S. Maken, *Thin Solid Films,* 158, 245 (1988).
222. R. M. Torresi and C. P. De Pauli, *Thin Solid Films,* 162:353 (1988).
223. I. Farnan, R. dupree, Y. Jeong, G. E. Thompson, G. C. Wood, and A. J. Forty, *Thin Solid Films,* 173:209 (1989).
224. S. B. Saidman, J. R. Vilche, and A. J. Arvia, *Thin Solid Films,* 182:185 (1989).
225. G. Mende, E. Hensel, F. Fenske, and H. Flietner, *Thin Solid Films,* 168:51 (1989).
226. I. Montero, O. Sanchez, J. M. Albella, and J. C. Pivin, *Thin Solid Films,* 175:49 (1989).
227. M. A. Mohammed and D. V. Morgan, *Thin Solid Films,* 176:45 (1989).
228. K. C. Kalra, P. Katyal, and K. C. Singh, *Thin Solid Films,* 177:35 (1989).
229. J. Van de Ven, J. J. M. Binsma, and N. M. A. deWild, *J. Appl. Phys.,* 67:7568 (1990).

7

Epitaxial Film Deposition Techniques

All mankind is divided into three classes: those who are immovable, those who are movable, and those who move.

Benjamin Franklin

Epitaxy is the oriented or single-crystalline growth of one substance over another having crystallographic relations between the deposit and the substrate. Epitaxy comes from the Greek words *epi* and *taxis*, the former meaning "upon" and the latter meaning "arrangement." When an epitaxial film grows on a material of the same kind, the phenomenon is called homoepitaxy; film growth on a different material is termed heteroepitaxy. Epitaxy is now used to grow thin crystalline layers of elemental and semiconductor compounds and alloys with better control of purity and perfection, as well as doping level. The different epitaxial growth techniques and the basic principles involved in epitaxy—thermodynamics, kinetics of mass transport, and surface processes—are discussed in a review paper by Stringfellow [1].

Epitaxial film deposition may be achieved by a variety of techniques from solution as well as from vapor. This chapter discusses the epitaxial techniques commonly used, such as liquid phase epitaxy (LPE), molecular beam epitaxy (MBE), hot wall epitaxy (HWE), and metal organic chemical vapor deposition (MOCVD). For epitaxial films prepared by other vapor deposition methods, see Chapter 6, under CVD.

7.1   MOLECULAR BEAM EPITAXY (MBE)

MBE is a sophisticated epitaxial growth technique and is in essence a development of the original idea of Gunther known as the "three-temperature method" [2], performed in an ultra high vacuum system. This therefore is basically an ultra high vacuum evaporation process. Here beams of atoms or molecules of the element or constituent elements of the compound and dopants of the epitaxial layer are created by thermal evaporation in ultra high vacuum using heated crucibles (effusion ovens or cells). These are directed onto clean heated substrates to form the element/compound film. The intensities of the beams in the substrate are controlled by the temperature of the effusion ovens, and these temperatures are chosen to provide the necessary flux of the various elements arriving at the surface. The composition of the film and its doping level depend on the relative arrival rates of the constituent element or elements, which in turn depend on the evaporation rates from the appropriate sources. Thus by varying the source temperature during growth, gradual changes in composition of the film can be achieved. Simple shutters kept in front of the evaporation sources interrupt the beam fluxes to start and stop deposition and also doping. Good quality epitaxial films of GaAs and related compounds from Groups III–V were first achieved by MBE by Chang et al. [3] and Cho and Arthur [4]. If, for example, the growth of GaAs at a given substrate temperature is considered, generally only as many As atoms adhere to the growing surface as collide with with the excess of Ga atoms on the surface and react to form GaAs. That is, one As atom for each Ga atom sticks to the surface to form the epitaxial film. It is seen then that the growth rate of GaAs is determined by the Ga arrival rate and stoichiometry of most of compounds in Groups III–V is self-regulating as long as there is excess flux of Group V molecules available for film formation. For a detailed growth model based on kinetic data, readers are referred to Foxon and Joyce [5, 6].

A basic MBE system consists of the UHV chamber (background vacuums in the 10^{-11} torr range), the substrate heating block, molecular effusion cells provided with individual shutters and liquid nitrogen cooling shroud, reactive gas inlet line for controlled admission of gases, and sample exchange load-lock system. In addition, the growth chamber incorporates components of several analytical techniques such as RHEEDS, SIMS, and AES, for in situ studies. The analytical facilities are used for the initial characterization of the starting substrate surface and are used continuously in situ, during molecular beam epitaxy, to optimize the conditions for the successful preparation of high quality epitaxial films. Apart from the use of high purity elemental sources resulting in epitaxial layers of high purity, and the provision for in situ analytical tools to control composition and to monitor crystal structures, MBE has the unique advantage that the growth occurs under ultra high vacuum conditions. Thus the background concentration of gases such as O_2, H_2O, and CO is very low. Also the very high degree of control of growth rate and composition can lead to very abrupt changes in composition, doping level, or both.

An integral part of the MBE system is the effusion sources for the evaporation of the film materials. An ideal effusion cell for use in MBE appropriate the Knudsen cell [7], although practical conditions do not approximate a true Knudsen design. In the designing and construction of effusion cells, many factors such as rapid thermal response, low outgassing rate for the cell material, the evaporation material to be used in the cell (i.e., negligible reaction between the source charge and the cell material), and uniform heating are important, and all these factors pose problems. Both high purity graphite and pyrolytic boron nitride (PBN) are used as cell materials. Low cost and ready machinability to any design are plus points for graphite, but graphite is more reactive chemically than PBN. PBN is generally used in spite of its high cost.

Designs of MBE effusion cells able to provide sufficiently high and reasonably stable fluxes have been reported in the literature [4, 8-10a]. Most of the cells generally used are resistance heated. MBE growth of high quality semiconductor films requires good thermal isolation of the effusion cells from the rest of the UHV chamber to minimize outgassing from the chamber walls. To achieve this, a liquid nitrogen cooled shroud surrounds the effusion cells.

For the evaporation of low melting point materials from effusion cells, flux stability is obtained by accurate temperature control. For high melting point materials evaporated with electron gun evaporators in MBE systems, the rate of control possible with a constant electric power source is not adequate for accurate fabrication of MBE-grown devices. The high degree of rate control needed for electron gun evaporation and the design criteria for rate controllers have been discussed recently by Schellingerhout et al. [10b].

A design of an effusion cell made from 99.999% pure boron nitride has been reported recently by Khadim [11a] for the successful growth of GaAs by MBE. The crucible, 50 mm long by 10 mm diameter and 2 mm wide, is resistively heated using 99.999% purity molybdenum wire. These windings, made on the outer surface very closely and running up to the top, ensure uniform heating and prevent crust formation near the opening. Three layers of radiation shields of molybdenum foils (99.999% pure, 0.1 mm thick) are used for thermal shielding and independent temperature control. A Pt/Rh thermocouple for temperature measurement of the cell is embedded in the bottom of the cells. For the As_4 cells, a temperature regulation of $\pm 2°C$ has been provided. According to Khadim [11a], these cells have been successfully used in a UHV MBE system to grow 30 GaAs films ranging from 0.5 to 3 μm in thickness.

A single filament effusion cell with reduced thermal gradient, which can generate reproducible and uniform beams of elemental molecules, has been reported by Mattord et al. [11b]. The filament design is made to conform to the shape of pyrolytic boron nitride crucibles used in MBE systems.

Careful control of beam flux is important in MBE. In GaAs, since Group III element flux determines the rate of growth of these materials, measuring this flux precisely is important. The same is true of the Ga:Al flux ratio in $Ga_xAl_{1-x}As$ and the Ga:In flux ratio in $Ga_xIn_{1-x}As$. Therefore careful control of the Group III beam fluxes is required for satisfactory MBE films. Beam flux can ultimately be calibrated from the measurements of the film thickness and growth rate. But the calibration must be repeated often because of the drift in the characteristics of the effusion cells. Another alternative is to have in situ ionization gages and mass spectrometers. These instruments, however, must be placed in the molecular beam near the substrate for precise measurements, hence usually cannot be used during film growth. To overcome this drawback, McClintock and Wilson [12] developed an optical technique for measuring Group III beam fluxes using optical resonant fluorescence. An elemental hollow cathode lamp serves as the light source, and the fluorescent signal is detected by a filter and a photomultiplier tube. the measured signal is linearly proportional to the Ga flux. This technique can be used during the growth process and requires no equipment inside the vacuum system.

Preparation of the substrate surface for epitaxial growth from the initial polishing stage to the final in situ cleaning stage is very important for MBE, as for any epitaxial film growth. Fronius et al. [13] have developed a new method for substrate preparation of GaAs. Substrate heating is generally done by mounting the substrate with indium solder to an internally heated Mo block [14], but several other methods, including direct radiative heating [15, 16] and direct resistance heating [17], have been reported. Direct radiative heating can achieve good temperature uniformity over large wafers without introducing mechanical stress in the substrate.

Since the early work of Cho and Arthur [14], and Chang et al. [3] on MBE-grown films of GaAs and related compounds from Groups III–V, rapid development has taken place in the field of molecular beam epitaxy, and dramatical progress has been made in the application of the MBE technique to many other materials, including compounds from Groups II–VI and IV–VI. MBE is now a well-established epitaxial process of major importance in the development of optoelectronic and microelectronic devices. Early work on the growth and application of epitaxial films by MBE was by Smith [18] and Ploog [19]. In the past decade, innumerable reports have been published, and more recent, extensive reviews on MBE have been published [20].

Standard commercial MBE systems are now available for growing epitaxial films, and MBE has experienced very rapid growth for the past few years. Quite a large number of reports have been published during this period, and we discuss here only a few special examples of molecular beam epitaxial growth.

The first report of a novel molecular beam epitaxy system designed with possibilities of using this high performance technology in the commercial production of compound semiconductor devices was made by Ueda et al. [21]. Many high quality thin films (e.g., AlGaAs, GaAs) have been grown in this system, which consists of five chambers for deposition, preparation/analysis, transfer, loading, and unloading. The deposition and preparation/analysis chambers can be evacuated to a level less than 5×10^{-11} torr, and the transfer chamber to less than 5×10^{-10} torr. These levels of vacuum were obtained by a high performance turbomolecular pump; the chambers were built of high quality stainless steel, with the best surface treatment to reduce the outgassing rate. A horizontal effusion cell geometry was used and could accommodate a maximum of 10 cells. A substrate up to 3 cm could be used, and the substrate transfer was controlled by a personal computer.

A modified MBE technique proposed by Horikoshi et al. [22] makes use of the rapid migration that is characteristic of metal atoms on metastable surfaces: for example, when Ga or Al atoms impinge on a clean GaAs surface in an As-free atmosphere, they are very mobile and migrate on the growing surface rapidly, even at low temperatures [23]. This characteristic has been utilized for growing high quality GaAs and AlAs layers at temperature below 300°C. Therefore in this modified MBE growth system, the interruption of the arsenic supply to the growing surface during the supply time of Ga or Al is essential for the enhancement of the surface migration of the surface atoms. This purpose is served with the experimental setup [24] in which Ga or Al atoms and As_4 molecules were alternately supplied to a (001) GaAs substrate. With this method the investigators have been able to lower the substrate temperature during GaAs and AlAs growth, and GaAs layers and AlAs-GaAs quantum well structures with reasonable photoluminescence characteristics were grown at 200 and 300°C, respectively.

In a modification of the MBE system [25], the arsenic beam intensity can be precisely regulated, as well as GaAs and AlGaAs epitaxial growth, to get good quality heterojunctions and good control of the thickness of the layers. Here GaAs layers were grown in a high vacuum MBE system with the As molecules transported by purified hydrogen as carrier gas. The conventional MBE system was modified by adding a turbomolecular pump; the high vacuum chamber used is shown in Figure 7.1. The minimum background pressure in the system (2×10^{-10} torr) was attained with the cryogenic pump. The hydrogen gas flow was adjusted to a background hydrogen gas pressure range of 1×10^{-6} to 1×10^{-3} torr. A special arsenic effusion cell has also been designed for this purpose. A large amount of hydrogen molecules was introduced through the heated arsenic cell, inducing frequent collisions of the arsenic molecules over the arsenic ingot with the hydrogen, thus effecting the purification of As. The investigators had shown that arsenic beam intensity could be regulated by way of the hydrogen flow rate.

A variant of the conventional solid source MBE is termed chemical beam epitaxy (CBE) by Tsang [26] and has been used for the epitaxial growth of GaAs, AlGaAs, InP, and InGaAs. CBE has many of the advantages of the MBE and MOCVD epitaxial techniques. In CBE, unlike MBE, where molecular beams of say, Al, Ga, or In evaporated at high temperatures from elemental sources are used, all the sources were gaseous Group III organometallics and Group V hydrides. Gas sources were first used in molecular beam epitaxial techniques by Panish [27] to make possible the well-controlled use of P and P plus As for preparing epitaxial compound semiconductor layers. In gas source MBE [27–29], the gas sources used were the Group V hydrides (arsine, phosphine), while solid elemental Group III sources were used as the evaporants. The following are the main advantages of CBE [30] over MBE.

1. The use of room temperature gaseous Group III organometallics enables easy multiwafer scale-up.
2. Semi-infinite source supply incorporates precision electronic mass flow controllers.
3. Single Group III beam automatically guarantees composition uniformity [31].
4. No oval defects found, even at high growth rates.
5. Could have high growth rates if desired.

Compared with OMCVD, the CBE technique has the following advantages.

1. Multiwafer scale-up is possible because no flow pattern problem is encountered.
2. Very abrupt heterojunctions and ultrathin layers can be produced conveniently.
3. Growth environment is very clean.
4. In situ surface diagnostic analysis instrumentation can be easily incorporated.

The growth system for CBE consisted of a gas handling system similar to that employed in MOCVD with precision electronic mass flow controllers for regulating the flow rates of the various gases admitted into the growth chamber. For transporting the low vapor pressure Group III materials, hydrogen gas was used as a carrier gas. Separate gas inlets were used for the organometallics and the hydrides. A low pressure Group V hydride cracker [28, 32] with reduced input pressure of

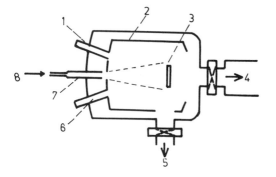

Figure 7.1 Modified MBE system featuring precise regulation of the As beam intensity: 1, Ga cell; 2, cryopanel; 3, substrate holder; 4, to turbomolecular pump; 5, to cryogenic pump; 6, Al cell; 7, As cell; 8, purified H_2. (Data from Ref. 25.)

about 200 torr was maintained on the high pressure side of the mass flow controller. The cracking temperature was about 920°C. Complete decomposition into arsenic/ phosphorus and hydrogen was routinely achieved. Group III alkyl molecules impinge directly (line of sight) onto the heated substrate surface. The growth chamber was a converted molecular beam epitaxial chamber. The gas handling system and the growth chamber are described in References 30 and 33.

In this technique, the growth kinetics differ completely from the kinetics of conventional MBE and in some respects from MOCVD [30]. Because Group III alkyl molecules impinge directly onto it, the heated substrate surface can, depending on the substrate temperature and arrival rate of the organometallics, either acquire enough thermal energy from the heated substrate to dissociate (leaving the Group III atom on the surface) or reevaporate (undissociated or partially dissociated). At a high substrate temperature, the growth rate is determined by the rate of arrival of the Group III alkyls; at lower substrate temperatures, the growth rate is limited by the rate of surface pyrolysis. The Group V material is always supplied in excess.

In earlier reports [26, 31], Group V alkyls were used because they are safer than hydrides; their purity, however, is rather poor. Epitaxial growth of GaAs on similar lines has been reported using trimethyl- or triethyl-gallium (TMG or TEG) and AsH_3 in a commercial MBE machine modified for metal–organic molecular beam epitaxy [34]. The same group have recently reported a comparative study of the epitaxial growth of GaAs by means of molecular beams of TMG or TEG and AsH_3 [35a].

Tsang et al. [35b] have reported the preparation of Fe-doped InP layers by CBE using a thermal atomic Fe beam, following the growth procedures and the CBE system discussed above [26]. Trimethylindium (TMIn) and thermally decomposed phosphine (PH_3) were used as vapor sources, and a thermal atomic beam generated by evaporating high purity elemental Fe in a standard Knudsen cell served as the doping source; S-doped (001) InP wafers were used as substrates. By using a standard MBE shutter for interrupting the Fe beam, very abrupt and ultrathin doping profiles were achieved here also, as with a metal–organic dopant beam. The thermal atomic beam has the advantage of starting with 99.9999% purity. The dopant concentration depends on the Fe beam intensity and the growth rate, the beam intensity being controlled by the effusion cell temperature. These investigators used a growth rate of $1.6 \mu m/h$ and a substrate temperature of 550°C. The sticking coefficient of Fe at this growth temperature is unity, and they estimated the Fe flux intensity from the effusion cell temperature and the distance between the sample and effusion cell, ranging from about 3×10^{11} to 1.5×10^{13} atoms/cm²s equivalent to Fe concentration ranging from about 7×10^{18} to 3×10^{20} cm⁻³ at the growth rate used. Epilayers with high resistivities ($\geq 10^7 \Omega \cdot cm$) were obtained for a wide range of Fe concentrations.

Recently $Si_{1-x}Ge_x$ heteroepitaxial layers have been grown on Si (100) surfaces by gas source MBE, where pure Si_2H_6 and pure GeH_4 were used as the Si and Ge gas sources respectively [35c]. The gas sources were mixed in a subchamber, which was used as a mixing chamber and a buffer chamber, and this subchamber was independently evacuated by a turbomolecular pump. The Si_2H_6 flow rate to the subchamber was kept constant at 6.7 sccm, while the GeH_4 flow rate was varied between 0 and 3 sccm to control the Ge mole fraction x in the $Si_{1-x}Ge_x$ layer that was grown. Part of the gas mixture was supplied to the 4 in. Si (100) substrate in the growth chamber through a gas source cell. The growth rate of the epitaxial layer decreased gradually with increasing GeH_4 flow rate.

During growth, the system pressure was in the range of 2×10^{-6} to 1×10^{-4} torr, this pressure increasing with increasing GeH_4 flow rate. The increase of system pressure with increasing GeH_4 flow rate was thought to be due to the low evacuation speed for GeH_4 of the turbomolecular pump and the liquid nitrogen shroud, which evacuated the growth chamber during growth. The flow rate of the mixture gas was about 2 sccm.

Partly SiO_2 patterned Si (100) wafers were also used as substrates, and selective epitaxial was successfully carried out.

The substrate temperature during growth was 630°C.

MBE has also been used to prepare epitaxial superconducting thin films.

Kwo et al. [35d] have reported the in situ preparation of highly oriented epitaxial $YBa_2Cu_3O_{7-x}$ thin films on MgO (100) by molecular beam epitaxy at a substrate temperature of 550-600°C. The in situ low temperature growth was achieved by using a combination of MBE and a reactive oxygen source generated from a microwave discharge in a flow tube design. The system used was a versatile ultra high vacuum molecular beam epitaxy system reported previously [35e].

The three metal sources (Y and Cu from a heated electron beam and the Ba from an effusion cell oven) were coevaporated. Shuttered growth of individual sources was not used. The growth temperature extended over a finite range of 550-600°C, the substrate temperature being calibrated by the optical pyrometer method. The overall growth rate was 0.5 Å/s (total thickness, ~ 1000 Å). A beam of activated oxygen was directed at the substrate to enhance the oxidation of the metal species in the growing film. The activated oxygen species (predominantly excited molecular oxygen and atomic oxygen) were generated by flowing molecular oxygen through a discharge contained in a flow tube reactor made of an untreated quartz tube (10 mm i.d.); the discharge was excited using a microwave power source of 120 W at 2.45 GHz. The gas flow was sampled with a 1.6 mm diameter hole located 60 cm from the discharge and directed at the substrate.

The pressure near the discharge region was kept at 400 mtorr, which produces a flux of 2×10^{17} species/cm²s impinging on the substrate placed approximately 2 cm from the opening. This flux corresponds to a pressure of 6×10^{-4} torr (i.e., 2 orders of magnitude over the pressure of 5×10^{-6} torr maintained in the growth chamber). Kwo et al. had estimated the flux of the activated oxygen at the substrate to be about 6×10^{15}/cm²s, about an order of magnitude over the amount of oxygen required at the given growth rate. They had confirmed the epitaxial (100) orientation by X-ray diffraction and ion channeling. In situ RHEEDS showed that a layer-by-layer growth had produced a well-ordered, atomically smooth surface in the as-grown tetragonal phase with oxygen stoichiometry of 6.2–6.3. Annealing the as-grown films in 1 atmosphere O_2 at 500°C converted the oxygen content to 6.7–6.8. The typical $YBa_2Cu_3O_{1-x}$ films exhibited a T_c (onset) of 92 K and a T_c (R = 0) of 82 K.

7.2   LIQUID PHASE EPITAXY (LPE)

Liquid phase epitaxy offers a rapid and simple method for the preparation of high purity films of semiconductor compounds and alloys. Indeed, the quality of the material obtained is often superior to the best obtained by vapor phase epitaxy or MBE. The morphology is difficult to control, however, and the quality of the surface is often far from what is desired. In some cases the thermodynamics of the system makes the use of this method difficult.

LPE in principle is growth from solution. A solvent of the material is needed in LPE, and the material from such a solution, when cooled, is grown onto an underlying substrate. For the preparation of films by LPE, the solution and the substrate are kept apart in the growth apparatus. The solution is saturated with the film material at an appropriate growth temperature. The solution/substrate is then brought in contact with the substrate surface/solution and allowed to cool at a suitable rate and time interval to get the desired film. Also dopants can be easily introduced into the films.

LPE was developed beginning with the work of Nelson [36] for the growth of GaAs and since the 1960s this has been developed into a useful technique for the preparation of thin films. There are three basic growth techniques in LPE.

1.   Using a tipping furnace as developed by Nelson [36]. Here the saturated or nearly saturated solution, with the growth material at a particular temperature, is allowed to come in contact with the substrate for a defined temperature interval by tipping the boat containing the solution (Figure 7.2). During cooling, the growth material precipitates from the solution and forms films on the substrate surface. The tube is then tipped back to its original position, the solution flows away from the substrate, and the remnants of the solution adhering to the surface are removed by wiping and dissolution in a suitable solvent.

2.   Using a dipping technique (Figure 7.3), first described by Ruprecht [37]. In this vertical growth system, the substrate is immersed in the solution at the desired temperature, and growth is terminated by withdrawing the substrate from the solution at the appropriate temperature. That is, the vertical motion of the substrate controls the melt–substrate contact.

3.   Using the slider system first described by Panish et al. [38]. Although similar in its principles of operation and cooling to the dipper system, the slider system

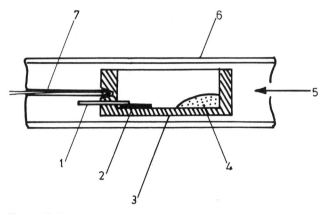

Figure 7.2   Schematic diagram of the tipper system for the growth of GaAs by LPE: 1, clamp; 2, substrate; 3, graphite boat; 4, solution; 5, H_2 flow; 6, quartz furnace tube; 7, thermocouple. (Data from Ref. 36.)

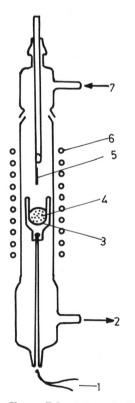

Figure 7.3   Schematic diagram of a dipper system for vertical growth by LPE: 1, thermocouple; 2, gas outlet; 3, crucible; 4, melt; 5, substrate; 6, furnace; 7, gas inlet. (Data from Ref. 37.)

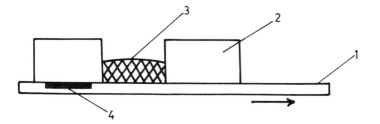

Figure 7.4 Schematic diagram of a simple slider system for LPE film growth: 1, slider; 2, boat; 3, melt; 4, substrate.

has a different method for controlling the contact between the melt and substrate. In a simple system (Figure 7.4), the melt is contained in a reservoir in a graphite boat with a movable slide, the upper surface of which constitutes the bottom of the reservoir. The substrate is placed in a recessed area of the slide outside the reservoir. After the growth conditions have been established in a horizontal furnace, the slide can be moved to place the substrate beneath the solution. In the multibin technique of Nelson [39], with several solution reservoirs provided in the graphite boat, the graphite slide can be moved to place the substrate sequentially beneath the solution in different reservoirs; the whole system is inserted into a quartz furnace, as in Figure 7.2. Thus by selecting appropriate solutions, dopants, and temperature schedules, films of different types, with controlled electrical and optical properties as well as thickness, can be deposited sequentially onto the substrate surface.

Different versions of these basic growth systems have since been developed [40–42], and the slider technique is now more commonly used [43–45].

LPE has now been developed into a useful technique for the preparation of the films of a wide variety of materials and is frequently used for the fabrication of thin layers of compounds and alloys from Groups III-V. Although other growth techniques are now available, LPE remains the dominant technology when high quality material is desired. The ability to carefully control alloy composition, carrier concentration, and the thickness of individual epitaxial layers is of utmost importance in device fabrication.

7.3 HOT WALL EPITAXY (HWE)

Hot wall epitaxy is a vacuum deposition technique in which epitaxial films are grown under conditions as near as possible to thermodynamic equilibrium [46]. This is achieved by having a heated wall between the source and the substrate, as shown schematically in Figure 7.5. The three resistance heaters (one each for source, wall, and substrate) are independent. The substrate acts as a lid to close the tube (sometimes the substrate is kept at a distance above the mouth of the tube), and

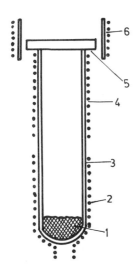

Figure 7.5 Schematic diagram of a simple hot wall system: 1, source; 2, source furnace; 3, quartz tube; 4, wall furnace; 5, substrate; 6, substrate furnace.

the whole apparatus is kept in a vacuum. The heated wall serves to direct the evaporating molecules onto the substrate. This system has the following advantages.

1. Loss of the evaporating material is kept at a minimum.
2. A clean environment is obtained within the growth tube.
3. Relatively high pressure can be maintained inside the tube.
4. The difference between the source and substrate temperature can be very much reduced, as a result of the high pressure inside the tube.

Many variations of the simple hot wall system above have been used by different authors to prepare different semiconductor compound films. Figure 7.6 shows a typical hot wall apparatus with a compensating source for the growth of PbTe films, developed by Lopez-Otero [47]. Early works in hot wall epitaxial growth of compounds from Groups II–VI, IV–VI, and III–V have been reported in detail, with the different variations of the hot wall system [48, 49]. Table 7.1 summarizes a selected list of materials prepared by HWE.

A few interesting examples of HWE published during the past few years, are described next.

Sadeghi et al. [61] were the first to report that the HWE technique can be used to prepare epitaxial GaAs films with controlled stoichiometry on SrF2 substrates. A schematic diagram of the system is shown in Figure 7.7. A polycrystalline GaAs source is kept at the bottom of the wide tube. Pure Ga in the boat on the upper half of the wide tube is for adjusting the stoichiometry and the growth rate of the GaAs films. The source for doping is kept in the inner tube. The substrate, consisting of

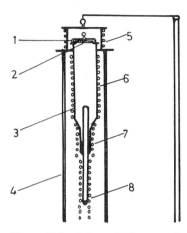

Figure 7.6 Schematic diagram of a variation of the hot wall apparatus including a compensating source for the growth of PbTe films: 1, substrate holder; 2, substrate; 3, quartz tube; 4, radiation shield; 5, substrate furnace; 6, wall furnace; 7, PbTe furnace; 8, Te furnace. (Data from Ref. 47.)

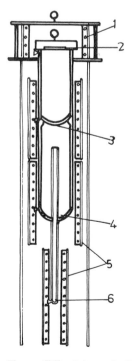

Figure 7.7 Schematic diagram of a hot wall system for the preparation of epitaxial GaAs films on SrF$_2$ surfaces: 1, substrate heater; 2, substrate; 3, Ga source; 4, GaAs source; 5, heaters; 6, doping source. (Data from Ref. 61.)

Table 7.1 Growth Details of Epitaxial Films Prepared by Hot Wall Epitaxy

		Temperature (°C)				
Film	Substrate	Source	Wall	Substrate	Growth rate	Remarks, if any	Ref.
Bi_2S_3	NaCl			30–250	2 nm/s	Films were grown on freshly cleaved NaCl. Hot wall was held at same temperature as substrate. At room temperature (30°C), amorphous films formed with Bi 29.39 and S 70.61 at. %, respectively. At 220°C films were polycrystalline and stoichiometric; at 250°C, polycrystalline and sulfur deficient.	50
CdTe, CdTe:In	Cleaved (111) oriented or mechanically polished (311) oriented BaF_2	500		430–435		Substrates were prebaked to 430 or 550°C before introducing into the growth tube. There were two independent sources for CdTe and In; the CdTe source contained 2% excess of Cd to maintain the stoichiometry. For doping, the In source kept at 540°C. Whole system was kept under a vacuum of 10^{-6} torr.	51a

PbTe	(111) oriented BaF$_2$	545	560	250–500	BaF$_2$ was cleaved in air and preheated in vacuum before bringing to the growth position. Epitaxial growth was studied by SEM and X-ray diffractometry. Temperature of the substrate was controlled by radiant heat from the substrate furnace and by heat conduction, as well as by radiation from the wall and source furnaces.	51b
(PbSn)Te	(111) BaF$_2$	500	500	500	Pregrowth substrate bakeout step at 500°C for 90 minutes was used in some growth runs. Source charge, polycrystalline Pb$_{0.8}$Sn$_{0.2}$Te, vacuum during growth, 10^{-6}–10^{-7} torr range. Thin films (70–2500 Å) were prepared for TEM by evaporating thin carbon backing onto the sample. TEM study was made of nucleation and growth. Using dark field technique and microdiffraction, the recrystallization process and degree of epitaxy were studied as a function of overgrowth thickness.	52

Table 7.1 (Continued)

Film	Substrate	Temperature (°C)			Growth rate	Remarks, if any	Ref.
		Source	Wall	Substrate			
Cd	Glass, air-cleaved NaCl (100)	375–450	365–435	60–80	30 μm/h	Substrate was initially kept away from tube. When the vacuum in the chamber attained a value of 3×10^{-5} torr, the substrate was brought down and made to rest on the mouth of the tube, closing it completely.	53
Zn_3P_2	Freshly cleaved mica or KCl	550	550	250–350	6 μm/h	Working chamber had a base pressure of 2×10^{-6} torr. Substrate was prebaked to about 450°C for 30 minutes. Homogeneity of the synthesized source material was analyzed by X-ray diffractometry. Single crystallinity and surface morphology of the films were examined by X-ray and optical microscopy. Lattice constants calculated were in close agreement with ASTM data. Films showed high resistivity of the order of 10^4 $\Omega \cdot$cm at room temperature.	54

ZnSe, ZnSe:In, ZnS-ZnSe	GaAs (100)	300		ZnSe, 0.63 μm/h; ZnS 0.33 μm/h	HWE system was used with two independent crucibles, one for Zn and the other for S or Se. For the preparation of ZnS-ZnSe superlatticess, two sets of HWE systems were used. Substrate slid over these two systems, and ZnS and ZnSe films were grown alternately. For In doping, a hot wall chamber with In source was used in addition.	55
CdTe	Semi-insulating GaAs doped with Cr <001> oriented	400	500	2-3 μm/h	Polycrystalline CdTe (stoichiometric) was used as source material. HWE was used for the first time for the growth of high quality epitaxial films on semi-insulating GaAs. Substrate was thermally cleaned at 600°C in a separate preheating oven before transfer into the growth reactor, mounted in the same vacuum chamber. Epitaxial layers obtained were (100) oriented. Quality of films was investigated by photoluminescence.	56

Table 7.1 (Continued)

Film	Substrate	Temperature (°C)			Growth rate	Remarks, if any	Ref.
		Source	Wall	Substrate			
$Cd_{1-x}Mn_xTe$ $x \leq 0.02$	(001) Rock salt					Prereacted binary compound CdTe and pure Mn were used as source materials. Stoichiometry of the films was controlled by a reservoir chamber containing one of the pure elements. Each source material was placed inside a separately heated zone. Substrates were outgassed prior to growth at about 400°C for 1 hour. With shadowed replicas, TEM examination of the films was carried out. Most preferred growth orientation is the (001) one parallel to the (001) substrate plane.	57
PbTe	KCl	PbTe: 550 Te reservoir: 380	550	450		TEM analysis of the films has been made and the structural perfection of the film investigated.	58
	BaF$_2$	PbTe: 530 Te reservoir: 220	530	470			
Cd$_3$P$_2$	Mica			150–300		Films were characterized by X-ray diffraction and SEM. Films were n type, as indicated	59

Material	Substrate						Ref
PbI$_2$	(001) face of CdI$_2$ single crystal	250	150	75–120	0.005 nm/s	by thermal probe. Optical absorption measurements gave a bandgap of 0.58 eV. Excellent quality films were obtained. Band edge absorption spectra were investigated for films grown under different conditions.	60
Zn$_3$P$_2$	Freshly cleaved mica and chemically cleaned glass plates	630	650	200–350		Source material was synthesized from high purity zinc and red phosphorus. Powder was confirmed as Zn$_3$P$_2$ by X-ray analysis. Transport properties of the films grown at different substrate temperatures (200–350°C) were studied from room temperature to 600°C.	92
Pb$_{1-x}$Eu$_x$Te	Cleaved BaF$_2$ (111)			350	10 Å/s	Temperatures of the PbTe, Eu, and Te sources during growth were 500, 460–500, and 370–400°C, respectively. Good crystalline films were obtained by preparing in excess Te vapor. All the films prepared were single crystalline, as evidenced by Laue photographs or reflection electron diffraction. Electrical properties of the films were examined.	93

Table 7.1 (Continued)

Film	Substrate	Temperature (°C)			Growth rate	Remarks, if any	Ref.
		Source	Wall	Substrate			
$Pb_{1-x}Eu_xS$	Cleaved BaF_2 (111)			300	10 Å/s	Source materials PbS, Eu, and S were deposited at 550, 450–500, and 110–130°C, respectively. Lattice constants of the films very close to that of PbS. Energy bandgap increases very rapidly with EuS content up to $x = 0.05$.	94
Zn_3P_2	Corning 7059 glass with 1 μm layer of vapor deposited Ag	800–850	400–450	300–320	1 μm/min	Thermal evaporation of Zn_3P_2 powder from a heated carbon crucible. Device grade low resistivity (100 $\Omega \cdot$cm) poly-crystalline films were obtained, to produce low leakage diodes by evaporating magnesium or aluminum dots onto the Zn_3P_2 films.	95
Zn_3P_2	Pyrex glass and (100) GaAs	420–540	420–540	240–380		Source temperature dependence on the deposition rate of both substrates was studied, keeping the substrates at 280 and 360°C; also investigated was deposition rate as a function of substrate	96

CdTe	(100) GaAs	495–510		390–420	3–5 μm/h	temperature for different source temperatures (480 and 540°C). Growth rate, structural properties, and electrical resistivity of the films were compared. Higher deposition rates were obtained for GaAs crystalline substrates at the same source and substrate temperature than for glass substrates.	97
						X-Ray rocking curves of (100) and (111) oriented epitaxial layers grown on (100) GaAs have been measured. Number of extended defects increased with thickness in (111) CdTe layers but decreased with thickness in (100) CdTe layers.	
CdTe	GaAs (100)	430–480	430–480	300	2 Å/s	Polycrystalline CdTe was used as the source. Strain relaxation of the CdTe (100) layers grown on GaAs (100) substrates was investigated by measurement of optical properties and by X-ray studies and TEM. Results indicated that the grown layers were of high quality.	98

(111) oriented SrF2 crystals cleaved in air immediately before deposition, serves as a semiclosed lid. The following ranges of temperatures were used:

GaAs source   900–925°C
Ga source   850–950°C
Substrate   560–610°C

Several radiation shields were used to prevent overheating of the walls of the bell jar that contained the whole apparatus.

For the preparation of $Pb_{1-x}Eu_xTe$ single-crystalline epitaxial film, Krost et al. [62] used a modified hot wall technique. Specifically, the growth tube of the standard hot wall system for Groups IV–VI epitaxy [47] was modified and a three-source quartz system was used (Figure 7.8). After preheating, the growth tube was closed by the substrate, BaF2 (111) or KCl (100) oriented crystal. To ensure a homogeneous temperature across the substrate surface, the substrate was mounted on a copper plate. Eight substrate furnaces were mounted on a push–pull rotary feed through, to enable eight samples to be prepared during each run. During evaporation, the Eu vapor partly reacts in the furnace, producing EuTe, thus reducing the vapor pressure. By elongating the inner tube (containing Te), unwanted reactions were eliminated. The following were the typical growth conditions employed, beginning with a total pressure in the vacuum system before evaporation of about 0.75×10^{-7} torr.

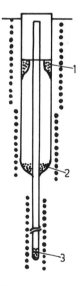

Figure 7.8   Three-source quartz system for the HWE preparation of $Pb_{1-x}Eu_xTe$ films: 1, PbTe; 2, Eu; 3, Te. (Data from Ref. 62.)

T_{wall}	560–660°C
T_{PbTe}	520–560°C
T_{Te}	300–350°C
$T_{substrate}$	440–460°C

The films obtained were single crystalline, as evidence by Laue photographs. In addition, the substitution of Pb by Eu ions was found to increase the lattice constant nonlinearly with Eu composition as well as the gap. The effective masses were increased accordingly.

The growth of CdTe films from vapor is usually complicated because of the different sticking coefficients of Cd and Te. The first use of HWE for the preparation of (100) oriented epitaxial films of CdTe on (100) oriented GaAs has been reported by Schikora et al. [56]. Recently a simplified hot wall setup has been used [63] to grow CdTe layers on GaAs substrates with excellent surface morphology. The deposition details are as follows.

Substrate	Undoped semi-insulating polished GaAs wafers (7 × 7 mm²)
Substrate temperature	390–420°C
Source temperature	495–510°C
Growth rate	2–5 μm/h
Film obtained	(100) oriented CdTe

To assess the quality of the films, the 1.8 K near-band-edge photoluminescence was measured. Again growth of (111) and (100) CdTe films of high quality on (100) GaAs substrates by HWE has been demonstrated by Korenstein and MacLeod [64]. Their setup is shown in Figure 7.9. The two resistance-heated furnaces were contained in a Pyrex bell jar maintained at a base pressure of 5×10^{-8} torr. The substrate could be isolated from the Cd and Te₂ flux by a shutter, which remained closed during the pregrowth bakeout. The source temperature was maintained between 500 and 550°C. The growth temperature varied from 340 down to 220°C. The (111) CdTe films obtained in the work were of comparable quality to MBE (111) CdTe films.

Using a novel hot wall system shown in Figure 7.10, Chaudhuri et al. [65] prepared CdSe film to glass, NaCl, and KCl substrates. CdSe powder (99.999%) was evaporated from a quartz crucible heated by a cylindrical graphite heater. There is a constriction at the neck of the crucible to hold a quartz wool plug, which was very effective in preventing spattering of CdSe. To ensure adequate mixing of the vapor before ejection, to form the film, there are appropriate buffers in the form of small dimples inside the quartz tube. A chimney serves as the hot wall. (A similar system was used earlier by this group to prepare CdS films [66].) The optimum values of the source and the hot wall were 802 and 202°C, respectively. The films were deposited at different substrate temperatures. Film deposited onto NaCl and KCl showed fully oriented epitaxial film at and above a substrate temperatures of 162°C; this was not true, however, for film on glass.

7.4 METAL-ORGANIC CHEMICAL VAPOR DEPOSITION (MOCVD)

Metal-organic chemical vapor deposition is a recently developed technique that is increasingly used for preparing high quality epitaxial films for applications in optoelectronics and microwave devices. Since the first demonstration of the deposition onto insulator substrates of single-crystal films of GaAs by MOCVD [67] using trimethylgallium (TMG) and AsH_3 as Ga and As sources, respectively, many reports have been published on the growth of compound semiconductor films by this vapor phase materials technology. Duchemin et al. [68] and Chang et al. [69] were first to report on the growth of GaAs thin films by low pressure MOCVD. Duchemin et al. used TMG and AsH_3 as the sources, while Chang et al. used TEG

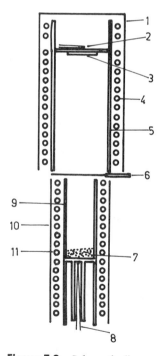

Figure 7.9 Schematic diagram of the hot wall epitaxial setup for the growth of high quality films of CdTe on GaAs substrates: 1, radiation shield; 2, thermocouple; 3, substrate; 4, heater; 5, quartz liner; 6, shutter; 7, CdTe source; 8, thermocouple; 9, quartz crucible; 10, radiation shield; 11, heater. (Data from Ref. 64.)

and AsH3. Duchemin at al. [70] have used the low pressure MOCVD technique successfully to grow a wide range of materials in Groups III-V: GaAs, GaAl AsInP, Ga0.47In0.53As. They have also shown that it is possible to grow heterojunctions with abrupt composition changes and low interfacial recombination velocities.

Manasevit [71], in his "Recollections and Reflections of MOCVD," discussed in general the development of MOCVD techniques for epitaxial growth and listed semiconductors grown by MOCVD from compounds of Groups III-V, II-VI, and IV-VI. Mullin et al. [72] have reviewed the developments in the MOCVD of compounds in Groups II-VI, discussing as well several compound films. The advantages and disadvantages of organometallic chemical vapor depositions are compared with those of conventional vapor phase epitaxial techniques. Stringfellow [73] has given a brief review of the development of MOCVD for the epitaxial growth of $Al_xGa_{1-x}As$. The development of the MOCVD technique and the problems involved in obtaining good quality semiconductor epitaxial films has been reported by Manasevit [74]. A critical appraisal examining the fundamental aspects of the growth mechanisms involved in this epitaxial deposition technology has been offered by Stringfellow [75]. Recently Schumaker et al. [76] have reported

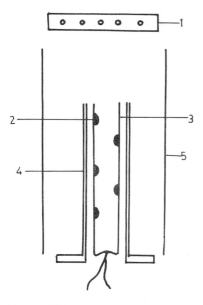

Figure 7.10 Schematic diagram of a hot wall apparatus for the preparation of CdSe films on glass, NaCl, or KCl substrates: 1, substrate; 2, buffers; 3, quartz crucible; 4, cylindrical graphite heater; 5, hot wall. (Data from Ref. 65.)

the potentials and developmental status of MOCVD as a commercial thin film deposition technique.

Quite a large number of reports have been published on the preparation of epitaxial films of several semiconductor compounds and alloys by the MOCVD technique. A few typical examples are described below.

Both vertical [77] and horizontal reactors at low pressure and at atmospheric pressure are used for the MOCVD reactor designs. In a typical horizontal system reported by Wright and Cockayne [78], single-crystal ZnSe, ZnSe, ZnS_xSe_{1-x} films have been grown on a variety of substrates in a horizontal reactor at atmospheric pressure, by direct reaction of dimethyl zinc (DMZ), hydrogen sulfide, and/ or hydrogen selenide, using hydrogen as the carrier gas. This apparatus represents a development of the equipment designed by Bass [79] for the MOCVD at atmospheric pressure of compounds in Groups III–V. The reactor consists of a water-cooled silica envelope containing an SiC-coated graphite pedestal (for substrate mounting) heated by a high frequency inductor. The reactor, mounted horizontally, can be evacuated by a rotary pump, and the graphite can be baked to high temperature ($\approx 1100°C$) in vacuum prior to use. The reactant gases are controlled by mass flow controllers. DMZ must be cooled to below –5 and –10°C; H_2S and H_2Se are supplied as 5% mixtures in high purity hydrogen. Purified H_2 is used to transport the reactants. The position of the inlet nozzle for the alkyl-plus-H_2 stream was critical in the sense that this short silica tube is positioned to prevent premature mixing with H_2S and H_2Se in the narrow inlet tube of the reactor. Careful positioning of this nozzle reduces undesirable reactions and consequent deposition downstream from the susceptor. The basic reaction for ZnSe formation is:

$$(CH_3)_2Zn + H_2Se \rightarrow ZnSe + 2CH_4$$

DMZ and H_2/Se can react at room temperature to give an undesired homogeneous gas phase reaction. Therefore a low concentration of the reactants [5×10^{-5} mole fraction of $(CH_3)_2Zn$ and 2×10^{-4} mole fraction of H_2Se together with a high flow rate: 4.5 L/min of the carrier gas (H_2)] were used, and Wright and Cockanye were able to grow thin single-crystal ZnSe films on GaAs and Ge (both providing a close lattice match to ZnSe).

Bhat [80] has developed a new growth technique for the epitaxial growth of GaAs and AlGaAs using TMG and TMA and a solid elemental or compound source for the Group V element, instead of the Group V hydrides traditionally used. Solid arsenic was used to generate arsenic vapor. GaAs polycrystals have also been used in the preliminary studies. GaAs and AlGaAs films were deposited on Cr-doped and Si-doped GaAs substrates. GaAs epitaxial films with excellent photoluminescence efficiency have been obtained. Carbon was the only acceptor observed. Epitaxial layers of $InAs_{1-x}Sb_x$ have been prepared by MOCVD [81] in a vertical atmospheric quartz reaction chamber using TMG, TMSb, and arsine in H_2, as source materials. Although several investigators reported earlier the growth of $InAs_{1-x}Sb_x$ by MOCVD [82], it is only here that a method of controlling the alloy composition by use of a thermodynamic model proposed by Stringfellow [83] has been used. This model predicts that the more stable of the compounds from Groups III and V will control the composition. For the $InAs_{1-x}Sb_x$ system, when the III/V molar ratio is less than 1.00 in the vapor phase, As preferentially incorporates into the epitaxial layer because InAs is more stable than InSb in the growth temperature range. For a III/V ratio close to or greater than 1, As and Sb are incorporated equally.

Biefeld [81] used a vertical reactor growth system with a quartz reaction chamber on a stainless steel handling system. The TMIn bubbler was held at 20°C and a TMSb bubbler at -25°C, and AsH_3 (5% in H_2) was run continuously either into the vent or into the reaction chamber. Pd-diffused H_2 was the carrier gas. The best surface morphology and the sharpest X-ray diffraction pattern were obtained with a growth temperature of 475°C with a TMIn/(TMSb + AsH_3) molar ratio of 1.00 and a growth rate of less than 1.0 μm/h.

Another interesting material grown by MOCVD that is useful in optoelectronic devices is $InP_{1-x}Sb_x$, whose bandgap covers the range from 1.35 eV (InP bandgap) to 0.17 eV (InSb bandgap at 27°C). This work by Jou et al. [84] provides the first evidence that single-phase epitaxial films of $InP_{1-x}Sb_x$ can be prepared by MOCVD on InP, InAs, and InSb substrates over the entire composition range including the region of solid immiscibility. The experiments were performed in atmospheric pressure using a horizontal infrared heated reactor [85]. The reactants (TMIn and TMSb) were kept in temperature-controlled baths 11 and 5°C, respectively.

The deposition parameters reported were as follows.

Ambient gas	Pd-diffused H_2
Flow rate	2 L/min
Flow rates of the reactants	
TMIn	4.33 μmol/min
TMSb	5.49–20.3 μmol/min
PH$_3$	20.3–203 μmol/min
Growth temperature	480–600°C
Substrates (100) semi-insulating InP: the layers formed generally have good morphologies and gave sharp X-ray diffraction peaks	

Khan and O'Brien [86] have reported the use of zinc acetate as a novel precursor for the direct growth of ZnO in a simple hot wall, low pressure MOCVD reactor. This consisted of a continuously pumped horizontal Pyrex tube that passed through a two-zone furnace; the system was evacuated to a pressure of 10^{-4} torr, and the temperature of the growth zone maintained around 350–420°C, while the source was placed in the collar zone (remote from the pump), about 150°C cooler. Zinc acetate was dried before use for about 2 hours at 80°C under reduced pressure (10^{-2} torr). The substrates were InP, Si, GaAs, and microscope slides. In a typical experiment, the growth region was kept at 380°C and the growth was for 1 hour. The films obtained were featureless and polycrystalline, as revealed by SEM. The interesting feature of this method is the direct growth of zinc oxide from zinc acetate at relatively low temperatures.

Hanna et al. [87] have shown that high mobility (peak mobilities as high as 200,000 cm^2/V·s at 50 K) GaAs can be grown using TMG and arsine in an atmospheric pressure organometallic vapor phase epitaxial reactor. The highest mobility GaAs has been obtained using TEG and arsine under reduced pressure [88]. The use of TEG imposes the need to operate at reduced pressure, thereby making necessary the use of complex systems; the most common precursors for the organometallic growth of GaAs are TMG and arsine [89, 90].

Hanna et al. [87] grew the GaAs layers simultaneously on undoped semi-insulating GaAs substrates oriented 2° off (100) toward a (100) direction and on (100) n^+-GaAs substrates in a conventional horizontal quartz reactor operating at atmospheric pressure (650 torr). The substrates were mounted on a SiC-coated graphite susceptor heated by infrared lamp. These substrates were cleaned first ultrasonically and then in a 3% ammonium hydroxide solution, to remove the native oxide; finally they were rinsed in deionized water and blown dry, with nitrogen and immediately loaded into the reactor.

The deposition details reported were as follows.

TMG molar flow rate	9.4×10^{-5} mol/min
Growth rate	1000 Å/min
Layer thickness	9.12 μm
Arsenic flow rate	1.9×10^{-3} to 7.5×10^{-3} mol/min, corresponding to an arsine/TMG (V/III) ratio of 20:80, respectively
Substrate temperature	580–660°C
Total hydrogen (Pd-diffused) flow into the reactor	5 standard liters/min (giving an average gas velocity of 30 cm/s above the substrate)

An oilless vacuum pump was used to remove any air that was introduced during the loading of the substrate. The single TMG and arsine source used in this study had the highest purity TMG and 100% arsine. Sometimes the arsine source passed through a type 3A molecular sieve during the process.

The layers grown were characterized by Hall effect studies and 10 K photoluminescence studies, to examine the dependence of the incorporation of residual impurities on the growth temperature and arsine partial pressure. With the optimum growth conditions, the amount of residual carbon densities was evaluated as less than 5×10^{13} cm^{-3}.

Kanehori et al. [91] have grown superconducting YBa$_2$Cu$_3$O$_{7-x}$ thin films with T_c (onset) of 86 K and T_c ($R = 0$) of 83 K by MOCVD on SrTiO$_3$ (100) substrates. Yttrium, barium, and copper tetramethyl heptanedione complexes were used as source materials. Oxygen was used as the oxidation gas. The reactor was a quartz tube heated by a conventional electric furnace. Every metal complex was sublimated to an argon carrier gas and then introduced into the reactor. The oxygen gas was mixed with the source gas just in front of the reactor. The films were grown at a substrate temperature of 700°C, the deposition pressure ranging from 10 to 1.5 mm Hg. The grown film was cooled in an O$_2$ atmosphere (760 mm Hg) at an average cooling rate of 20°C/min.

The films obtained were polycrystalline and had a crystallographic orientation with an axis perpendicular to the substrate plane.

REFERENCES

1. G. B. Stringfellow, *Rep. Prog. Phys.*, 45: 469 (1982).
2. K. G. Gunther, in *The Use of Thin Films in Physical Investigations* (J. C. Anderson, Ed.), Academic Press, New York, 1966, p. 213.
3. L. L. Chang, L. Eskai, W. E. Howard, and R. Ludeke, *J. Vac. Sci. Technol.*, 10: 11 (1973).
4. A. Y. Cho and J. R. Arthur, *Prog. Solid State Chem.*, 10: 157 (1975).
5. C. T. Foxon and B. A. Joyce, *J. Cryst. Growth*, 44: 75 (1978).
6. C. T. Foxon and B. A. Joyce, in *Current Topics in Materials Science*, Vol. 7 (E. Kaldis, Ed.), North Holland, Amsterdam, 1981, p. 1.
7. M. Knudsen, *Ann. Phys.*, 28: 999 (1909).
8. C. T. Foxon, M. R. Boundry, and B. A. Joyce, *Surf. Sci.*, 44: 69 (1974).
9. D. L. Smith and V. Y. Pickhardt, *J. Appl. Phys.*, 46: 2366 (1975).
10. (a) K. G. Wagner, *Vacuum*, 34: 7 (1984). (b) A. J. G. Schellingerhout, M. A. Janocko, T. M. Klapwijk, and J. E. Mooij, *Rev. Sci. Instrum.*, 60: 1177 (1989).
11. (a) N. J. Khadim, *Vacuum*, 38: 189 (1988). (b) T. J. Mattord, V. P. Kesan, D. P. Neikark, and B. G. Streetmen, *J. Vac. Sci. Technol.*, B7: 214 (1989).
12. J. A. McClintock and R. A. Wilson, *J. Cryst. Growth*, 81: 177 (1987).
13. H. Fronius, A. Fischer, and K. Ploog, *J. Cryst. Growth*, 81: 169 (1986).
14. A. Y. Cho, *J. Appl. Phys.*, 46: 1733 (1975).
15. E. S. Hellman, P. H. Pitner, A. Harwit, D. Liu, G. W. Yoffe, Y. S. Harris, Jr., B. Cafee, and T. Hierl, *J. Vac. Sci. Technol.*, B4: 574 (1986).
16. K. Oe and Y. Imamura, *Jpn. J. Appl. Phys.*, 24: 779 (1985).
17. C. T. Foxon, J. A. Harvey, and B. A. Joyce, *J. Phys. Chem. Solids*, 34: 1693 (1973).
18. D. L. Smith, *Prog. Cryst. Growth Charact.*, 2: 33 (1979).

19. K. Ploog, in *Crystals—Growth, Properties and Applications*, Springer—Verlag, New York, 1980, p. 73.
20. (a) E. H. C. Parker, *The Technology and Physics of Molecular Beam Epitaxy*, Plenum Press, New York, 1985. (b) B. A. Joyce, *Rep. Prog. Phys.*, 48: 1637 (1985).
21. S. Ueda, H. Kamohara, Y. Ishikawa, N. Tamura, S. Katoo, and Y. Shiraki, *J. Vac. Sci. Technol.*, A4: 602 (1986).
22. Y. Horikoshi, M. Kawashima, and H. Yamaguchi, *Jpn. J. Appl. Phys.*, 25: L868 (1986).
23. J. Arthur, *J. Appl. Phys.*, 37: 3057 (1966).
24. H. Sugiura, M. Kawashima, and Y. Horikoshi, *Jpn. J. Appl. Phys.*, 25: 847 (1986).
25. Y. Horikoshi, N. Kobayashi, and H. Sugiura, Two dimensional systems and new devices, "*Proc. Int. Winter School (Austria)*, 1986, p. 2.
26. W. T. Tsang, *Appl. Phys. Lett.*, 45: 1234 (1984).
27. M. B. Panish, *J. Electrochem. Soc.*, 127: 2729 (1980).
28. A. R. Calawa, *Appl. Phys. Lett.*, 38: 701 (1981).
29. M. B. Panish, *J. Cryst. Growth*, 81: 249 (1987).
30. W. T. Tsang, *J. Cryst. Growth*, 81: 261 (1987).
31. W. T. Tsang, *J. Appl. Phys.*, 58: 1415 (1985).
32. F. J. Morris and H. Fukui, *J. Vac. Sci. Technol.*, 11: 506 (1974).
33. W. T. Tsang and R. C. Miller, *J. Cryst. Growth*, 77: 55 (1986).
34. N. Putz, E. Veuhoff, H. Heinecke, M. Heyen, H. Luth, and P. Balk, *J. Vac. Sci. Technol.*, B3: 671 (1985).
35. (a) N. Putz, H. Heinecke, M. Heyen, P. Balk, M. Meyers, and H. Luth, *J. Cryst. Growth*, 74: 292 (1986). (b) W. T. Tsang, A. S. Sudbo, L. Yang, R. Camarda, and R. E. Leibenguth, *Appl. Phys. Lett.*, 54: 2336 (1989). (c) H. Hirayama, M. Hiroi, K. Koyama, and T. Tatsumi, *Appl. Phys. Lett.*, 56: 1107 (1990). (d) J. Kwo, M. Hong, D. J. Trevor, R. M. Fleming, A. E. White, R. C. Farrow, A. R. Kortan, and K. T. Short, *Appl. Phys. Lett.*, 53: 2683 (1988).
36. H. Nelson, *RCA Rev.*, 24: 603 (1963).
37. H. Rupprecht, in *Gallium Arsenide*, Institute of Physics Conference Vol. 3 (J. Frankel, Ed.), Bristol Institute of Physics, 1967, p. 57.
38. M. B. Panish, I. Hayashi, and S. Sumski, *Appl. Phys. Lett.*, 16: 326 (1970).
39. H. Nelson, U.S. Patent 3,565,702 (1971).
40. S. R. Sashital, *J. Cryst. Growth*, 74: 203 (1986).
41. M. Panck, M. Ratuszek, and M. Tlaczala, *J. Cryst. Growth*, 74: 568 (1986).
42. (a) C. F. Wan, *J. Cryst. Growth*, 80: 270 (1987). (b) M. Yoshikawa, *J. Appl. Phys.*, 63: 1533 (1988). (c) Y. R. Lee, R. G. Alonso, E. K. Suh, A. R. Ramdas, L. X. Li, and J. K. Furdyna, *J. Appl. Phys.*, 68: 1023 (1990). (d) R. F. Belt, J. Ings, and G. Diercks, *Appl. Phys. Lett.*, 56: 1805 (1990).
43. E. Lendvay, L. Petras, and V. A. Gevorkyan, *J. Cryst. Growth*, 71: 317 (1985).
44. Y. Sasai, M. Ogura, and T. Kajiwara, *J. Cryst. Growth*, 78: 461 (1986).
45. (a) F. Fiedler, H. H. Wehmann, and A. Schlachetzki, *J. Cryst. Growth*, 74: 27 (1986). (b) M. Ohshima and N. Hirayama, *J. Cryst. Growth*, 92: 461 (1988). (c) G. Sarusi, Z. Zemel, and D. Eger, *J. Appl. Phys.*, 65: 672 (1989). (d) P. Koppel and K. Owens, *J. Appl. Phys.*, 67: 6886 (1990).
46. J. S. Blakemore, *J. Appl. Phys.*, 53: R123 (1982).
47. A. Lopez-Otero, *J. Appl. Phys.*, 48: 446 (1977).

48. A. Lopez-Otero, *Thin Solid Films*, 49: 3 (1978).
49. H. Holloway, in *Physics of This Films*, Vol. 11 (G. Hass and M. H. Francombe, Eds.), Academic Press, New York, 1980, p. 105.
50. P. A. Krishnamoorthy, *J. Mater. Sci. Lett.*, 3: 551 (1984).
51. (a) H. Sitter and D. Dchikora, *Thin Solid Films*, 116: 137 (1984). (b) H. Clemens, E. J. Fantner, W. Ruhus, and G. Bauer, *J. Cryst. Growth*, 66: 251 (1984).
52. H. C. Snyman, G. J. Gouws, and R. J. Muller, *J. Cryst. Growth*, 70: 373 (1984).
53. A. V. Kulkarni, S. Bose, and R. Pradap, *Thin Solid Films*, 120: L73 (1984).
54. K. R. Murali, P. R. Vaya, and J. Sobhanadri, *J. Cryst. Growth*, 73: 196 (1985).
55. H. Kuwabara, H. Fujiyasu, H. Shimizu, A. Sasaki, and S. Yamada, *J. Cryst. Growth*, 72: 299 (1985).
56. D. Schikora, H. Sitter, J. Humenberger, and K. Lischka, *Appl. Phys. Lett.*, 48: 1276 (1986).
57. S. Miotkowska, I. Miotkowski, and T. Warminski, *J. Cryst. Growth*, 78: 457 (1986).
58. P. Pongratz and H. Sitter, *J. Cryst. Growth*, 80: 73 (1987).
59. K. R. Murali and B. S. V. Gopalam, *Acta. Phys. Hung.*, 61: 157 (1987).
60. Y. Nagamune, S. Takeyama, N. Miura, T. Minagawa, and A. Misu, *Appl. Phys. Lett.*, 50: 1337 (1987).
61. M. Sadeghi, H. Sitter, and H. Gruber, *J. Cryst. Growth*, 70: 103 (1984).
62. A. Krost, B. Harbecke, R. Faymonville, H. Schlegel, E. J. Fantner, K. E. Ambrosch, and G. Bauer, *J. Phys. C, Solid State Phys.*, 18: 2119 (1985).
63. H. Sitter, K. Lischka, W. Faschinger, J. Wolfrum, H. Pascher, and J. L. Pautrat, *J. Cryst. Growth*, 86: 377 (1988).
64. R. Korenstein and B. MacLeod, *J. Cryst. Growth*, 86: 382 (1988).
65. S. Chaudhuri, A. Mondal, and A. K. Pal, *J. Mater. Sci. Lett.*, 6: 366 (1987).
66. S. Chaudhuri, J. Battacharyya, D. De, and A. K. Pal, *Solar Energy Mater.*, 10: 223 (1984).
67. H. M. Manasevit, *Appl. Phys. Lett.*, 12: 156 (1968).
68. J. P. Duchemin, M. Bonnet, F. Koelesch, and D. Huyghe, *J. Cryst. Growth*, 45: 181 (1978).
69. C. Y. Chang, Y. K. Su, M. K. Lee, L. G. Chen, and M. P. Young, *J. Cryst. Growth*, 55: 24 (1981).
70. J. P. Duchemin, J. P. Hirtz, M. Razeghi, M. Bonnet, and S. D. Hersee, *J. Cryst. Growth*, 55: 64 (1981).
71. H. M. Manasevit, *J. Cryst. Growth*, 55: 1 (1981).
72. J. B. Mullin, S. J. C. Irwine, and D. J. Aschen, *J. Cryst. Growth*, 55: 92 (1981).
73. G. B. Stringfellow, *J. Cryst. Growth*, 55: 43 (1981).
74. H. M. Manasevit, *Proc. SPIE Int. Soc. Opt. Eng.*, 323: 94 (1982).
75. G. B. Stringfellow, *J. Cryst. Growth*, 68: 111 (1984).
76. N. E. Schumaker, R. A. Stall, W. R. Wagner, and L. G. Polgar, *J. Met.*, 38: 41 (1986).
77. P. Blanconnier, M. Cerclet, P. Henoc, and A. M. Jean-Louis, *Thin Solid Films*, 55: 375 (1978).
78. P. J. Wright and B. Cockayne, *J. Cryst. Growth*, 59: 148 (1982).
79. S. J. Bass, *J. Cryst. Growth*, 31: 172 (1975).
80. R. Bhat, *J. Electron. Mater.*, 14: 433 (1985).
81. R. M. Biefeld, *J. Cryst. Growth*, 75: 255 (1986).

82. (a) G. Nataf and C. Verie, *J. Cryst. Growth*, 55: 87 (1981). (b) P. K. Chiang and S. M. Bedair, *J. Electrochem. Soc.*, 131: 2422 (1984).

83. (a) G. B. Stringfellow, *J. Cryst. Growth*, 62: 225 (1983). (b) G. B. Stringfellow, *J. Cryst. Growth*, 70: 133 (1984).

84. M. J. Jou, Y. T. Cherng, and G. B. Stringfellow, *J. Appl. Phys.*, 64: 1472 (1988).

85. C. P. Kuo, R. M. Cohen, and G. B. Stringfellow, *J. Cryst. Growth*, 64: 461 (1983).

86. O. F. Z. Khan and P. O'Brien, *Thin Solid Films*, 173: 95 (1989).

87. M. C. Hanna, Z. H. Lu, E. G. Oh, E. Mao, and A. Majerfeld, *Appl. Phys. Lett.*, 57: 1120 (1990).

88. M. Razeghi, F. Omnes, J. Nagle, M. Defour, O. Acher, and P. Bove, *Appl. Phys. Lett.*, 55: 1677 (1989).

89. T. F. Kuech and E. Veuhoff, *J. Cryst. Growth*, 68: 148 (1984).

90. J. Van de Ven, H. G. Schoot, and L. J. Giling, *J. Appl. Phys.*, 60: 1648 (1986).

91. K. Kanehori, N. Sughii, T. Fukazawa, and K. Miyauchi, *Thin Solid Films*, 182: 265 (1989).

92. V. S. Babu, P. R. Vaya, and J. Sobhanadri, *J. Appl. Phys.*, 64: 1922 (1988).

93. A. Ishida, S. Matsuura, M. Mizuno, Y. Sase, and H. Fujiyasu, *J. Appl. Phys.*, 63: 4572 (1988).

94. A. Ishida, N. Nakahara, T. Okamura, Y. Sase, and H. Fujiyasu, *Appl. Phys. Lett.*, 53: 274 (1988).

95. A. Haque, T. Dinh, N. Christoforou, D. E. Brodic, and J. D. Leslie, *Thin Solid Films*, 176: 13 (1989).

96. S. Fuke, T. Imai, K. Kawasaki, and K. Kuwahara, *J. Appl. Phys.*, 65: 564 (1989).

97. K. Lischka, E. J. Fantner, T. W. Ryan, and H. Sitter, *Appl. Phys. Lett.*, 53: 1309 (1989).

98. H. Tatsuoka, H. Kuwabara, Y. Nakanishi, and H. Fujiasu, *J. Appl. Phys.*, 67: 6860 (1990).

8

Other Methods of Film Deposition

The secret of success is constancy of purpose.
Benjamin Disraeli

This chapter discusses certain other important techniques for the preparation of useful, good quality thin films that do not fall directly under the classifications given earlier.

8.1 LANGMUIR–BLODGETT TECHNIQUE

The technique of building up molecular layers on substrates from spread insoluble monolayers on liquid, discovered by Katharine Blodgett and Irving Langmuir in 1933, is known as the Langmuir–Blodgett (LB) technique. A brief discussion on the history of LB films (also called Langmuir films) is given by Gaines [1]. High-quality, ordered monomolecular layers or multilayers with high dielectric strength can be produced using this technique, and the possible applications of these films extend from electronic devices using the thinness and perfection of the insulating films to solar energy conversion systems. This field has undergone tremendous progress in the past few years, and quite a number of materials—fatty acids or other long chain aliphatic materials, aromatics substituted with only very short aliphatic chains, and a host of other similar molecules—can be formed into high quality LB films. Quite a number of reports have been published on the preparation and characterization of single-layer and multilayered LB films during the past few years. It is not intended here to discuss the preparation of the materials that have been studied in LB form. However the general principles and the technique involved in the preparation of LB films are treated below.

Molecules to be deposited using this technique must have carefully balanced hydrophilic (affinity for water) and hydrophobic (no affinity for water) regions if the initial monolayer and subsequent multilayers are to be formed. That is, the long chain should be hydrophilic (e.g., COOH) at one end and hydrophobic (e.g., CH3) at the other. The fatty acid molecular structure is suitable for LB deposition; for example, stearic acid CH3(CH2)16COOH has 16 CH2 groups forming the long chain body with CH3 at one end and COOH at the other.

In Langmuir's original method [2], a clean hydrophilic substrate is dipped into the water before the monolayer to be deposited is spread. Next the monolayer is spread and is maintained at a constant surface pressure. The substrate is then slowly withdrawn across the water surface and the monolayer is transferred to it. The deposition process is in principle, simple [3,4]. The substance is dissolved in a volatile solvent (e.g., chloroform, petroleum ether, toluene), and this solution is spread on a water surface, termed the subphase. The solvent evaporates, and the insoluble molecules float on the surface with their long axes randomly oriented (Figure 8.1a). By application of an appropriate constant surface pressure, the molecules are compacted and the molecules orient with their long axis normal to the water surface (Figure 8.1b). Because of the delicate nature of these films, the compression must be carried out in a very sensitive manner to avoid collapse of the film on the subphase; in addition, the film should be continuously compressed as the film is removed, to maintain the original uniformity. The whole system should also be vibration free. The details of the various techniques for obtaining constant surface pressure are described in Reference 3. The metallic salt of the molecules can be obtained by introducing metallic ions into the water. For example, to get LB films of manganese stearate, Pomerantz and Segmuller [5] used water containing divalent ion Mn^{2+} at a concentration of 10^{-3} M. When a clean solid substrate was inserted through the surface and withdrawn, the monolayer deposited on the surface. The films so formed can thus adhere to a hydrophilic (e.g., Al_2O_3, MgO, SiO_2) or a hydrophobic substrate (pure metals such as Au, Ag, or Ge). The types of built-up film are X, Y, and Z. If the layers are deposited only during the lowering of the substrate, such a deposition or built-up film is termed X type. If the deposition takes place upon both lowering and withdrawing the substrate, the film built up thus is called Y type. This is the type of film that has been most commonly studied.

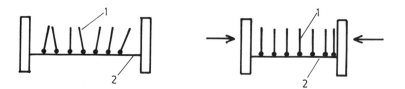

Figure 8.1 Dispersed molecules on a water surface (subphase): 1, molecules; 2, water surface. Molecules oriented (a) at random and (b) vertically after compression.

When the deposition takes place only upon withdrawal of the slide (substrate) across the monolayer, the built-up film is termed Z type. This mode of deposition is rather uncommon.

The process of building up Y films is as follows. First a hydrophobic substrate is introduced through the molecules monolayer into the water. The monomolecular layer folds itself in the direction of motion of the substrate (the advance angle is obtuse), and the first layer is transferred on its downward trip (Figure 8.2a). Upon withdrawal, the layer curves in the direction of motion of the substrate and causes the deposition of the second layer (Figure 8.2b). The next downward journey deposits the third layer (Figure 8.2c), and so on. It can be seen that the slide finally removed contains an even number of layers, and both the upper and lower surfaces of the films are made up of hydrophobic methyl groups.

A hydrophilic substrate wets when immersed in water and forms a meniscus as shown in Figure 8.3a. For deposition to take place, the meniscus should curve in the same direction as the motion of the substrate. Therefore, during the initial immersion of the substrate, no deposit will be formed, and during the withdrawal, the film will be deposited on the substrate (the meniscus curves—the layer folds itself in the direction of motion of the substrates, as in Figure 8.3b), the hydrophilic sites of the molecules attaching to the hydrophilic sites on the substrate surface. It can be seen that the substrate surface now becomes hydrophobic. During the next (second) immersion, film deposition will take place, the resulting surface becoming hydrophilic. The process, when repeated, ends in a multilayer films of odd number of layers (Figure 8.3c). Here it can be seen that the film is not deposited in the first dipping; the film deposited upon withdrawal and the subsequent deposition, however, is of type Y.

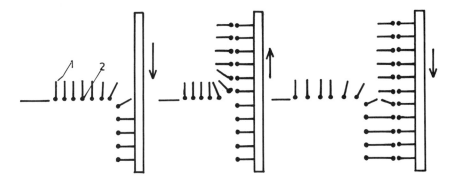

Figure 8.2 Langmuir film deposition of Y-type multilayered films on a hydrophobic substrate: 1, methyl groups; 2, carboxyl groups. (a) First downward motion of the hydrophobic substrate, (b) withdrawal of the substrate, and (c) next downward motion.

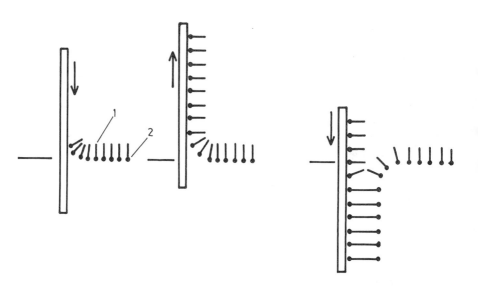

Figure 8.3 Langmuir films: deposition of Y-type multilayer films on a hydrophilic substrate: 1, methyl groups; 2, carboxyl groups, (a) First immersion of the substrate, (b) withdrawal of the substrate, and (c) second immersion of the substrate.

The subphase plays a very important role in determining the quality of the deposited films. Ultrapure water is the most favored liquid because of its exceptionally high surface tension value. The nature of the deposited film is also controlled by the pH and the temperature of the subphase. The surface quality and the chemical composition of the substrate influence the deposited layer. The speed of dipping and the age of the floating monolayer are also important. Many such factors involved in the preparation of LB films offer the advantage of versatility. Unfortunately a great effort is required to ensure that the different factors are varied in a controlled manner or held constant to produce reproducible films.

Surface pressure is a crucial factor in the preparation of high quality LB films. Various techniques have been reported for obtaining constant surface pressure and for measuring the surface pressure accurately. For details of the early film balances, the reader is referred to Reference 3. Single moving barrier [6], rotating barrier [7], constant perimeter barrier [8], and other systems to deposit LB films have since been reported. A review by Roberts [9a] describes the preparation and characterization of LB films, summarizing their importance in basic and applied sciences. The LB method is also described in Reference 9b, which makes useful reading. The resurgence of recent interest in LB films focused on electronic and nonlinear optical applications has led to the development of highly sophisticated LB deposition systems [10–12].

8.2 SPRAY METHOD

The spray method differs from other chemical solution deposition techniques in that the film is formed on a substrate kept outside the solution, which is sprayed onto the heated substrate kept outside to produce the final film, either by pyrolytic or hydrolytic chemical reaction of the liquid droplets. This has been a useful, simple, and inexpensive technique for the preparation of uniform coatings over large areas. Deposition rates and film thickness can be easily controlled, and doping can be conveniently done by simply dissolving the dopants in the spray solution in the required quantity. Although this technique has the advantage of low equipment cost compared with, say, a vacuum deposition technique, the overall cost of the materials is high when coating involves large areas. This is because only a small fraction of the material supplied is deposited on the substrate, a large fraction of the droplets being carried out of the coating region. Also a good number of droplets contained in the spray, especially the small ones, vaporize before reaching the substrate and do not contribute to the formation of the required film.

8.2.1 Spray Hydrolysis

In the method of film preparation based on the surface hydrolysis of metal halides on a heated substrate, the substrate is preheated to the deposition temperature and is then sprayed with the starting reagents, using, say, compressed argon gas as a carrier, in open atmosphere. The substrate is kept horizontal below the nozzle or at an angle.

The basic reaction for example can be represented as follows:

$$SnCl_4 + 2H_2O \rightarrow SnO_2 + 4HCl$$

Normally an alcoholic solution (or other organic acid) of the reagents are made. The amount of water introduced, from ambient humidity and in the strating reagents, largely determines how the reaction will take place and is not as simple as the basic reaction might indicate. High substrate temperatures also increase the deposition rate [13].

Several conducting transparent films, doped and undoped (e.g., SnO_2, SnO_2:Cd, In_2O_3) have been prepared by spray hydrolysis. A typical example of thin film preparation by spray hydrolysis is cited below [14].

In_2O_3 films doped with tellurium were deposited by spraying an alcoholic solution of $InCl_3$ and $TeCl_4$ onto heated (380–550°C) borosilicate glass slides or polished silicon plates in an open air atmosphere. The atomic ratio of tellurium to indium in the solution was varied from 0.01 to 0.05.

Spraying of the solution was done with a standard sprayer using compressed argon gas as a carrier gas. The lowest resistivity and highest transparency obtained exceeded 90% in the visible range, with the following parameters.

Substrate temperature	480–500°C
Spraying rate	0.08 cm³/s
Distance between nozzle and substrate	50 cm
Tellurium-to-indium atomic ratio in the solution	0.028
Angle of spraying	30–45°

The effects of the spraying angle and the deposition temperature on the grain size of the films obtained have also been investigated by Ratcheva et al. [14].

8.2.2 Spray Pyrolysis

The spray pyrolysis method was first described by Chamberlin and Skarman in 1966 for preparing CdS and certain other sulfide and selenide films [15]. Spray pyrolysis has been developed extensively since then. For example, Pamplin and Feigelson [16] have reported the spray pyrolysis of thin films of $A^I B^{III} C_2^{VI}$ semiconductor (A is copper; C^{VI} either sulfhur or selenium) and of quaternary and quinary alloys based on these compounds.

Spray pyrolysis involves a thermally stimulated chemical reaction between fine droplets of different chemical species. In this technique of film preparation, a solution (usually aqueous) containing soluble salts of the constituent atoms of the compound is sprayed onto a heated substrate in the form of fine droplets by a nozzle sprayer with help of a carrier gas, as in spray hydrolysis. Upon reaching the hot surface, these droplets undergo pyrolytic decomposition to form a film on the substrate surface. The hot substrate provides the thermal energy for the decomposition and subsequent recombination. The carrier gas may or may not play an active role in the pyrolytic reaction process.

The general setup for spray pyrolysis deposition of thin films is the same as that used for spray hydrolysis. The nozzle and the substrate are generally kept in a closed chamber provided with an exhaust to ensure a stable flow pattern. Sometimes a continuous flow of inert gas is maintained in the chamber. Spray pyrolysis has been conveniently used for large-area polycrystalline thin films. These films generally also show good adherence to the substrates. Doping can be easily accomplished by simply dissolving the dopants in the required quantities in the spray solution. Spray pyrolysis has often been used for preparing CdS films from a solution of cadmium- and sulfur-containing compounds. To cite a typical example, Krishnakumar et al. [17a] have deposited thin films of indium-doped CdS by spray pyrolysis using a spray solution of 0.09 M $CdCl_2$ and 0.1 M thiourea with $InCl_3$ added for doping. The deposition was carried out simultaneously on SnO_2-covered and plain glass substrates at a substrate temperature of 320 ±10°C, using an all-glass atomizer. The substrates were kept on a hot plate vertically below the nozzle at a distance of 40 cm. Each spray lasted for about 4 seconds. The temperature dropped 20°C during the spraying operation, and sufficient time was allowed between consecutive sprays to allow the substrate temperature to reach the preset value. After the substrates had attained the desired thickness, they were allowed to cool to room temperature.

Unlike the conventional spray pyrolysis method, the pyrolysis of an aerosol produced by ultrasonic spraying, an original deposition process named the pyrosol process [17b], has been used to deposit films of yttrium/iron/garnet (YIG) by Deschanvres et al. [17c]. The schematic of a pyrosol process apparatus is shown in Figure 8.4. Yttrium and iron acetylacetonates dissolved in butanol at 0.03 mol^{-1} served as the source solution to be sprayed. The bottom of the glass vessel was fitted with a piezoelectric transducer. The excitation of piezoelectric ceramic (at 800 kHz) resulted in the formation above the surface of the liquid of an aerosol (microscopic droplets), which was then transported by a carrier gas (air) and sprayed onto the heated substrates, where pyrolysis occurred and a thin film formed.

The substrates were fused quartz and gadolinium-gallium-garnet (GGG) single crystals oriented in the [111] direction. The substrate temperature ranged from 520 to 620°C. The as-deposited layers were amorphous and annealing at 700°C in air; epitaxial YIG layers were obtained on GGG single crystals, and high quality polycrystalline films on fused quartz.

DeSisto and Henry [17d] have reported recently the deposition of (100) oriented

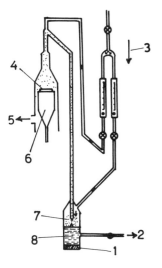

Figure 8.4 Schematic diagram of the pyrosol process apparatus for the deposition of YIG films: 1, transducer; 2, to constant level system; 3, carrier gas; 4, substrate; 5, exhaust; 6, furnace; 7, aerosol; 8, source solution. (Data from Ref. 17c.)

thin MgO films on sapphire by a novel spray pyrolysis method (Figure 8.5). The substrates used were (100) Si wafers (1.2 × 1.2 cm), fused silica (2.5 × 2.5), and sapphire (2.5 cm diameter), and these were cleaned by degreasing in hot trichloroethylene, followed by rinsing in acetone, deionized water, and methanol just prior to deposition. The substrates were heated to temperatures between 400 and 550°C. The spray solution was ultrasonically nebulized and sprayed, using the ultrasonic nozzle, and transported in flowing oxygen to the substrate, to be thermally decomposed on the hot substrates (i.e., to deposit the film). The operating frequency of the nozzle was 60 kHz, giving a droplet size of 30 μm. The spray rate and the oxygen flow rate were 10 cm^3/h and 400 cm^3/min, respectively. The films obtained were from 0.1 to 0.5 μm thick, with a typical deposition rate of 1500 Å/h, and were optically transparent and smooth. The as-deposited MgO films were of poor crystalline quality, but the films on sapphire crystallized with strong (100) orientation after annealing in flowing oxygen at 700 and 930°C.

Extensive work has been carried out on the deposition of thin films by the spray technique, and the literature on the early works can be had from these [18–20] reviews. Table 8.1 summarizes the later work done in this area.

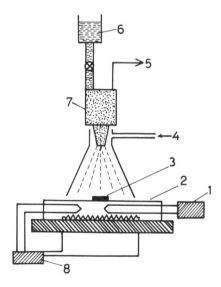

Figure 8.5 Schematic diagram of the spray pyrolysis setup for the preparation of thin MgO films on sapphire: 1, digital thermometer; 2, substrate heater; 3, substrate; 4, oxygen; 5, to power supply; 6, liquid feed; 7, ultrasonic nozzle; 8, temperature controller. (Data from Ref. 17d).

Table 8.1 Thin Films Deposited by the Spray Method

Film	Spray solution	Substrate and substrate temperature (°C)	Remarks, if any	Ref.
SnO_2:F	HCl solution of $SnCl_4$ with 1–2 wt % NH_4F	(100) oriented Si wafers 500	Polished silicon surface and texturized surface was used. Semiconductor–insulator–semiconductor solar cells were prepared.	21
Cd_2SnO_4	Aqueous solution of $CdCl_2$ plus $SnCl_4$ (0.1 M $SnCl_4$ and 0.2 M $CdCl_2$ mixed in equal proportion)	Pyrex glass 370–450	Solution pH was adjusted to 1.5 with HCl to obtain complete dissolution of $SnCl_4$. Temperature of substrate was varied at 20°C intervals. Structural, optical, and electrical properties of the films were studied.	22
CdS_xSe_{1-x} (0 < x < 1)	0.01 M aqueous solution of $CdCl_2$, 0.01 M aqueous solution of selenourea, and 0.01 aqueous solution of thiourea	Glass 425–450	After deposition, the films were cooled rapidly to room temperature. Complete solid solutions formed over the whole composition range. Films were of hexagonal structure. Lattice constants varied with composition linearly, while the variation of optical bandgap with composition was slightly nonlinear.	23
$Cd_yZn_{1-y}S$	$ZnCl_2$, $CdCl_2$, and thiourea	ITO-coated glass ≈ 300	Films were prepared for use in solar cells. Zinc was added in the form of $ZnCl_2$ to give the right weight proportion. It was assumed that the stoichiometry of the films obtained would be proportional to the fraction of Zn added in relation to Cd. Whole setup was enclosed in nitrogen atmosphere. Carrier density varied differently with Zn concentration and film thickness.	24

Table 8.1 (continued)

Film	Spray solution	Substrate and substrate temperature (°C)	Remarks, if any	Ref.
$CuInS_2$	Aqueous solutions of cupric acetate, indium trichloride, and thiourea mixed in appropriate stoichiometric proportions; solution concentration, 0.05 M	Glass 200–500	Films with chalcopyrite structure were prepared. Excess sulphur varying from 5 to 30% (in the form of thiourea) was added to the mixture to compensate for the loss of sulphur in the deposited films. Best results were obtained at 350°C. Films were characterized by X-ray, TEM, SEM, optical transmission, and the electrical measurements. Films were found to be p-type.	25
$Cd_{0.8}Zn_{0.2}S$	320 cm³ of 0.2 M $CdCl_2$ + 80 cm³ of 0.2 M thiourea solution	Glass 300–500 with interval of 50°C	Fast cooling used at termination of spray. Effect of temperature on the transport and optical properties was studied. Optical adsorption revealed that there is not much change in bandgap with substrate temperature.	26
$CuInSe_2$	Aqueous solutions indium trichloride, cuprous chloride, and selenourea (Cu/In/Se 1:1:3); concentrations of indium trichlorides and selenourea, 0.05 M; cuprous chloride, 0.01 M	Glass 350	Excess quantity of selenourea to counter the loss of selenourea and to maintain the p-type nature. Spraying was done in the dark to prevent selenourea from reacting with light. X-Ray diffraction confirmed that films were $CuInSe_2$ with chalcopyrite structure. p-type conductivity was ascertained by hot probe. Resistivities of the films were controlled by varying Cu:In ratio. Optical and structural properties were studied.	27
SnO_2:F	Alcoholic solutions of $SnCl_4 \cdot nH_2O$ (200 g $SnCl_4 \cdot nH_2O$, 90 mL	Glass 430	Spraying interval was 1–2 seconds, with pauses of 60 seconds between. Experimentally optimized values of the different parameters: L (distance	28

Material	Solution	Substrate, temperature (°C)	Comments	Ref.
	C_2H_5OH, 10 mL HCl and 2.8 mL 50% HF; $[F]/[Sn] = 0.147$)		between nozzle and substrate) $\approx$ 50 cm; spraying angle–45°. Spraying was done in open air atmosphere. Electrical and optical properties were studied. High reflectivity was reported over a wide interval in the IR region.	29
$CuInSe_2$	Solution of 0.01 M solution of $CuCl_2$, N,N-dimethyl selenourea and InCl	CdS 300 $\pm$ 5, Mo-coated glass 300 and 250 Glass 300–375	Films were grown on Mo-coated glass. First 100 mL was sprayed at 275°C to take account of the emissivity of molybdenum; then the temperature was reduced to 250°C. Study involved the kinetics of film formation using spray pyrolysis. To determine rthe phases present, the films were analyzed by X-ray diffraction.	30
$CdS:O_2$	Solution of thiourea with either cadmium chloride or cadmium acetate with varying mole ratio	Glass or sapphire, both bare and coated, with either ITO or SnO_2 250–350	Substrate temperature was maintained at $\mp$ 5°C by a liquid metal bath. Films subsequently were heat treated in oxygen, then annealed in O_2 for 60 minutes at 180–280°C in 20°C increments. Nozzle-to-substrate distance was 33 cm.	31
$SnO_2:Sb$ $In_2O_3:Sn$	$SnCl_2$, $SbCl_3$, HCl, and C_2H_5OH with constant Sb/Sn mass ratio of 0.015. In_2O_3, $SnCl_4 \cdot 5H_2O$, HCl, and C_2H_5OH, with constant Sn/In mass ratio of 2:1	Lime–alkali metal silicate glass 500	Solutions were mixed together (2:1) for spraying. Films were obtained from spraying 0.1–0.6 cm^3 of the mixture. Electrical and optical properties of the film were investigated.	32
SnO_2	$SnCl_4 \cdot 5H_2O$ (3.5 g) in ethyl alcohol (50mL)	Oxidized Si, quartz 270–605	Carrier gas was nitrogen. First, the substrate was heated to the desired temperature under alcohol spray. Then solution was sprayed in the required quantity at constant flow rate. Growth rate was thermally activated with substrate temperature and proportional to the rate of flow of spray solution.	

Table 8.1 (continued)

Film	Spray solution	Substrate and substrate temperature (°C)	Remarks, if any	Ref.
ZnO and ZnO:In	0.1 M solution of isopropyl alcohol and water (volume ratio, 3:1) plus InCl₃ for doping	Glass 402	Pure ZnO and ZnO were doped with 1.0–3.0 at. % indium. Films were deposited by spray pyrolysis. ZnO films with 2–3 at % indium exhibited excellent transparent conductivity and associated spectral reflectivity in the IR region. Optical properties of ZnO were studied in the UV, visible, and IR regions.	33
SnO₂	0.5 M SnCl₄ in ethyl acetate	Carbon film and Tempax glass 340–500	Films were amorphous at substrate temperatures of 340–350°C, partially crystallized at 360–410°C, and polycrystalline at 420–500°C. Deposition rate varied from 5 to 35 Å/s.	34
ZnO	0.1 M aqueous zinc chloride and zinc acetate solutions; hydrochloride and acetic acids added at 0.1–0.2 vol % to adjust the pH to 2.4–5.8.	(111) oriented single crystal and zinc doped p-type	Filtered compressed air was used as spray carrier gas. Complete dissolution of the zinc compounds was obtained by adding hydrochloric and acetic acids. After spraying, the ZnO films were rinsed in isopropanol and deionized water and then blown dry.	35
Al₂O₃	Aluminum sec-butoxide in toluene; concentration, 70 vol %	Soda lime glass 550	Substrate was heated for 5 minutes, followed by spraying as soon as the sample was removed from the furnace. Sample was then returned to the furnace and annealed for 5 minutes. Whole process was repeated to get the desired thickness. X–Ray diffraction analysis showed that the films were amorphous. Adherent, pinhole-free, and dense films were obtained.	36

Al_2O_3-ZrO_2	Aluminum sec-butoxide and n-butylzirconate mixed at 20 and 80 vol %, respectively	Soda lime glass 550	Spray procedure was same as above. Coating obtained was amorphous, as observed by X-ray analysis.	36
SnO_2:Sb	1 M $SnCl_4$ in ethanol + 0.1 M Sb_2S_3 in HCl	Glass 300–600 ± 5	Antimony-doped SnO_2 films were prepared by spray pyrolysis; 0.1 M Sb_2S_3 in HCl was added in varying quantities to make up different concentrations of antimony in the final composition. Spraying period was 3 seconds with 3-minute intervals. Structural, electrical, and optical properties were investigated by varying concentration of antimony, substrate temperature, and film thickness. Systematic optimization of the parameters, made it possible to obtain films with an average transmission of 88% in the visible region and an IR reflectance of 76%.	37
Pyrite FeS_2	Aqueous solution of $FeCl_4$ and thiourea ($FeCl_4$/thiourea/H_2O = 0.15:0.07:0.78	Quartz 550	Solution was immediately sprayed with a double-nozzle sprayer. This composition gave higher optical transparency. Layers were clear, transparent, and yellow, very stable, and had good adhesive properties. Purity and structure were confirmed by X-ray diffraction. Absorption edge measurements were carried out.	38
CdS	Solution of $CdCl_2$ and $(NH_2)_2CS$; solution concentrations, 0.10 M; solution ratio of Cd to S: 1:1	Glass 235, 285, 335, 400, and 450	Jet was positioned directly over the heated substrate; nozzle-to-substrate distance, 30 cm. Temperature and time of deposition were varied (time, 5, 10, 15, and 20 min). Nucleation, growth, and microstructure of the films were studied.	39

Table 8.1 (continued)

Film	Spray solution	Substrate and substrate temperature (°C)	Remarks, if any	Ref.
$YBa_2Cu_3O_{7-\delta}$	Mixed nitrate powder of Y, Ba, and Cu (Y/Ba/Cu = 1:2:3); dilute solution of 1–2 wt % prepared by adding water.	(100) single crystals of MgO, ZrO_2 with 9% Y_2O_3, and $SrTiO_3$ 180–250	Mixed nitrate powder was prepared by mixing stoichiometric amounts of Y_2O_3, $BaCo_3$, and (Y/Ba/Cu = 1:2:3) with nitric acid and evaporated to dryness. Nitrogen was used as the carrier gas. Addition of up to 25% ethanol to the solution allowed deposition at lower temperatures without precipitation. Typical growth procedure: spraying the solution (T_s = 180°C), prebaking films in air (20 min at 500°C), oven annealing under flowing O_2 (900–950°C), and slow cooling to 200°C in about 3 hours. Best films (T_c = 87 K) were obtained in the YSZ substrate.	40
In_2O_3:Tb	0.1 M $InCl_3$ in 3 parts iso-propyl alcohol and 1 part deionized water, $TbCl_3$ added with doping concentration of 10 at. %	Pyrex glass slide 270, 300, and 330	$TbCl_3$-doped In_2O_3 films were prepared by spray pyrolysis using air as carrier gas. Substrate-to-nozzle distance, 30 cm. Spraying time was 10 minutes. Intrinsic films were also prepared; both showed a cubic crystalline structure.	41
Bi-Sr-Ca-Cu-O	Bismuth oxide, strontium carbonate, and cupric oxide in concentrated nitric acid. Metal stoichiometry, 2:2:1:2 and 4:3:3:4 (Bi/Sr/Ca/Cu). Solution diluted to get an	MgO and BeO substrate held on 600°C heated block	After spray deposition, the films were heated in air in a single-zone furnace, then densified by melt quenching/flux sintering. These dense, superconducting films exhibited a high degree of crystallographic orientation, with c-axis normal to the substrate surface. Films were characterized by X-ray diffraction, SEM, and four-probe resistance	42

Material	Solution	Substrate	Remarks	Ref
	aqueous solution that was 1.6 M in nitric acid and a total metal ion concentration of 0.5 M		measurements.	
SnO_2	Solutions based on dibutyltin and tributyltin trifluoroacetate	Hot glass substrate	Electrical measurements were made and the free electron concentration and mobility as a function of temperature in the range 80–600 K determined. Solution based on tributyltin tri-fluoroacetate gave very good spectrally selective layers.	43
$YBa_2Cu_3O_{7-x}$	$YBa_2Cu_3O_{7-x}$ powder dissolved with a total concentration of about 0.04 mol in a solution of 90% propionic acid and 10% deionized water	Single crystals of MgO Substrate surface maintained at 120°C	Microscopic droplets formed at the surface of the liquid (formed by the excitation of a piezoelectric transducer fitted at the bottom), transported by airflow and sprayed onto the substrate. Films were annealed in air, heat treated in oxygen (850–1000°C), and slowly cooled to room temperature.	44
Cr_2O_3 and CrO_3	$Cr(NO_3)_3 \cdot 9H_2O$ solution in distilled water	Glass plates ≈ 420	Films obtained when sprayed onto glass plates at this temperature were reddish. When the temperature was raised to near 500°C, the films became greenish. X–Ray analysis showed that reddish films are CrO_3, polycrsytalline, and orthorhombic in structure, and the greenish films are Cr_2O_3, polycrystalline, and hexagonal.	45
Ag-doped $YBa_2Cu_3O_{7-x}$	0.05 mol/L of $YBa_2Cu_3O_7$ + 0.2 mol/L of $AgNo_3$	MgO 350–850	Spray solution was made by dissolving Y_2O_3, $BaCO_3$, and CuO in dilute nitric acid; silver doping done by adding $AgNO_3$ to this solution so that the final solution contained up to 0.2 mol/L of $AgNO_3$. Oxygen used as the propellant gas. With various amounts of $AgNO_3$ in solution, the films obtained were characterized.	46

Table 8.1 (continued)

Film	Spray solution	Substrate and substrate temperature (°C)	Remarks, if any	Ref.
$YBa_2Cu_3O_{7-x}$	Appropriate amounts of metal nitrates $Y(NO_3)_3 \cdot 6H_2O$, $Ba(NO_3)_2$, and $Cu(NO_3)_2 \cdot 3H_2O$, dissolved in ultrapure water, aqueous solution mixed into Y/Ba/Cu (= 1:2:3 molar ratio) while adding ethanol up to 20% to the solution	(100) YSZ and (100) $SrTiO_3$ (surface chemically etched to mirror finish) 250–280	After heating in air at 500°C for 30 minutes, the films were introduced into the furnace within 1 minute and fired rapidly at 1000–1020°C in flowing helium atmosphere, then annealed in oxygen. Melt textured films had aligned grain structure with large platelike grains over a large area. The best J_c of the films fired at 1010°C was 4080 A/cm^2 at 77 K in zero magnetic field.	47
SnO_2	Tin chloride ($SnCl_4 \cdot 5H_2O$, 2 g), 99.9% purity, dissolved in 10 cm^3 ethyl alcohol	Soda lime glass 280–440	Oxygen and nitrogen (99.9% purity) were used as carried gases. Dependence of optical and electrical properties on substrate temperature and substrate–nozzle distance was studied.	48
Sb_2O_5	0.5 M $SbCl_3$ in ethyl acetate	Polished glassy carbon disks, and glass 400	Glassy carbon substrates were used for electrochemical investigations and glass for conductivity measurements. Water necessary for the hydrolysis reaction was supplied as vapor saturated in air or nitrogen. Effect of thickness of a film on its physical properties was studied.	49

REFERENCES

1. G. L. Gaines, *Thin Solid Films,* 99:IX (1983).
2. I. Langmuir, *Trans. Faraday Soc.,* 15:62 (1920).
3. G. L. Gaines, *Insoluble Monolayers at the Liquid–Gas Interface,* (Wiley, Interscience) New York, 1966.
4. V. K. Srivastava, in *Physics of Thin Films,* Vol. 7 (G. Hass, M. H. Francombe, and R. W. Hoffman, Eds.), Academic Press, New York, 1973, p. 311.
5. M. Pomerantz and A. Segmuller, *Thin Solid Films,* 68:33 (1980).
6. C. W. Pitt and L. M. Walpita, *Thin Solid Films,* 68:101 (1980).
7. P. Fromherz, *Rev. Sci. Instrum.,* 46:1380 (1980).
8. G. G. Roberts, W. A. Barlon, and P. S. Vincett, *Phys. Technol.,* 12:69 (1981).
9. (a) G. G. Roberts, *Adv. Phys.,* 34:475 (1985). (b) A. Barraud and M. Vandevyver, in *Non-linear Properties of Organic Molecules and Crystals,* Vol. 1 (D. S. Schemia and J. Zyss, Eds.), Academic Press, New York, 1987, Ch. II.5.
10. M. F. Daniel, J. C. Dolphin, A. J. Grant, K. E. N. Kerr, and G. W. Smith, *Thin Solid Films,* 133:235 (1985).
11. G. Hunger, L. Lorrain, G. Gagne, and R. M. Leblanc, *Rev. Sci. Instrum.,* 58:285 (1987).
12. (a) K. Miyano and T. Maeda, *Rev. Sci. Instrum.,* 58:428 (1987). (b) T. Kasuga, H. Kumehara, T. Watanabe, and S. Miyata, *Thin Solid Films,* 178:183 (1989). (c) B. R. Malcolm, *Thin Solid Films,* 178:191 (1989).
13. J. A. Aboaf, V. C. Marcotte, and N.J. Chou, *J. Electrochem. Soc.,* 120:701 (1973).
14. T. M. Ratcheva, M. D. Nanova, L. V. Vassilev, and M. G. Mikhailov, *Thin Solid Films,* 139:189 (1986).
15. R. R. Chamberlin and J. S. Skarman, *J. Electrochem. Soc.,* 113:86 (1966).
16. B. R. Pamplin and R. S. Feigelson, *Thin Solid Films,* 60:141 (1979); *Mater. Res. Bull.,* 14:1 (1979).
17. (a) R. Krishnakumar, Y. Ramaprakash, V. Subramonian, K. Chandrasekharapillai, and A. S. Lakshmanan, *SPIE Opt. Mater. Tech. Energy Effic. Solar Energy Convers.,* 562:187 (1985). (b) J. Spitz and J. C. Viguie, French Patent 2,110,622 (1972). (c) J. L. Deschanvres, M. Langlet, and J. C. Joubert, *Thi Solid Films,* 175:281 (1989). (d) W. J. DeSisto and R. L. Henry, *Appl. Phys. Lett.,* 56:2522 (1990).
18. J. L. Vossen, in *Physics of Thin Films,* Vol. 9 (G. Hass, M. H. Francombe, and R. W. Hoffman, Eds.), Academic Press, New York, 1977, p. 1.
19. K. L. Chora, R. C. Kainthla, D. K. Pandya, and A. P. Thakoor, in *Physics of Thin Films,* Vol. 12 (G. Hass, M. H. Francombe, and R. W. Hoffman, Eds.), Academic Press, New York, 1982, p. 167.
20. B. Mooney and S. B. Radding, *Annu. Rev. Mater. Sci.,* 12:81 (1982).
21. F. J. Gracia, J. Muci, and M. S. Tomar, *Thin Solid Films,* 97:47 (1982).
22. R. A. Ortiz, *J. Vac. Sci. Technol.,* 20:7 (1982).
23. A. Raturi, R. Thangaraj, A. K. Sharma, and O. P. Agnithotri, *J. Phys. C. Solid State Phys.,* 15:4933 (1982).
24. H. H. L. Kwok, M. Y. Leung, and Y. W. Lam, *J. Cryst. Growth,* 59:421 (1982).
25. P. Rajaram, R. Thankaraj, A. K. Sharma, A. Raza, and O. P. Agnihotra, *Thin Solid Films,* 100:111 (1983).
26. M. D. Uplane and S. H. Pawar, *Solid State Commun.,* 46:847 (1983).

27. O. P. Agnihotra, P. Rajaram, R. Thankaraj, A. K. Sharma, and A. Raturi, *Thin Solid Films*, 102:291 (1983).
28. G. Mavrodiev, H. Gajdardziska, and N. Novkovski, *Thin Solid Films*, 133:93 (1984).
29. C. R. Abernathy, C. W. Bates, Jr., A. A. Anani, and B. Haba, *Thin Solid Films*, 115:L41 (1984).
30. D. Richards, A. M. El-Korashy, R. J. Stirn, and P. C. Karulkar, *J. Vac. Sci. Technol.*, A2:332 (1984).
31. K. Kulaszewicz, W. Jarmoc, and K. Turowska, *Thin Solid Films*, 112:313 (1984).
32. J. Mimila-Arroya, J. A. Reynoso, E. Saucedo, and J. C. Bourgoin, *J. Cryst. Growth*, 68:671 (1984).
33. S. Major, A. Banerjee, and K. L. Chopra, *Thin Solid Films*, 125:179 (1985).
34. J. Melsheimer and B. Tesche, *Thin Solid Films*, 138:71 (1986).
35. C. Eberspacher, A. L. Fahrenbruch, and R. H. Bube, *Thin Solid Films*, 136:1 (1986).
36. D. Nguyen, M. vanRoode, and S. Johar, *Thin Solid Films*, 135:L19 (1986).
37. I. S. Mulla, H. S. Soni, V. J. Rao, and A. P. B. Sinha, *J. Mater. Sci.*, 21:1280 (1986).
38. A. K. Abass, Z. A. Ahmed, and R. E. Tahir, *J. Appl. Phys.*, 61:2339 (1987).
39. D. S. Albin and S. H. Risbud, *Thin Solid Films*, 147:203 (1987).
40. A. Gupta, G. Koren, E. A. Giess, N. R. Moore, E. J. M. O'Sullivan, and E. I. Cooper, *Appl. Phys. Lett.*, 52:163 (1988).
41. A. Ortiz, M. Garcia, S. Lopez, and C. Falcony, *Thin Solid Films*, 165:249 (1988).
42. H. M. Hsu, I. Yee, J. De Luca, C. Hilbert, R. F. Miracky, and L. N. Smith, *Appl. Phys. Lett.*, 54:957 (1989).
43. J. J. Ph. Elich, E. C. Boslooper, and H. Haitjema, *Thin Solid Films*, 177:17 (1989).
44. M. Langlet, E. Senet, J. L. Deschanvres, G. Delabouglise, F. Weiss, and J. C. Joubert, *Thin Solid Films*, 174:263 (1989).
45. R. H. Misho, W. A. Murad, and G. H. Fattahallah, *Thin Solid Films*, 169:235 (1989).
46. E. J. Cukauskas, L. H. Allen, H. S. Newman, R. L. Henry, and P. K. Van Damme, *J. Appl. Phys.*, 67:6946 (1990).
47. E. Ban, Y. Matsuoka, and H. Ogawa, *J. Appl. Phys.*, 67:4367 (1990).
48. V. Vasu and A. Subramoniam, *Thin Solid Films*, 189:217 (1990).
49. W. A. Badaway, *Thin Solid Films*, 186:59 (1990).

9

Summary

Piling up knowledge is as bad as piling up money. You have to begin sometime to use what you know.

Robert Frost

This chapter gives a summary overview of the various thin film preparation techniques covered in the book, discussing basic principles and salient features, with selected literature references.

Thin films play an important role in present-day technological development, and techniques of film deposition offer a major key to the fabrication of solid state microelectronic devices. Thin film deposition can be broadly classified as either physical or chemical. Under physical methods we have vacuum evaporation and sputtering, where the deposition takes place after the material to be deposited has been transferred to a gaseous state either by evaporation or an impact process. Under chemical methods we have the gas phase chemical processes such as conventional chemical vapor deposition, photo CVD, and plasma-enhanced CVD. Thermal oxidation is a chemical thin film process in which the substrate itself provides the source for the constituent of the oxide. Liquid phase chemical techniques include electrolytic deposition, electroless deposition, electrolytic anodization, spray pyrolysis, and liquid phase epitaxy. Molecular beam epitaxy (MBE) is a sophisticated, finely controlled evaporation technique performed in ultra high vacuum. There are other methods based on the application of an ion or ionized cluster source, and these can be treated as variants of the physical methods of film deposition. Reactive deposition makes use of a reactive component for the deposition of compound films.

The broad classification is outlined in Figure 9.1.

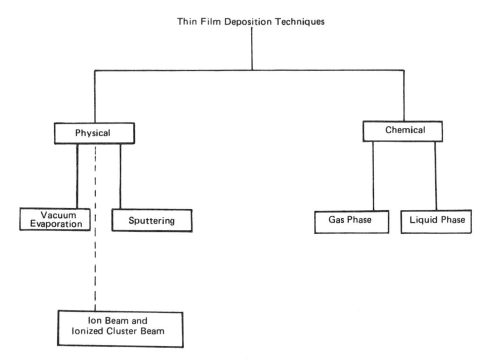

Figure 9.1 Broad classification of thin film deposition techniques.

9.1 VACUUM EVAPORATION

Vacuum evaporation, although one of the oldest techniques, is still widely used both in the laboratory and in industry. It is a very simple and convenient technique.

The three basic steps involved in the deposition are as follows.

1. Generation of vapor from the condensed phase, solid or liquid.
2. Transfer of the vapor from the source to the substrate.
3. Condensation of the vapor on the substrate surface to form the solid film.

The pressure used for normal evaporation is about 10^{-5} torr. When evaporation is made in vacuum, the evaporation temperature is considerably lowered and the incorporation of oxides and other impurities is very much reduced. Also this ensures a straight line path for the vapor emitted from the source for the usual working source-to-substrate distance of 10–50 cm. Excellent treatments of the subject of vacuum evaporation are given in Holland [1] and Maissel and Glang [2].

There are different methods for evaporating the material to be deposited. The oldest and simplest one, which is still commonly used, is evaporation from a boat or wire of a refractory metal—tungsten, tantalum, or molybdenum, heated by electric current. (The refractory metals have high melting points and very low vapor pressures.) Vapor sources (support materials) of different types—hairpin, wire helix, dimpled foil, dimpled foil with oxide coating, wire basket, indirectly heated crucible made from Al_2O_3 or other oxide—are used for the evaporation depending on the evaporant material. Vapor sources of different designs have been described by many authors [3-6].

In another technique, flash evaporation, small quantities of the material to be evaporated are dropped in powder form onto a boat that is hot enough to ensure that the material is evaporated instantaneously. The method is useful in the preparation of multicomponent alloys or compounds that tend to distill fractionally. Since the feed of the material is continuous, the instantaneous, discrete evaporation of the material results in a vapor stream of the same composition as that of the source material. Thus stoichiometric thin films of compound or alloy are deposited on the substrate. Different methods (mechanical, ultrasonic, electromagnetic, etc.) are used for feeding (dropping) the powder onto the heated support boat. A serious drawback of this technique is the difficulty in preoutgassing (for several hours) the evaporant powder. Unless this outgassing is performed the pressure will increase during evaporation as a result of the release of the gases from the powder. Also, expanding gases can cause "spitting" during evaporation.

Electron beam bombardment is another method of accomplishing evaporation. Here a stream of electrons accelerated by a voltage of 5-10 kV is focused onto the surface of the material, which is contained in a water-cooled crucible. The kinetic energy of the electrons is transferred very rapidly into thermal energy, and the material melts at the surface and evaporates. The material in contact with the crucible remains unmelted, so that the molten mass is contained in a crucible of itself and reactions with the crucible walls are avoided completely. Any material can be evaporated with this method of evaporation, and the rate of evaporation is significantly higher than that of resistance heating. Electron beam evaporator devices of various types (with axial gun, magnetic focusing, etc.) are now commercially available to produce thin films of a wide variety of materials.

Laser evaporation is another method of evaporation for the deposition of thin films. Lasers are used as the thermal source, and the laser source, kept outside the vacuum system, is focused onto the target material; the ejected material is deposited on the substrate, placed in front of the target material inside the vacuum. The ejection process is still not very clearly understood. Target materials may be ejected by laser pulses (collisional sputtering, normal thermal evaporation, etc.), and the process probably depends on the power of the laser and also on the type of material ejected [7].

Laser evaporation has several advantages: congruent evaporation of the target material, for example, and the ability to maintain the original purity of the source material (eliminating crucible contamination). With the high power density obtained by focusing the laser beams, high melting point materials can be evaporated at high deposition rates. Simultaneous and sequential evaporation using multiple sources can be easily done by directing the laser beam with external mirrors.

In another method, arc evaporation, an arc is struck between two electrodes, producing a temperature high enough to evaporate the materials of the conducting electrodes. The process, which is not easily reproducible, is used to prepare films of refractory metal.

In the rf heating method, by suitable arrangement of the rf coils, levitation and evaporation can be accomplished. This method of deposition is not commonly used because of the difficulty in positioning the coil and the sample properly, for effective coupling, in a vacuum system.

9.2 SPUTTERING

Bombardment of a surface with high velocity positive ions causes the surface atoms to be ejected. [8]. This ejection of atoms form the surface due to bombardment by positive ions, usually inert, is commonly known as (cathode) sputtering. When the ejected atoms are made to condense on a substrate, thin film deposition takes place.

Various theories have been put forward to account for the mechanism of cathode sputtering. Several works on the subject of sputtering and sputtering processes of thin film deposition are available [9-13].

The simplest arrangement for the deposition of thin films by sputtering is the glow discharge dc sputtering system. Here the plate of the material to be deposited is connected to a negative dc voltage supply (1-5 kV) and the substrate is mounted on the anode facing the target. A neutral gas such as argon is introduced into the vacuum chamber until the pressure reaches 10^{-1} to 10^{-2} torr. When the electric field is applied, a glow discharge is formed. Here the positive ions strike the target plate, removing from the surface mainly neutral atoms, which eventually condense on the substrate as a thin film.

The operation of a sputtering system calls for a self-sustained glow discharge. In a practical sputtering system, however, a self-sustained dc discharge at pressures below 10 mtorr is impossible because there are too few ionizing collisions. The most serious drawback here is contamination of the deposited film by the inert gas used to produce the discharge. This can be minimized if the sputtering is done at low pressures. To increase the ionization and sustain the discharge at low pressures, therefore, additional electrons from a source other than the target must be supplied.

In triode sputtering, the electrons are provided in the discharge from an independent source. A hot cathode (heated tungsten filament), which emits electrons through thermionic emission, is used to inject electrons into the discharge system. A supported discharge system like this allows operation at much lower pressures (10^{-3} torr) than are possible in a conventional diode glow discharge system. The main limitation lies in the difficulty in producing uniform sputtering from flat targets of large size. Also the supported discharge is difficult to control for reproducible results.

A depositing film is an active getter of impurities, and a sputtering gas, when made to pass over an area of freshly deposited films, cleanses the new film of its impurities. This principle is used in the design of the getter sputtering systems. The residual gas molecules impinging on an atomically clean metal surface are chemisorbed or form a metal compound, and chemical cleanup occurs. Rare gases are merely physisorbed and not readily trapped.

In a conventional dc sputtering system, if the metal target is replaced by an insulator, a surface positive charge is built up on the front surface of the insulator during ion bombardment and prevents further ion bombardment. It has been shown [14] that this charge buildup can be prevented by simultaneously bombarding the insulator with both ion and electron beams. For sputter depositing insulator films, Davidse and Maissel [15] developed a practical rf sputtering system in which an rf potential is applied to the metal electrode placed behind the dielectric plate target. Here the electrons oscillating in the alternating field have sufficient energies to cause ionizing collisions and to keep the discharge self-sustained. The high voltage at the cathode required in dc sputtering for the generation of secondary electrons is not necessary here to maintain the discharge. Since the electrons have much higher mobilities than ions, far more electrons will reach the dielectric target surface during the positive half-cycles than ions during the negative half-cycles, and the target becomes self-biased negatively. The negative dc potential on the dielectric target surface repels electrons from the vicinity of this surface, creating a sheath enriched in ions in front of the target. These ions bombard the target, and sputtering is achieved. At frequencies less than 10 kHz, such an ion sheath is not formed; 13.56 MHz is generally used for rf sputtering. metals can also be sputtered if the rf power supply is coupled capacitively to a metal electrode.

Another variant of the sputtering process is magnetron sputtering, a magnetically enhanced sputtering technique discovered by Penning [16] and subsequently developed. For a simple planar magnetic system, a planar cathode is backed by permanent magnets that provide a torodial field with field lines forming a closed path over the target (cathode) surface. The secondary electrons generated are trapped in cycloidal orbits near the target and prevent self-heating of the substrate. The confinement of the plasma and the resultant intense plasma allow magnetron sputtering systems to operate at much lower pressures and lower target voltages than are possible for rf diode sputtering. Also here the deposition rates are relatively higher and cover large deposition areas. Low substrate heating allows the use of a variety of substrates for a wide variety of applications. Several magnetron sputtering systems have been developed, and the technology in general has been reviewed [17,18a].

Ion beam sputtering is another useful film deposition technique affording independent control of the ion beam energy, as well as the current density of the bombarding ions. Here the ion beam generated at an ion source is extracted into a high vacuum chamber and directed at the target material, which is sputtered and deposited on a nearby substrate. The directionality of the beam allows the angle of incidence (target) and angle of deposition (substrate) to be varied. Other advantages over conventional sputtering include the low background pressure and the greater isolation of the substrate from the ion production process.

9.3 ION-ASSISTED DEPOSITION

Ion-assisted techniques include conventional ion plating [18b] and its variants, cathodic arc deposition, and hot hollow cathode gun evaporation, where a significant percentage of the vapor of the source material is ionized. Again there is concurrent ion bombardment deposition, with a separate ion source irradiating the substrate during deposition in a controlled manner. So also there is direct ion beam deposition: ion beams consisting of the desired film material are extracted and deposited directly onto the substrate at low energy.

In conventional ion plating the material is evaporated from a resistively heated crucible serving as the anode and the substrate is made the cathode of a diode dc discharge (2–5 kV) normally used in diode sputtering. The variants of ion plating include triode ion plating, hollow cathode discharge ion plating, and magnetron sputter ion plating.

In cathodic arc deposition the material is made the cathode in an arc circuit and the material evaporated by the action of vacuum arcs, the arc occurring in regions a few micrometers in size and carrying very high current densities.

Hot hollow cathode (HHC) is an arc-generating device; by extracting electrons to form an electron beam, it can be used as a heating device for the preparation of thin films, where the evaporated material is substantially ionized.

In all these ion-assisted deposition methods, the properties of the deposited films depend largely on the available energy per condensing atom. Ion-assisted deposition has been discussed in two recent articles [19,20], and the reader is referred to these for more information.

9.4 REACTIVE DEPOSITION

Compound thin films (nitrides, oxides, carbides, etc.) are prepared by reactive deposition techniques, entailing the presence of a reactive gas or component. The material is converted physically from the condensed phase to the vapor phase by using, for example, thermal energy (evaporation) or momentum transfer (sputtering), whereupon a chemical reaction take place to form the compound films. For example, BeO films have been prepared by evaporating pure beryllium from a resistance-heated tantalum boat in an oxygen atmosphere at reduced pressure [21].

Not all compounds can be deposited when evaporated in the presence of reactive gas. For certain oxides, nitrides, and other materials, the high activation energy that is necessary if chemical reaction is to take place is achieved in the activated reactive evaporation (ARE) process developed by Bunshah and Raghuram [22]. Here a low voltage electrode (biased positively around 100 V) attracts electrons from the molten pool of the source, which is heated by an electron beam, and these electrons ionize the reactive gas to form a thick plasma. Because of the presence of this plasma, the chemical reaction rate is increased, leading to compound film formation and a higher deposition rate. Several modifications of ARE have been developed [23,24].

In reactive arc evaporation, the basic principle is the arc evaporation of the metal in the respective reactive gas. The ionization efficiency in cathodic arc deposition processes is as high as 30–50%, and this high degree of ionization in the plasma and the high energies of the ions increase the reaction efficiency, resulting in the formation of compound films exhibiting enhanced adhesion and density.

In reactive sputtering, the sputtered metal from the target reacts with the reactive gas present to form the compound. Sometimes reactive gas is added to make up for a constituent in the film (e.g., a compound target that chemically decomposes during sputtering), and this process is also treated as a case of reactive sputtering. Sometimes "target poisoning" occurs: that is, a reactive gas reacts with the target during sputtering and a compound is formed on the surface of the target, causing the sputtering rate to drop considerably as a result of the exposure to ion bombardment of the compound, rather than the target material. The effect of target poisoning on sputter deposition depends on the particular metal–reactive gas combination and the properties of the compound surface layer formed. Several models have been developed for reactive sputter deposition [25–27a]. The principle of reactive sputter deposition of the preparation of compound films is used in the different variants of sputtering systems.

In reactive ion plating, a useful development of basic ion plating, the residual atmosphere in the vacuum system contains a reactive gas. Ionization can be improved by the use of various techniques such as hollow discharge and magnetron sputtering.

Reactive ion beam sputter deposition is a modification of the ion beam sputter deposition technique and differs from other reactive deposition methods in that the reactive ion can be introduced in two ways.

1. *As ion beams*: the compound film is formed by the sputtering of the compound or target solid component of the binary compound with the reactive ion beam or mixture of inert and reactive ions. Alternatively, when sputtering is done by one ion source and the reactive ion beam from another source is directed onto the growing film on the substrate, we have dual ion beam deposition.

2. *As a gaseous phase*: film formation is accomplished with the inert gas bom-

barding the target in the presence of the reactive gas, which is admitted in the vicinity of the substrate.

9.5   IONIZED CLUSTER BEAM (ICB)

Ionized cluster beam deposition is a recently developed thin film deposition technique that has been widely used for the preparation of high quality films of a wide variety of materials [27b]. Epitaxial films of many materials on a variety of substrates are obtained at low temperatures. The material to be deposited is vaporized and ejected from the nozzle of a special crucible to expand into a high vacuum environment. A supersaturated state is formed by adiabatic expansion. The atoms lose their energy by collision, and nucleation starts. The nuclei grow to form clusters of up to a thousand atoms held together by weak interatomic forces. These clusters are ionized by electron bombardment (impact ionization), whereupon they are accelerated toward the substrate and impinge on it, finally forming the film. The energy of the ionized clusters striking the substrate depends on the potential difference between the substrate and the crucible. In the reactive ionized cluster beam (R-ICB) method, the reactive gas is introduced in the vicinity of the substrate in an ICB system. In the dual ion beam deposition process using ionized clusters, the ionized clusters of the source material from the ICB system and the reactive gas ions from another source are incident simultaneously on the substrate to form the required films. Brief reviews of the ICB method of film deposition have been published recently [27c,d].

9.6   CHEMICAL METHODS

Another major technology for depositing thin films and coatings consists of the chemical methods, including thin film formation by chemical processes in the gas or vapor phase as well as from liquid phases.

Thermal growth is a gas phase reaction method of thin film formation. Thermal growth is not a commonly used method, but thermally grown oxides of metals and semiconductors, where the substrate itself provides the metal or semiconductor constituent of the oxide, have been widely investigated. All metals except gold react with oxygen, and thermal oxidation is usually carried out by conventional furnace oxidation, although other reports different from conventional oxidation are available [30]. Thermal oxidation of silicon to form SiO_2 has been extensively studied, since SiO_2 has very important applications in silicon device technology. Mott [31] has reviewed some current theories on silicon oxidation.

Chemical vapor deposition (CVD) is an important and popular technique in which films of high purity and quality, with required composition and doping levels, can be prepared. Here constituents of the vapor phase react to form a solid film on the substrate surface, which is maintained at a suitable temperature. The chemical reaction itself is an important characteristic of all CVD processes, and reactions basic to CVD include thermal decomposition (pyrolysis), reduction, oxidation, nitride and carbide formation, and chemical transfer reactions.

Flow rate, input concentration, deposition temperature, pressure, and reactor geometry are process variables controlling the deposition of films. Thus a basic CVD system for the deposition of thin films involves three fundamental processes: transportation of the reactants to the site of reaction, provision of activation energy for reaction, and removal of by-product gases and vapors.

Thin films that can be prepared by CVD cover a very wide range of materials, elements as well as compounds, on various substrates. There are reactor systems for atmospheric pressure CVD (APCVD), low pressure (0.1–10 torr) CVD (LPCVD), and different designs and shapes for low temperature (< 600°C), medium temperature (600–900°C), and high temperature (900–1500°C) systems. A good and detailed account of the different aspects of CVD, reactor system types, and early work on this mode of deposition has been presented in Reference 32. In conventional CVD, the reaction is thermally activated and the reactive gases pass over the heated substrate, with a thermal reaction taking place to deposit the film on the substrate.

Laser chemical vapor deposition (LCVD) is a recently developed CVD technique in which a laser source is used to activate the chemical reaction. Laser CVD can be divided into two broad categories, pyrolysis and photolysis, depending on the mechanism that initiates the chemical reaction. In pyrolysis the laser beam is utilized as a heat source for localized heating, and pyrolytic decomposition occurs as a result of the reaction of the ambient gases with the locally heated substrate. In photolysis, photons of sufficient energy are absorbed to activate gaseous reactant atoms or molecules, resulting in chemical reactions to form the films on the adjacent substrate. It should however be borne in mind that thermal effects are very often present in LCVD reactions, and so these reactants can be more generally said to be mainly photolytic or pyrolytic. Pyrolytic laser decomposition will be termed as LCVD, as distinguished from photo CVD. Although the reactant systems in LCVD is similar to conventional CVD, the film growth characteristics will be different [33,34].

Photo CVD relies on high energy photons for selective absorption; the commonly used sources are low pressure mercury discharge lamps and excimer lasers. Photo CVD is relatively new and is an attractive low temperature deposition technique for the preparation of high quality films for many interesting applications.

Plasma-enhanced CVD is yet another versatile technique for depositing a variety of materials for micro-electronic, photovoltaic, and many other applications. The primary role of the plasma is to promote chemical reactions. The electrons are the most energetic species in the plasma, and the average electron energy, varying from 1 to 20 eV, is sufficient to ionize and dissociate most types of gas molecule. Rf discharges are more common, though dc and microwave fields have also been used.

Many reactor systems have been reported, and for each system one must examine the role of electrons, ions, photons, and other excited species in the deposition of thin films. The disadvantage of exposing the substrate to energetic species and charged particles is eliminated by using microwave-excited plasma, where the discharge region can be separated from the reaction region [35,36]. In microwave electron cyclotron resonance plasma CVD [37], highly excited plasma is easily obtained at low gas pressures of $10^{-5}-10^{-3}$ torr, and films of excellent quality can be produced.

Several reviews on plasma CVD have been published [38,39].

Electrodeposition is an electrochemical process in which the anode and a cathode are immersed in a suitable electrolyte and the passage of electric current serves to deposit the material on the cathode. A variety of insulator, semiconductor, and alloy films have been prepared by electrodeposition.

Anodization, another electrochemical process, relies on anodic reactions. Here when the electric current is passed an oxide coating forms on the surface of the metallic anode. In other words, the anode reacts with negative ions from the electrolyte and forms an oxide coating. Aluminum, tantalum, silicon, and zirconium are a few of the metals that give useful films by this simple method, with aluminum by far the most important. In anodic sulfurization, native sulfide films instead of native oxides terminate the lattice and passivate the surface [40].

Electroless deposition is a film deposition process in which no electrode potential is applied, unlike electrodeposition. The chemical reduction can be without a catalyst, as in silver coating on glass using a reducing agent for silver nitrate, or with a catalyst, as in the case of reduction of $NiCl_2$, by sodium hypophosphite, the metal being deposited on nickel itself. Here the metal acts as a catalyst. Electroless deposition is simple, and large-area deposition is possible with this technique.

9.7   EPITAXIAL DEPOSITION

Oriented growth of one material over another is commonly called epitaxy, and epitaxial growth is of particular interest in many semiconductor applications.

Molecular beam epitaxy (MBE) is a sophisticated and finely controlled epitaxial growth technique [41,42] that is basically an ultra high vacuum evaporation process. Here atomic or molecular beams of the element or constituent elements of the compound are created (by slowly evaporating them from separate Knudsen effusion cells) and directed onto clean, heated single-crystal substrates to form the film. The beam intensities are controlled by the temperature of the effusion ovens, chosen to provide the necessary flux of the various elements arriving at the surface. In many cases the composition is controlled by the flux ratio, with the fluxes controlled via the source temperature. Fast shutters, introduced between the sources and the substrate, interrupt the beam fluxes, and by controlling the shutters, one can grow layers of precisely controlled characteristics (thickness, dopant profile, etc.)

Liquid phase epitaxy (LPE) is a thermally controlled technique for the preparation of high purity epitaxial films of semiconductor compounds and alloys. This is a relatively old technique, developed in the 1960s beginning with the work of Nelson [43]. Although the method is simple, in some cases, the thermodynamics of the system makes its use difficult. Other growth techniques are available, but LPE is still used when high quality material is required. As indicated, however, the morphology is difficult to control and, in addition, the quality of the surface is poor compared to that obtained by MBE.

Hot wall epitaxy (HWE) is a vacuum deposition technique in which epitaxial films are grown under conditions as near as possible to thermodynamic equilibrium [44]. This is achieved by having a heated wall between the source and the substrate, to direct the evaporating molecules onto the substrate. In one system, three resistance heaters are used—one each for the substrate, the wall, and the source. Many variations of this simple hot wall system have been used by different authors to prepare different semiconductor compounds [45,46].

Metal–organic chemical vapor deposition (MOCVD) is a recently developed technique for the growth of thin layers of compound semiconductor materials for use in optoelectronic and microwave devices. Here various combinations of organometallic compounds and hydrides are used for the growth of the epitaxial layers. Compounds and alloys from Groups III–V and II–VI are commonly prepared by using this technique, GaAs and AlGaAs being most extensively studied because of their technological applications. GaAs/GaAlAs heterostructures grown by MOCVD are the basic materials in the fabrication of sophisticated high powered laser devices [47].

One advantage of the MOCVD technique is the ready availability of relatively pure organometallic compounds for most of the elements that are used for the epitaxial growth of the semiconductor compounds. In principle, for MOCVD of epitaxial films, the source reactants—organometallic compounds such as dimethylcadmium, trimethylgallium, and trimethylindium, and hydrides such as AsH_3, and PH_3—are introduced as vapor phase constituents into the reaction chamber, where they participate in the elevated temperature reaction (thermal decomposition) of the hot susceptor and substrate, the desired films being formed on the substrate. The need for careful construction and system design imposes constraints on the use of MOCVD as an epitaxial growth technique for certain device applications.

The development of the MOCVD technique and problems involved in obtaining good quality films [48], the fundamental aspects of the growth mechanism [49], and process potentials and developmental status [50] have been reported. For an excellent review of the MOCVD process, the reader is referred to Reference 51.

9.8 OTHER METHODS

The Langmuir-Blodgett (LB) technique makes use of molecules that have carefully balanced hydrophilic (affinity for water) and hydrophobic (no affinity for water) regions to form a monomolecular film on the surface of water and transfer it to a solid substrate. High quality monolayers, bilayers, or multilayers can be produced using this technique. LB films developed recently include electrically conducting films [52] and magnetic LB films [53], and the versatility and simplicity of the method for adjusting the properties of molecular assemblies have led to growing interest in LB techniques.

The spray method of depositing thin films is quite simple; it uses inexpensive equipment to make coatings over large areas. Unlike the other chemical solution deposition techniques, the film is formed on a substrate kept outside the solution. The solution is sprayed onto a heated substrate, where the film is formed either by pyrolytic or hydrolytic chemical reaction of the liquid droplets, the hot substrate providing the thermal energy. Different techniques for spraying that have been employed include standard sprayers using compressed argon gas as a carrier gas and ultrasonic spraying using nozzle sprayers. For early work, the reader is referred to Reference 54.

REFERENCES

1. L. Holland, *Vacuum Deposition of Thin Films*, Chapman Hall, London, 1961.
2. L. I. Maissel and R. Glang, Eds., *Handbook of thin film Technology*, McGraw-Hill, New York, 1970, Ch. 1.
3. C. E. Drunheller, *Transactions of the 7th AVS Symposium*, Pergamon Press, Oxford, 1960, p. 306.
4. W. G. Vergara, H. M. Greenhouse, and N. C. Nicholas, *Rev. Sci. Instrum.*, 34:520 (1963).
5. L. I. Maissel and R. Glang, Eds., *Handbook of Thin Film Technology*, McGraw-Hill, New York, 1970, p. 1.43.
6. Z. Chen, K. Yang, and J. Wang, *Thin Solid Films*, 162:305 (1988).
7. K. Reichelt and X. Jiang, *Thin Solid Films*, 191:91 (1990).
8. W. R. Grove, *Phil. Trans. R. Soc. (London)*, 142:87 (1952).
9. K. L. Chopra, *Thin Film Phenomena*, McGraw-Hill, New York, 1969, p. 23.
10. L. I. Maissel and R. Glang, Eds., *Handbook of Thin Film Technology*, McGraw-Hill, New York, 1970.
11. G. N. Jackson, *Thin Solid Films*, 5:209 (1970).
12. J. L. Vossen, *J. Vac. Sci. Technol.*, 8:512 (1971).
13. R. Behrisch, Ed., *Topics in Applied Physics*, Vol. 47, *Sputtering by Particle Bombardment*, Springer-Verlag, Berlin, 1981.
14. R. L. Hines and R. Wallor, *J. Appl. Phys.*, 32:202 (1961).
15. P. D. Davidse and L. I. Maissel, *Transactions of the 3rd International Vacuum Congress*, Stuttgart, 1965; *J. Appl. Phys.*, 37:574 (1966).
16. F. M. Penning, *Physica (Utrecht)*, 3:873 (1936); U.S. Patent 2,146,025 (1939).

17. J. A. Thorton, *Z. Metallkd.*, 75:847 (1984).
18. (a) M. Wright and T. Beardow, *J. Vac. Sci. Technol.*, A4:388 (1986). (b) D. M. Mattox, *Electrochem. Technol.*, 2:295 (1964)/
19. N. Savvides, *Thin Solid Films*, 163:13 (1988),
20. H. Oechsner, *Thin Solid Films*, 175:119 (1989).
21. G. Tschulena, *Thin Solid Films*, 39:175 (1976)/
22. R. F. Bunshah and A. C. Raghuram, *J. Vac. Sci. Technol.*, 9:1389 (1982).
23. K. S. Joseph, Ph. D. Thesis, University of Cochin, 1983.
24. J. George, G. Pradeep, and K. S. Joseph, *Thin Solid Films*, 148:181 (1987).
25. D. K. Hohnke, D. J. Schmatz, and M. D. Hurley, *Thin Solid Films*, 118:301 (1984).
26. S. Berg, T. Larsson, C. Nender, and H. -O. Blom, *J. Appl. Phys.*, 63:887 (1988).
27. (a) T. Larsson, S. Berg, and H. -O. Blom, *Thin Solid Films*, 172:241 (1989). (b) T. Takagi, *Met. Res. Symp. Proc.*, 27:501 91984). (c) S. E. Huq, R. A. McMahon, and H. Ahmed, *Semicond. Sci. Technol.*, 5:771 (1990). (d) M. Sosnowski and I. Yamada, *Nuclar Instrum. Methods, Phys. Res.*, B46:397 (1990).
28. H. Takaoka, K. Fujino, J. Ishikawa, and T. Takagi, *Proceedings of the 11th I-SIAT Symposium, ISIAT '87*, Tokyo, p. 351.
29. H. Takaoka, J. Ishikawa, and T. Takagi, *Thin Solid Films*, 157;143 (1988).
30. J. George, B. Pradeep, and K. S. Joseph, *Thin Solid Films*, 144:255 (1986).
31. N. F. Mott, *Phil. Mag.*, B55:117 (1987).
32. W. Kern and V. S. Ban, in *thin Film Process W. Kern*, (J. L. Vossen, Eds.), Academic Press, New York, 1978, p. 257.
33. S. D. Allen, R. Y. Jan, S. M. Mazuk, and S. D. Vernon, *J. Appl. Phys.*, 58:327 (1985).
34. S. D. Allen, A. B. Trigubo, and R. Y. Jan, *Mater. Res. Soc. Symp. Proc.*, 17:207 (1983).
35. I. Kato, S. Wakana, S. Hara, and H. Kezuka, *Jpn. J. Appl. Phys.*, 21:L470 (1982).
36. S. Zaima, Y. Yasuda, S. Takashima, T. Nakamura, and A. Yoshida, *18th International Conference on Solid State Devices and Materials*, Tokyo, A-7-2:249 (1986).
37. S. Matsuo and M. Kiuchi, *Jpn. J. Appl Phys.*, 22:L210 (1983).
38. S. M. Ojha, in *Physics of Thin Films*, Vol. 12 (G. Hass, M. H. Francombe, and J. L. Vossen, Eds.), Academic Press, New York, 1982, p. 237.
39. S. V. Nguyen, *J. Vac. Sci. Technol.*, B4:1159 (1986).
40. Y. Nemirovsky and L. Burstein, *Appl. Phys. Lett.*, 44:443 (1984).
41. E. H. Parker, Ed., *The Technology and Physics of Molecular Beam Epitaxy*, Plenum Press, New York, 1985.
42. W. Knodle, *Res. Dev.*, 28:73 (1986).
43. H. Nelson, *RCA Rev.*, 24:603 (1963).
44. J. S. Blakemore, *J. Appl. Phys.*, 53:R123 (1982).
45. C. Geibel, H. Maier, and R. Schmitt, *J. Cryst. Growth*, 86:386 (1988).
46. K. Lischka, T. Schmidt, A. Pesek, and H. Sitter, *Appl. Phys. Lett.*, 55, 1220 (1989).
47. J. Luft and E. Wudy, *Thin Solid Films*, 175:213 (1989).
48. H. M. Manasevit, *Proc. SPIE Int. Soc. Opt. Eng.*, 323:94 (1982).
49. G. B. Stringfellow, *J. Cryst. Growth*, 68:111 (1984).
50. N. E. Schumaker, R. A. Stall, W. R. Wagner, and L. G. Polgar, *J. Met.*, 38; 41 (1986).
51 M. Ludowise, *J. Appl. Phys.*, 58:R31 (1985).
52. K. Hong and M. F. Rubner, *Thin Solid Films*, 160:187 (1988).

53. A. Ruaudel-Teixier, A. Barraud, P. Coronel, and O. Khan, *Thin Solid Films,* 160:107 (1988).
54. K. L. Chopra, R. C. Dainthla, D. K. Pandya, and A. P. Thakoor, in *Physics of Thin Films,* Vol. 12 (G. Hass, M. H. Francombe, and R. W. Hoffman, Eds.), Academic Press, New York, 1982, p. 167.

Index

a-C:H films:
  by ECR plasma CVD, 277–278
  by ion beam sputter deposition, 84, 85
ac sputtering (*see* Superconducting oxide
    films)
Activated reactive evaporation (ARE),
    145, 150–155, 359
Ag films (*see* Silver coatings)
Al films, (*see also* LCVD, Nozzle jet
    beam deposition):
  by hollow cathode discharge ion plat-
    ing, 114–115, 116
  by magnetron sputter ion plating
    (MSIP), 104–105
  by rf heating, 35
Al-Ag alloy thin films, by triode sputter-
    ing, 48
AlN films, preparation of (*see* Dual
    beam deposition)
Al₂O₃ films (*see also* Dual beam deposi-
    tion, Ceramic coatings, Metal
    oxide films):
  by direct current reactive magnetron
    sputtering, 184, 186–187
  by reactive sputtering, 163

[Al₂O₃ films]
  by rf sputtering, 54
  by R-ICB method, 217
  using a modified cathodic arc appara-
    tus, 160–161
Amorphous films:
  boron (*see* ECRCVD)
  boron nitride, by CVD, 229, 240
  carbon, using PEI system, 281
  Co-Zr, by ion beam sputtering, 86, 87
  GdCo, by getter sputtering, 52
  polyphosphide, 182
  prepared by flash evaporation, 11, 12
  silicon, by laser evaporation, 20
  Si, Ge:H alloys, 183
  silicon nitride, 280
  Si, Sn:H alloys, using dual magnetron
    sputtering system, 183–184
  Tb(FeCo), 52–53
Anodic deposition, 287–293, 362
  table, 288–293
Anodization (*see* Anodic deposition)
Arc evaporation (*see* Vacuum evapora-
    tion methods)

a-SiC:H films, by a novel rf PECVD
system, 281–282
a-Si films (*see also* Mercury sensitized
photo CVD):
using PEI system, 281–282
a-Si:H films (*see also* Mercury sensitized photo CVD, PECVD):
by compressed magnetic field magnetron sputtering, 77
by concurrent ion bombardment, 123
by direct photolysis, 247–248
by ECR plasma CVD, 277
using interdigital vertical electrode
deposition apparatus, 260
a-SiO$_x$ films, by laser photo CVD,
248–249

BaTiO$_3$ films, by laser photo CVD,
248–249
by ARE, 153–154
by laser evaporation, 28
BeO films, by reactive evaporation, 143,
358
Bi$_2$O$_3$ films, preparation of:
by ARE, 150–152
by thermal oxidation, 224–225
Boron nitride films:
by laser evaporation, 28
using rf sputtering, 54
Borophosphosilicate glass films, preparation of, 240–241

CaF$_2$ film deposition, by combined laser
evaporation and ion bombard-
ment, 127–128
Carbon-boron coatings, by vacuum arc
deposition, 120
Carbon films (*see also* Graphite):
by ion beam sputtering, 82–83
by laser evaporation, 19–20, 26
effect of hydrogen ion bombardment,
83–84
Catalytic CVD (cat-CVD), 242
Cathode spot, 116, 117

Cathode sputtering, 41
advantages of, 42
theories for the mechanism of, 41, 42
Cathodic arc plasma deposition,
115–120, 358
advantages of, 115, 116
alloy thin films by, 119
basic coating system for, 116–117
Cathodic arc source, 118
Cd$_3$As$_2$ films, preparation by PLE, 23,
24
Cd films, by laser evaporation, 27
CdS films, indium doped:
by ion assisted deposition, 112–113
by spray pyrolysis, 340–341
CdTe films:
by hot wall epitaxy (*see* Epitaxial
films)
by laser evaporation, 27
CeO$_2$ films, deposition of, 27
Ceramic coatings, by laser evaporation,
20, 28
Cermet films, preparation of, 12
Chemical beam epitaxy (CBE), 307–309
advantages of, 308
Fe doped InP layers by, 309
growth system for, 308
Chemical ferrite plating, 283, 287
Chemical methods of film deposition,
223–293, 360
Chemical vapor deposition (CVD),
225–282, 360–361
table, 230–239
technological applications of,
225–226
typical reactions involved in, 226–229
Chromium deposition, concurrent ion
bombardment, 122–123
Co-Cr thin films by FTS, 79
Co-ferrite films (*see* Chemical ferrite
plating)
Concurrent ion bombardment deposition,
122–123
Copper coatings:
by hollow cathode discharge ion plating, 114

[Copper coatings]
  by ion beam sputtering deposition, 84,
    86
  deposition by HHC gun evaporation,
    121-122
Copper molybdenum sulphide films, by
    reactive evaporation, 143
Cr-N films, preparation by RIP, 191
Cubic boron nitride films:
  by ARE, 150
  concurrent ion bombardment deposi-
    tion, 129-130
CuInSe$_2$ films, 11-12 (see also Epitaxial
    films)
CuS films, preparation by ARE, 145,
    151

DC sputtering, glow discharge, 42-46,
    356
  basic system for, 42-43
  table, 45-46
Diamond-like carbon films (see also i-C
    coating):
  by a hybrid ion beam technique,
    136-137
  by laser evaporation, 27
  by UM gun sputtering, 132
Diamond thin films, by DC plasma
    CVD, 278-279
Dual beam deposition, preparation of
    AlN and Al$_2$O$_3$ films by, 217,
    221

Electrodeposition, 282-283, 362
  table, 284-286
Electroless deposition, 283, 287, 362
  films prepared by, 283, 287
Electron beam evaporation (see Vacuum
    evaporation methods)
Electron beam guns, types of, 13-17,
    18-19
Electron cyclotron resonance (ECR)
    CVD, 276-278, 362
  amorphous boron films by, 278

Epitaxial film deposition techniques,
    303-331, 362-363
  hot wall epitaxy, 313-326, 363
    advantages of, 313-314
    compound films by, 314 (see also
      Epitaxial films grown)
    table, 316-323
  liquid phase epitaxy, 311-313, 363
  metal-organic chemical vapor deposi-
    tion (MOCVD), 326-331, 363
    compound films by, 327 (see also
      Epitaxial films grown)
  molecular beam epitaxy (MBE),
    304-311, 353, 362
    a basic system of, 304-305
    effusion cell for use in, 305-306
    substrate preparation in, 306
Epitaxial films grown (see also Flash
    evaporation, PLE, Superconduct-
    ing oxide films):
  AlGaAs, 306, 307, 327, 328
  AlAs, 307
  CdSe, by HWE, 325, 326
  Cd$_{1-x}$Mn$_x$Te, by PLE, 27
  CdTe, 88, 89, 325
  CuInSe$_2$, by flash evaporation, 12
  Fe, by ion beam sputtering, 87-88
  GaAlAsInP, 326-327
  GaAs, 304, 306, 307, 309, 314, 326,
    328, 330-331
  Ga$_{0.47}$In$_{0.43}$As, 327
  ^{74}Ge, by direct ion beam deposition,
    136
  HgCdTe, by ion sputter deposition,
    88, 89
  InAs$_{1-x}$Sb$_x$, by MOCVD, 328-329
  InGaAs, 307
  InP, 249-250, 307
  InP$_{1-x}$Sb$_x$ by MOCVD, 329
  Pb$_{1-x}$Eu$_x$Te by HWE, 324-325
  PbTe by HWE, 314, 315
  silicon by ion beam sputtering, 86, 87
  Si$_{1-x}$Ge$_x$ (see Gas source MBE)
  silver deposited during ion bombard-
    ment, 129

[Epitaxial films grown]
YSZ by ion beam sputter deposition,
91
ZnO, 329
ZnS, 328
ZnSe, 328
Zn S_xSe_{1-x}, 328
Epitaxy, 303
Evaporation (see Vacuum evaporation)

Facing targets sputtering (FTS), 78-81
films prepared by, 78-81 (see also
specific films)
Flash evaporation (see Vacuum evaporation methods)

GaAs films (see also Epitaxial films
grown):
using a supported plasma technique,
48-49
Gas source MBE, 308 (see also CBE)
$Si_{1-x}Ge_x$ layers by, 309-310
Germanium films, deposited under concurrent ion bombardment, 122
Getter sputtering, 52-53, 357
Gold films (see also LCVD):
by concurrent ion bombardment deposition, 123-125, 126
deposition by laser evaporation, 26
using ionized cluster assisted deposition, 215
Graphite, cathodic arc evaporation of,
119, 120

Heat mirror, 152
HfN coatings, 163
High-temperature superconducting
(HTS) oxide films (see super-
conducting oxide films)
Hollow cathode discharge ion plating,
114-115, 358
Hot hollow cathode (HHC) gun, 18-19,
120-121, 358

[Hot hollow cathode]
coatings using, 121-122 (see also
specific films)
Hydrogenated amorphous carbon films
(see a-C:H)
Hydrogenated amorphous Si films (see
a-Si:H)
Hydrogenated amorphous SiC films (see
a-SiC:H)

i-C coating, by ion beam deposition,
133, 136
$InAs_{1-x}Sb_x$ films (see Epitaxial films
grown)
Induction heated (IH) plasma assisted
CVD, 273-275
In_2O_3 films (see also Electroless deposition, Metal oxide films):
by reactive sputtering, 184
by rf reactive ion plating, 188-189
tellurium doped, 339-340
Inp layers, Fe doped (see CBE)
In_2S_3 films, preparation by ARE, 145,
151
Ion assisted deposition, 101, 102-132,
358
Ion beam deposition, 101, 132-137
table, 134-135
Ion beam sputtering, 81-92, 357-358
advantages and disadvantages of, 81,
82
ion beam sources used for, 82
simple system of, 81
survey of materials prepared by, 82
Ionization efficiency, 114
Ionized cluster assisted film deposition,
214-215
Ionized cluster beam deposition (ICBD),
207-214, 360
advantages of, 209
laboratory ICB system for, 210
table, 211-213
Ionized cluster source, basic configuration, 208
Ion plating, 102-115, 358

[Ion plating]
benefits of, 104
simple setup for, 102, 103
table, 106–107
Ion surface interactions, 102
IR dichroic filters, 129
Iron films, by facing target sputtering, 78, 79
Iron oxide films, deposition by laser evaporation, 25, 28

Knudsen cell, 305, 309
Knudsen's cosine law, 2

Langmuir-Blodgett (LB) technique, 335–338, 364
Langmuir expression, 2
Laser chemical vapor deposition (LCVD), 244–245, 361
films deposited by, 245
Laser evaporation (see Vacuum evaporation methods)
LB films, preparation of the different types of, 336–338

Magnetron sputtering, 55, 64–78, 357
basic principles of, 55
compressed magnetic field (CMF), 77
with a hollow cathode electron source, 75, 76, 77
with multiple targets, 73
table, 64–72
Magnetron sputter ion plating, 358 (see also Al films)
Manganine films, by ion beam sputtering (see Thin film sensors)
Mean free path, 3
Metal-organic chemical vapor deposition (see Epitaxial deposition techniques)
Metal oxide films, by pyrolytic decomposition, 240, 241

MgF$_2$ deposition, with ion bombardment, 126
MgO films, spray pyrolysis deposition of, 341–342
Molecular beam epitaxy (see Epitaxial deposition techniques)
Mullite films, using rf sputtering, 54
Multilayer films, deposition of (see also Ion beam deposition):
aluminum/aluminum oxide, 112
CuO/Al$_2$O$_3$ (see Metal oxide films)
Fe/Al, 88
Fe/Ti, 79, 81
Ta/Au, 77

Nb films, preparation of, 242
Nb$_3$Ge films:
deposition by halide CVD, 241–242
deposition by reactive evaporation, 143
Nb$_2$O$_5$, single crystal films by RIP, 191
Nickel based alloy coatings, using a dual beam ion system, 126–127
NiFe films, deposition by ion beam sputtering, 87
Nickel films, (see also Electroless deposition, LCVD):
by facing target sputtering, 78, 79
by magnetron sputtering, 74–75, 76
by triode sputtering, 48
Nozzle jet beam deposition, SiO$_2$ and Al films by, 210, 214

Pb$_{1-x}$Cd$_x$Se films, by laser evaporation, 23–25
Pb$_{1-x}$Eu$_x$Te films, by HWE (see Epitaxial films grown)
PbS films, by flash evaporation, 10
PbS-Ag films, by flash evaporation, 10
PbTe films:
by HWE (see Epitaxial films grown)
by laser evaporation, 27
doped, 27

Permalloy films, preparation of, 78-79, 80

Phosphosilicate glass (PSG) films, preparation of, 181-182

Photochemical vapor deposition (photo CVD), 245-251, 252-259, 361
mercury sensitized, 246, 247
table, 252-259

Plasma enhanced chemical vapor deposition (PECVD), 251, 260-282, 361-362
application of hollow cathodes for the deposition of a-Si:H, 279-280
materials prepared by, 251 (*see also* specific films)
table, 261-271

Platinum films, by ion sputter deposition (*see* Thin film sensors)

Polyphosphide films (*see* Amorphous films)

Pulsed laser evaporation (PLE), 21-32
films prepared using, 21-32 (*see also* specific films)

Pyrite films, by plasma assisted thermal reaction, 282

Quartz films:
by HHC gun evaporation, 121-122
by rf sputtering, 54

Radiofrequency heating (*see* Vacuum evaporation methods)

Radiofrequency sputtering, 53-55, 56-63, 357
thin films prepared by, 54 (*see also* Specific films)
table, 56-63

Reactive deposition techniques, 141-200, 358-360
reactive arc evaporation, 156-161, 359
advantages of, 156
and magnetron sputtering (*see* $Ti_xAl_{1-x}N$ coating)

[Reactive deposition techniques]
reactive evaporation, 141-145, 146-149, 358
advantages and drawbacks of, 142
table, 146-149
reactive ion beam sputtering (RIBS), 192, 195-200, 359-360
table, 197-200
thin film materials prepared by, 195
reactive ion plating (RIP), 188-192, 193-194, 359
table, 193-194
reactive sputtering, 161-188, 359
compound films by, 163
hysteresis curve for, 162-163
magnetron type of, 180-185, 187-188
models for, 163
nitride coating of tool steel by, 163
table, 164-179
target poisoning in, 162, 359
use of unbalanced magnetrons for (*see* In_2O_3 films)

Reactive ionized cluster beam (RICB) deposition, 216-217, 218-220, 360
table, 218-220

Refractory metal silicide films (*see* Silicide films)

Remote plasma enhanced CVD (RPECVD), 260, 272-273

Resistive heating (*see* Vacuum evaporation methods)

rf sputtering (*see* Radiofrequency sputtering)

Ruthenium oxide films, by ion beam sputtering, 84, 86

Sb_2S_3 films, by flash evaporation, 10-11

Self-sustained glow discharge, 44

Si films (*see also* Epitaxial films grown, LCVD):
phosphorous doped, deposition of, 73-74

Silicide films, deposition of, 74, 75, 250, 251
Silicon nitride films (*see also* Ion beam sputtering, Reactive ion beam sputtering, Reactive sputtering):
  by catCVD, 242-244
  by ECRCVD, 277
  by excimer laser photolysis, 249
  by RPECVD, 260, 272-273, 274
Silver coatings (*see also* Epitaxial films grown):
  by electroless deposition, 283
  by hollow cathode discharge ion plating, 114
  by hollow cathode electron beam, 122
SiN films, deposition of (*see also* Induction heated plasma assisted CVD):
  by ARE, 152
SiN$_x$ films, deposition using microwave excited plasma, 275-276
SiO$_2$ films (*see also* Nozzle jet beam deposition, Reactive sputtering):
  by ECRCVD, 277
  by PECVD, 260
  by RPECVD, 260, 272-273, 274
  by vacuum UV photo CVD, 246-247
SnO$_2$ films (*see also* Electroless deposition):
  arsenic doped, 242
  preparation by PLE, 21-22
SnS$_2$ films, by ARE, 145, 151
Spray method, 339-350, 364
  table, 343-350
Spray hydrolysis, 339-340
Spray pyrolysis, 340-342
Sputtering, 41-94, 104, 356-358 (*see also* Cathode sputtering)
Superconducting oxide films:
  by ac sputtering, 92-94
  by ARE, 152-153, 154
  by compressed magnetic field magnetron sputtering, 77-78
  by flash evaporation, 12
  by FTS method, 81

[Superconducting oxide films]
  by ion assisted laser deposition, 108-109
  by ion beam sputtering, 91-92
  by laser evaporation, 28-32
  by MBE, 310-311
  by MOCVD, 331
  by reactive magnetron multitarget sputtering, 187-188
  by reactive plasma evaporation, 154-155
  by resistive evaporation, 5
  concurrent ion bombarded, 130
  deposition by triode ion plating, 108-109

TaNfilms, by RIP, 188-189
Tantalum films, by triode sputtering, 50
Thermal growth, 223-225, 360
  high-temperature short time method for, 224
Thin film deposition techniques, classification of, 354
Thin film sensors, deposition of films for the fabrication of, 88, 90-91
Ti$_x$Al$_{1-x}$N coating, 159-160
TiC$_x$N$_{1-x}$ films, preparation of, 158-15
TiN films (*see* Titanium nitride films)
Tin oxide films, arsenic doped (*see* SnO$_2$)
Tin sulphide films, (*see also* SnS$_2$):
  preparation by reactive evaporation, 143-145
Titanium carbide films, deposition of:
  by reactive arc deposition, 156, 157, 158, 159
  by reactive ion plating, 189-191
Titanium carbonitride (Ti(C,N)) films, by ARE, 145, 150
Titanium films, by cathodic arc evaporation, 118-119
Titanium nitride films:
  by laser evaporation, 25-26

[Titanium nitride films]
by reactive arc deposition, 156–157, 158
by reactive ion plating, 188–189
by reactive sputtering, 163
single crystal films by magnetron sputtering, 180–181
Titanium oxide films, by reactive sputtering, 184
Titanium silicide films (*see* Silicide films)
Triode ion plating, 105, 108–114, 358
apparatus for coating small components, 110, 112
thermionically assisted, 109–110, 111
Triode sputtering, 47–51, 356–357
advantages and limitations of, 47
comparison with dc diode, 50
Tungsten films:
by triode sputtering, 47–48
selective deposition of, 244

UM gun, 131
sputtering system for thin film preparation, 131–132
Unbalanced magnetron assisted deposition, 131–132

Vacuum arc source, ion gun type, 156
Vacuum evaporation, 1–35, 354–356
Vacuum evaporation methods, 4–35
arc evaporation, 33–34, 356
refractory metal films, 33–34

[Vacuum evaporation methods]
electron beam evaporation, 13–19, 104, 355
crucibles for, 17
thin films prepared by, 18
flash evaporation, 10–13, 104, 355
drawbacks of, 12–13
thin films prepared by, 10–12 (*see also* Specific films)
laser evaporation, 19–32, 355–356
advantages of, 19, 355–356
radiofrequency heating, 35, 356
resistive heating, 4–10, 354–355
sources for, 4–5, 104
support materials used in, 4–9
Vanadium oxide (*see* RIBS)

WS_2 films, by CVD, 229, 240

Yttrium-iron garnet (YIG) films, preparation by spray pyrolysis, 341
Yttrium oxide (*see* RIBS)

Zirconium nitride films, by reactive arc evaporation, 157, 158
ZnO films:
by CO_2 laser evaporation, 22, 23
by MOCVD (*see* Epitaxial films grown)
ZnTe films, by ion beam sputtering, 84, 85, 86
$ZnTe_x$ films, by triode sputtering, 50
ZrO_2 films, by ion-assisted deposition, 125, 126